MICRO-CUTTING

Microsystem and Nanotechnology Series

Series Editors – Ron Pethig and Horacio Dante Espinosa

Micro-Cutting: Fundamentals and Applications
Cheng, Huo, August 2013

Nanoimprint Technology: Nanotransfer for Thermoplastic and Photocurable Polymer
Taniguchi, Ito, Mizuno and Saito, August 2013

Nano and Cell Mechanics: Fundamentals and Frontiers
Espinosa and Bao, January 2013

Digital Holography for MEMS and Microsystem Metrology
Asundi, July 2011

Multiscale Analysis of Deformation and Failure of Materials
Fan, December 2010

Fluid Properties at Nano/Meso Scale
Dyson et al., September 2008

Introduction to Microsystem Technology
Gerlach, March 2008

AC Electrokinetics: Colloids and Nanoparticles
Morgan and Green, January 2003

Microfluidic Technology and Applications
Koch et al., November 2000

MICRO-CUTTING
FUNDAMENTALS AND APPLICATIONS

Editors

Kai Cheng
Brunel University, UK

Dehong Huo
Newcastle University, UK

WILEY

This edition first published 2013

Registered Office
John Wiley & Sons, Ltd, The Atrium, Southern Gate, Chichester, West Sussex, PO19 8SQ, United Kingdom.

Editorial Offices
9600 Garsington Road, Oxford, OX4 2DQ, United Kingdom.
The Atrium, Southern Gate, Chichester, West Sussex, PO19 8SQ, United Kingdom.

For details of our global editorial offices, for customer services and for information about how to apply for permission to reuse the copyright material in this book please see our website at www.wiley.com/wiley-blackwell.

Library of Congress Cataloging-in-Publication Data

Micro cutting : fundamentals and applications / edited by Kai Cheng, Dehong Huo.
pages cm
Includes bibliographical references and index.
ISBN 978-0-470-97287-8 (cloth)
1. Micromachining. I. Cheng, K. (Kai), editor of compilation. II. Huo, Dehong, editor of compilation.
III. Title: Microcutting.
TJ1191.5.M4983 2013
671.3′5–dc23

2013015108

A catalogue record for this book is available from the British Library.

ISBN: 978-0-470-97287-8

Set in 10/12pt Times by SPi Publisher Services, Pondicherry, India
Printed and bound in Singapore by Markono Print Media Pte Ltd

1 2013

Contents

List of Contributors

Dr Yoshikazu Arai
Department of Nanomechanics
Tohoku University
Sendai, Japan

Dr Ampara Aramcharoen
Singapore Institute of Manufacturing Technology (SIMTech), A*STAR
Singapore

Dr Qingshun Bai
School of Mechanical and Electrical Engineering
Harbin Institute of Technology
Harbin, P R China

Professor Dr.-Ing Christian Brecher
Fraunhofer-Institut fuer Produktionstechnologie IPT
Werkzeugmaschinenlabor WZL
der RWTH Aachen, Germany

Dr Jiaxuan Chen
School of Mechanical and Electrical Engineering
Harbin Institute of Technology
Harbin, P R China

Professor Kai Cheng
School of Engineering and Design,
Brunel University
Uxbridge, Middlesex, UK

Professor Wei Gao
Department of Nanomechanics
Tohoku University
Sendai, Japan

Professor Han Huang
School of Mechanical & Mining Engineering
The University of Queensland
Queensland, Australia

Dr Dehong Huo
School of Mechanical and Systems Engineering,
Newcastle University
Newcastle Upon Tyne, UK

Professor Mark J. Jackson
Department of Mechanical Engineering Technology
College of Technology
Purdue University
West Lafayette, USA

Kang-Won Lee
Department of Nanomechanics
Tohoku University
Sendai, Japan

Professor Yingchun Liang
School of Mechanical and Electrical Engineering
Harbin Institute of Technology
Harbin, P R China

Dr Paul T. Mativenga
School of Mechanical, Aerospace and Civil Engineering (MACE)
The University of Manchester
Manchester, UK

Professor Shreyes N. Melkote
George W. Woodruff School of Mechanical Engineering
Georgia Institute of Technology
Atlanta, USA

K. Mosiman
Department of Mechanical Engineering Technology
College of Technology
Purdue University
West Lafayette, USA

Dr Young-Jin Noh
Department of Nanomechanics
Tohoku University
Aramaki Aoba 6-6-01, Aoba-ku
Sendai, Japan

T. Novakov
Department of Mechanical Engineering Technology
College of Technology
Purdue University
West Lafayette, USA

Dr Yuki Shimizu
Department of Nanomechanics
Tohoku University
Sendai, Japan

Dr Sathyan Subbiah
School of Mechanical and Aerospace Engineering
Nanyang Technological University
Singapore

Dr Ing Christian Wenzel
Oberingenieur
Fraunhofer-Institut fuer
Produktionstechnologie IPT
Steinbachstrasse, Germany

Dr Tao Wu
School of Engineering and Design,
Brunel University
Uxbridge, Middlesex, UK

Series Preface

This book series provides a thorough summary of the methods used in micro- and nano-technology research and shows how these advances are currently influencing many scientific fields of study and practical application. This contextual presentation ensures that the books are appropriate for readers with varied backgrounds, while being useful for self-study or as classroom materials. Readers of these books will learn the fundamental principles necessary to understand the topic and then explore examples that are representative of the application of these fundamental principles.

Micro- and nano-scale materials created by novel fabrication techniques and metrology methods are the basis for many modern technologies. Several books in this series provide a resource for building a thorough scientific understanding of the field. These include *Introduction to Microsystem Technology* by Gerlach and Dotzel, *Microfluidic Technology and Applications* edited by Koch, Evans, and Brunnschweiler, *Fluid Properties at Nano/Meso Scale* by Dyson, Ransing, P. Williams and R. Williams, and *Nanoimprint Technology: Nanotransfer for Thermoplastic and Photocurable Polymer* edited by Jan Taniguchi. Multiscale modeling, an important aspect of microsystem design, is extensively reviewed in *Multiscale Analysis of Deformation and Failure of Materials* by Jinghong Fan. Specific implementations and applications are presented in *AC Electrokinetics: Colloids and Nanoparticles* by Morgan and Green, *Digital Holography for MEMS and Microsystem Metrology* edited by Asundi. Topics of biological mechanics are discussed in *Nano and Cell Mechanics: Fundamentals and Frontiers* edited by Espinosa and Bao.

This book on micro-cutting, edited by Kai Cheng and Dehong Hou, presents technology that has been developed over the last two decades to bridge the manufacturing size-scale between precision and nanoscale manufacturing, i.e. feature sizes from a few millimeters to tens of micrometers. Featured here are the micro-cutting tool fundamentals, principal processes, and design-guiding modeling that have led to new applications for micro-cutting manufacturing. New micro-cutting tools, often developed via miniaturization of conventional machining tools, retain many of the advantages of conventional machining, not least of which is the capability to process many different materials such as metals, polymers, and ceramics. This flexibility establishes micro-cutting as an important manufacturing advancement and guarantees that the technology will find use in a wide array of practical applications.

Horacio D. Espinosa
Ronald Pethig

Preface

The motivation for manufacturing the smaller and smaller workpieces has been essentially the same since manufacturing was first established as an art/science – new applications, easily fabricated, less expensive, better performance and higher quality. The emergence of miniature and micro products in the last one or two decades is increasingly demanding the production of components and products with dimensions in the range of a few tens of nanometres to some few millimetres. Mechanical micro cutting is one of the key technologies to enable the realization of high accuracy complex micro products made from a variety of engineering materials including silicon, while a great many micro manufacturing processes have been developed.

Kinematically similar to conventional macro cutting, micro cutting is a mechanical material removal process to fabricate micro and miniature components using geometrically defined cutter edges, but the uncut chip thickness is normally a micrometer or less. As the unit removal size decreases, issues of cutting tool edge geometry, grain size and material micro structure, and so on, considered to have little or no influence at macro scale, become dominant factors with strong influences on the cutting mechanics and dynamics, and eventually result in machining accuracy, surface integrity and the quality of the machined component or product.

Micro cutting raises a great number of issues, mainly due to the size or scale associated with the process. When either the ratio of part size to be produced or size of the micro structure of the work material to the cutting tool dimension or cutting parameters becomes smaller, the size effects can change the whole aspect of the machining. Furthermore, scientific understanding of fundamentals and applications in micro machining is essential and much needed so as to address the underlying necessities for predictability, producibility, repeatability and productivity of manufacturing at micro and nano scales.

There are numerous books on MEMS and MEMS based micro manufacturing, which focus on lithography-based micro manufacturing processes. There are also several excellent books in non-MEMS micro manufacturing published in the last few years. But these books only briefly discussed certain aspects of mechanical micro machining normally within a single book chapter. Micro cutting as a new subject area in its own right with emerging cutting edge technologies is attracting international interest especially in the micro manufacturing community. This book is concerned with the state of the art research and engineering practices in micro cutting, including its concept and scope, enabling technologies, underlying theory, research methods, latest development and applications, and so on. It is the first book dedicated solely to the topic of micro cutting.

The book comprises two parts, Part One addresses fundamental aspects of micro cutting processes and Part Two, their applications. The first chapter overviews the micro cutting processes. The history and development process of micro cutting is presented, followed by conception and scope of micro cutting. It is difficult to give a single definition of micro cutting as it is a fast developing and timely subject area. Therefore, Chapter 1 attempts to characterize and define micro cutting by using a number of key features, namely, uncut chip thickness, micro part/feature dimensions and cutting tool geometry, underlying mechanics and application areas. Micro cutting mechanics are central to the progress of this subject area. Micro cutting mechanics aspects are therefore discussed in Chapter 2. Size effects related to micro cutting mechanics are categorized into three groups – cutting edge radius, grain size and material properties size effect. Influences of these size effects on cutting force, surface generation and burr formation in particular are reviewed. There are increasing demands on an industrially applicable micro cutting process for hard materials, scientific approaches to tackling micromachinability are therefore discussed. Chapter 3 focuses on the enabling technology for micro cutting – micro tooling design and fabrication – which is the key to the interface between micromachining machine tools and processes. The latest developments on micro tooling design, tool materials, coating and tool wear are presented and the chapter concludes with the smart micro tooling as a future trend. In addition to the process and material size effects, micro cutting process performance is strongly dependent on the development of machine tools which is another enabling technology for micro cutting. Ultra-precision and micro machine tools for micro cutting are the subject of Chapter 4. The state of the art machine tools suitable for micro cutting and characteristic machine components for both industrial and in-house research machine tools have been studied. Micro cutting is capable of processing a full range of materials; however micro cutting some engineering materials efficiently and effectively still remains a challenge. Chapter 5 investigates engineering material behaviours at the micro scale and provides an overview of machinability of various engineering materials that have been processed using micro cutting processes. It is usually difficult and expensive to conduct micro cutting experiments in particular to make in-process observations for condition monitoring and quality control purposes. Micro cutting mechanics research relies heavily on accurate modelling and simulation. Key modelling methods, including FE (Finite Elements), MD (Molecular Dynamics) and newly developed multi-scale simulation, are discussed in Chapter 6. The modelling and simulation are critically important for undertaking the investigation on micro cutting chip and surface generation, cutting temperature, defects and burrs formation, cutting optimization strategies, and so on, in a scientific and interactive manner.

The chapters in Part Two of the book are devoted to applying the fundamentals of micro cutting (presented in Part One) in a variety of machining processes including diamond turning and micro turning (Chapter 7), micro milling (Chapter 8), micro drilling (Chapter 9) and micro grinding (Chapter 10). Operation principles, machine equipment and micro tooling used material removal mechanisms specific to individual micro cutting processes, and application perspectives are discussed in these respective chapters. In-process micro/nano measurement is essential to the success of micro cutting research development and applications. In-process precise measurement of micro cutting force, cutting temperature, micro tool wear and micro surfaces are desirable for process monitoring, quality control and inspection purposes. This topic is fully explored in the final chapter (Chapter 11) of the book. The application chapters above are intended to reveal micro manufacturing researchers and practitioners with good exemplars of how to implement and apply micro cutting in precision and micro manufacturing routine practices, although more concrete detailed application examples may need to be provided.

The international interest in the subject is evident, with more than 20 esteemed authors coming from 11 institutions in seven countries on four continents. We are grateful to them all, for the benefit of their advice and expertise, and their patience in supplying us with their specialist chapters.

This book can be used as a textbook for a final year elective subject on manufacturing engineering, or as an introductory subject on advanced manufacturing methods at the postgraduate level. It can also be used as a textbook for teaching advanced manufacturing technology in general. The book can also serve as a useful reference for manufacturing engineers, production supervisors, tooling engineers, planning and application engineers, as well as machine tool designers.

At Brunel University and Newcastle University, we are indebted to colleagues Dr Richard Bateman, Dr Robin C. Wang, Dr Tim Minton, Dr Sarah Sun, Dr Khalid Nor, Dr Lei Zhou, Dr Najmil Aris, Dr Ibrahim Shidi, Mr Paul Yates, Dr Atanas Ivanov and Professor Kenneth Dalgarno for their assistance in checking many of the details of the chapters and for stimulating discussions. We have been appreciative of the support from Tom Carter, Anne Hunt, Debbie Cox and many others at the publisher, John Wiley and Sons Ltd., as the book has developed from its draft outline form through various stages of production.

Finally and most importantly, our greatest thanks have to be reserved for our respective darling wives, Lucy Lu and Jun Tian, for their steadfast support and interest throughout the preparation of the book.

Kai Cheng and Dehong Huo
London, UK
November 2012

Part One

Fundamentals

1

Overview of Micro Cutting

Dehong Huo[1] and Kai Cheng[2]
[1] *School of Mechanical and Systems Engineering, Newcastle University*
[2] *School of Engineering and Design, Brunel University*

1.1 Background and Scope

1.1.1 Micro Manufacturing

The increasing demands on micro and miniature parts, components and systems have led to the development of micro and nanotechnology. It is well-recognized that micro manufacturing has been a key enabling technology in industrially producing useful micro products and processes.

Micro Electric Mechanical Systems (MEMS) or micro system technology (MST) as known in Europe has been booming over the last two decades or so. Numerous MEMS products mainly micro sensors and micro actuators using silicon have been fabricated. These MEMS sensors and actuators have been widely used in various applications including medical engineering (e.g. pressure sensors), communications (high frequency resonators), inertial sensing (e.g. accelerometers and gyroscopes), to name a few. The processes employed to fabricate MEMS devices and other microelectronics products can be described as MEMS micro manufacturing or lithography-based micro manufacturing. Common techniques include photolithography, chemical-etching, plating and LIGA, and so on. Lithography-based micro manufacturing has existed and been developed for many years and is regarded as a mature large volume production process, hence the term micro manufacturing is familiar in the semiconductor or microelectronics fields.

In the past 20 years or so, high-accuracy complex shape micro and miniature components made from a range of engineering materials are increasingly in demand for various engineering industries. The geometry and functional requirements have led to the development of another category of micro manufacturing techniques termed as non-MEMS micro manufacturing or non-lithography-based micro manufacturing, which are fundamentally different from MEMS micro manufacturing in many aspects.

Micro-Cutting: Fundamentals and Applications, First Edition. Edited by Kai Cheng and Dehong Huo.

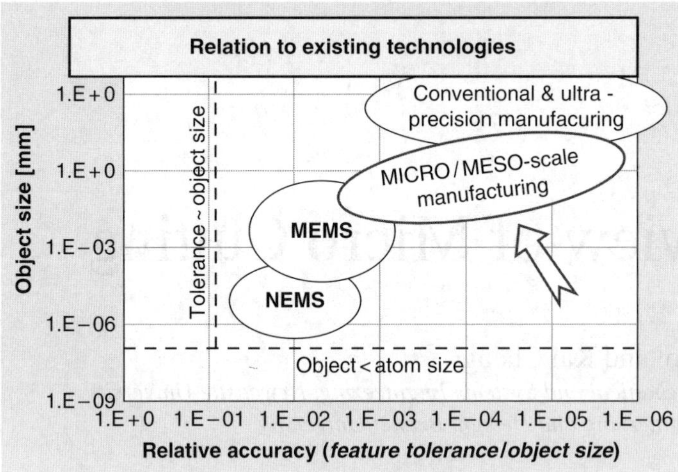

Figure 1.1 Micro manufacturing size/precision domains

Non-lithography-based micro manufacturing is a relatively new area, its concept, theories, processes and applications have been developed and formulated in the past around two decades. A report published by the WTEC panel on micro manufacturing describes non-lithography-based micro manufacturing as the creation of high-precision three dimensional products using a variety of materials and possessing features with sizes ranging from tens of micrometers to a few millimeters (WETC report). Figure 1.1 illustrates micro manufacturing size/precision domains. Micro manufacturing is normally used to produce part or feature size ranging from tens or hundreds of microns. Although micro manufacturing may not be capable of producing the smallest feature size as would be the case using MEMS and NEMS (Nano Electric Mechanical Systems) processes, it is a critical technology in bridging the gap between macro and nano domain [1]. It has many advantages over lithography-based micro manufacturing processes in terms of material choices, relative accuracy and the complexity of part geometry.

Typically non-lithography-based micro manufacturing includes micro EDM, micro mechanical-cutting, laser-cutting/patterning/drilling, micro-extrusion, micro-embossing, micro stamping, micro-injection moulding, and so on. These processes are based on different working principles and have their own respective characteristics in terms of production rate, attainable accuracy and surface finishes, and so on. But they are capable of producing 3D shape geometry micro parts over a wider range of engineering materials. This book will only focus on the micro mechanical-cutting process. Table 1.1 highlights the difference between MEMS micro manufacturing and non-MEMS micro manufacturing techniques (using mechanical micro machining as an example) to compare the fundamental differences between the two category micro manufacturing processes.

From Table 1.1 it can be found that micro mechanical machining has many advantages over MEMS-based process, such as wider materials choices, higher accuracy and capability of producing complex 3D geometry micro parts.

Recently, significant research efforts have been made on non-lithography-based micro manufacturing techniques. Europe has invested heavily in the research and development in

Table 1.1 Comparisons between MEMS-based process and micro machining

	MEMS-based process	Micro mechanical machining
Workpiece materials	Silicon, some metals	Metals, alloys, polymers, composite, technical ceramics
Component geometry	Planer or 2.5D	Complex 3D
Assembly methods	None or bonding	Fastening, welding, bonding
Relative accuracy	10^{-1}–10^{-3}	10^{-3}–10^{-5}
Process control	Feedforward	Feedback
Machine size	Macro	Macro to micro
Production volume	High	High or low
Production rate	High	Low
Initial investment	High	Intermediate or low
Applications	MEMS, microelectronics, some planner micro parts	Various applications requiring 3D micro components

micro manufacturing. In the past decade, dozens of EU large Framework projects have been initiated, such as MASMICRO, 4M, Launch-Micro, Production4ì, EUPASS, Hydromel, HYTI, NANOSAFE2, Manudirect, Napolyde, PRONANO, NaPa, CHARPAN, NANOIMPRINT, and so on. These projects cover all areas in micro and nano manufacturing, precision manufacturing and metrology [2]. In a study conducted by the UK Technology Strategy Board (TSB) on high value manufacturing in the UK, micro and nano manufacturing processes are identified as one of the most significant emerging manufacturing processes which would address challenges for the UK high value manufacturing industry [3].

1.1.2 History and Development Process of Micro Cutting

Micro cutting as an emerging subject area in its own right has attracted growing attention from both researchers and industry in the last two decades. Because mechanical cutting is a well-established area much knowledge from macro cutting has thus been adapted to study micro cutting processes. Some researchers from the conventional mechanical cutting research community shifted their research interests to micro-domain. Basically there are two research approaches being taken to study micro cutting. One approach is miniaturization of the conventional cutting process, tooling and equipment with an emphasis on their scaling down effect. The other approach can find its origin in ultra-precision machining, especially single point diamond turning (SPDT) with the emphasis on cutting mechanics, although the two approaches overlap in some areas and attempt to address similar issues, such as cutting tool edge size effect, minimum chip thickness, and so on.

The approach of miniaturization of the conventional cutting process tends to be process parameters centric. Macro-phenomena such as machining dynamics, chatters, cutting forces, and so on are directly translated into the micro-domain and the machine-tool interaction effect is well considered and modelled. Macro analytical, mechanistic and numerical cutting process models are adapted to micro cutting with consideration of the so called size effect.

The other approach utilizes research output from ultra-precision diamond cutting and tends to be cutting mechanics centric in nature. This approach is similar to diamond cutting research,

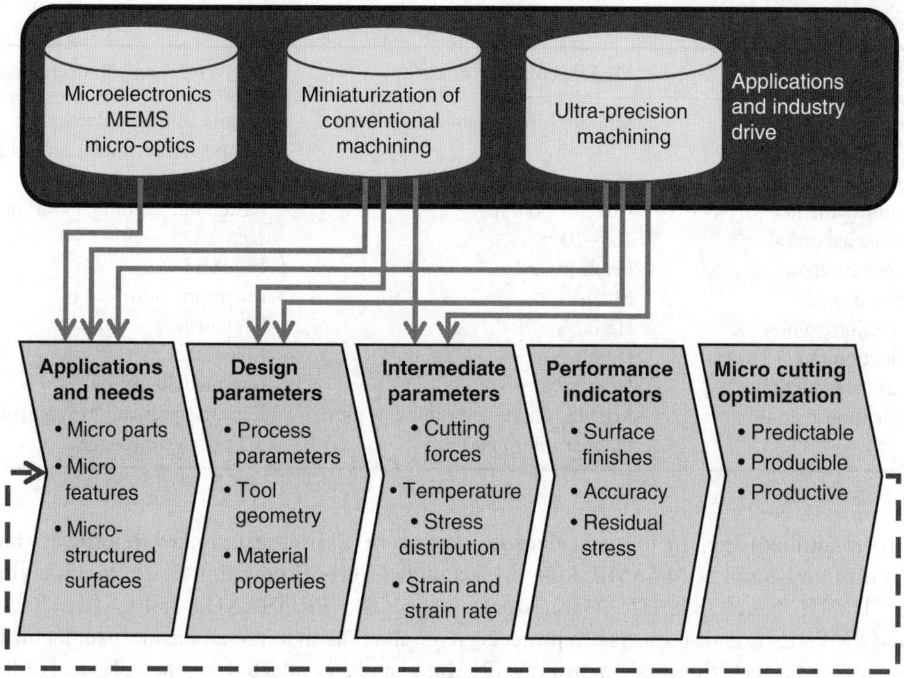

Figure 1.2 A typical micro cutting development process

but studies micro cutting, with more emphasis on tool geometries, material crystalline orientation and micro structures. Machine dynamics are often neglected as cutting forces are given very little consideration so that ultra-precision machines are treated as rigid and their effects do not appeared in the models. Atomic scale simulation or other numerical modelling considering micro structure and grain size effects are used for this approach and study.

As discussed above, traditionally MEMS and microelectronics use silicon materials-based micro manufacturing processes which are fundamentally different from mechanical micro cutting. With the increasing requirement on 3D complex shape MEMS devices, mechanical micro cutting will have great potential to fabricate micro parts for MEMS and microelectronics applications. On the other hand, hybrid micro manufacturing approaches, for example, a combination of micro cutting and etching processes to fabricate high precision 3D micro parts, is likely to be a promising method.

Figure 1.2 highlights a typical micro cutting development process, starting from applications and needs which come from microelectronics and MEMS, miniaturization of conventional machining and ultra-precision machining; a micro cutting development flow has a number of key stages. Design parameters including process parameters (e.g. cutting speed, uncut chip thickness, feedrate), tool geometry, material properties, and so on will be developed with the help of existing knowledge from both miniaturization of conventional machining and ultra-precision machining. Intermediate parameters such as cutting forces, temperature, stress distribution, and strain and strain rate are measured and analyzed during the micro cutting process. Depending on the applications, a number of performance indicators such as surface finishes, accuracy and residual stress can be chosen to evaluate the micro cutting performance towards the predictable, producible and highly productive manufacture of micro products.

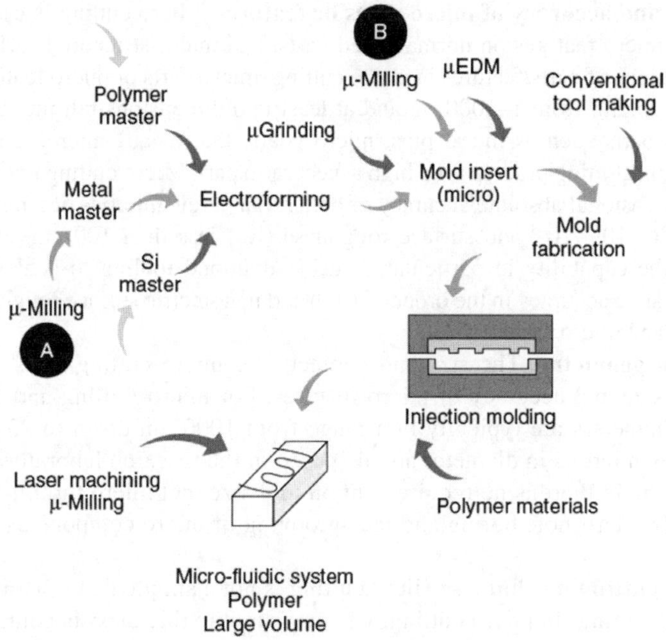

Figure 1.3 Micro cutting processes for micro-injection moulding. Reproduced with premission from [4]

Precision micro-structured surfaces or micro components are commonly directly machined by micro cutting, or through micro injection moulding or micro embossing with micro-cut micro moulds. Figure 1.3 shows that micro cutting, for example, micro milling, is used to fabricate micro moulds.

1.1.3 Definition and Scope of Micro Cutting

Micro cutting is kinematically similar to conventional cutting, but fundamentally different from conventional cutting in many aspects. It is important to define the scope and context of micro cutting, as the term may have different meanings for different people.

Micro cutting refers to mechanical micromachining (direct removal of materials) using geometrically defined cutter edge(s) carried out on conventional precision machines or micro machines. Micro cutting is normally used for machine high accuracy 3D components in a variety of engineering materials. A number of features can be used to characterize and define the scope of micro cutting as follows:

- **Uncut chip thickness**. Uncut chip thickness is the material layer being removed during the cutting process. Uncut chip thickness in micro cutting is different from that in conventional macro cutting. Masuzawa and Tonshoff [5] defined the micro-macro border as around 200 μm, while this border obviously changes according to the contemporary levels of conventional technologies. This borderline of uncut chip thickness decreases with advances in machining technologies. In the current state-of-the-art an uncut chip thickness less than tens of microns has been widely accepted by the micro machining community.

- **Dimensions and accuracy of micro parts or features**. Micro cutting is used to fabricate micro parts, micro features on normal-sized parts, and micro-structured surfaces. In terms of the dimensions of parts/features in micro cutting, micro parts or micro features must have dimensions ranging from 1–1000 μm and at least two dimensions fall into this range. For miniaturized parts such as micro pins, micro gears, that means micro cutting is a three dimensional machining process for a high aspect ratio part. Micro cutting normally achieves form and dimensional absolute accuracy of better than a few microns or a relative accuracy in the order of 10^{-3}–10^{-5} and surface roughness (Ra) less than 100 nm, although micro cutting has the capability in particular of using diamond tooling to achieve sub-micron accuracy (relative accuracy in the order of 10^{-6}) and nanometric surface roughness for micro components and micro structures.
- **Cutting tool geometry**. The size and geometry of micro cutting tools determine the limit of the size and accuracy of micro features. For micro milling and micro drilling tools, tool diameters are typically in a range from 1000 μm down to 25 μm, although tools of a few microns in diameter are also used in the research laboratories. For micro peripheral turning there is no requirement on tool size, but micro turning tools must be employed for micro-hole boring and face grooving of micro components with the high aspect ratio.
- **Underlying cutting mechanics**. Micro cutting is not a simple down scaling of conventional macro cutting. In micro cutting, when uncut chip thickness becomes comparable to the cutting edge radius of tools or grain size of workpiece materials, a number of critical issues, such as cutting edge radius effect, negative rake angle, tool-workpiece contact at the flank face, minimum chip thickness and micro structure effect, become prominent. These behaviours are known as size effects, which can influence underlying cutting mechanics in terms of micro cutting forces and specific cutting energy, chip formation process, surface generation, burr formation and tool wear mechanism. On the other hand, size scaling down of machine tools and cutting tools results in size effect on machining dynamics which in turn interacts with and affects cutting mechanics fundamentally.
- **Application area**. Micro cutting is capable of machining a broad range of engineering materials including metals, polymers, technical ceramics and composites, and also with achievable accuracy and surface roughness. Micro cutting has found applications in many areas requiring micro components.

Figure 1.4 shows some examples of high-accuracy micro components and micro structures manufactured by micro cutting. These examples illustrate that micro components having complex 3D geometries need to be made from a variety of materials and not just from silicon. Mechanical micro machining is an ideal method for producing complex 3D micro components with high accuracy.

1.1.4 Micro Cutting and Nanometric Cutting

There is no general agreement on the definition of nanometric cutting. But if the uncut chip thickness of mechanical cutting falls to the nanometric level, that is, less than tens of nanometers, the cutting process can be regarded as nanometric cutting. Some researchers have

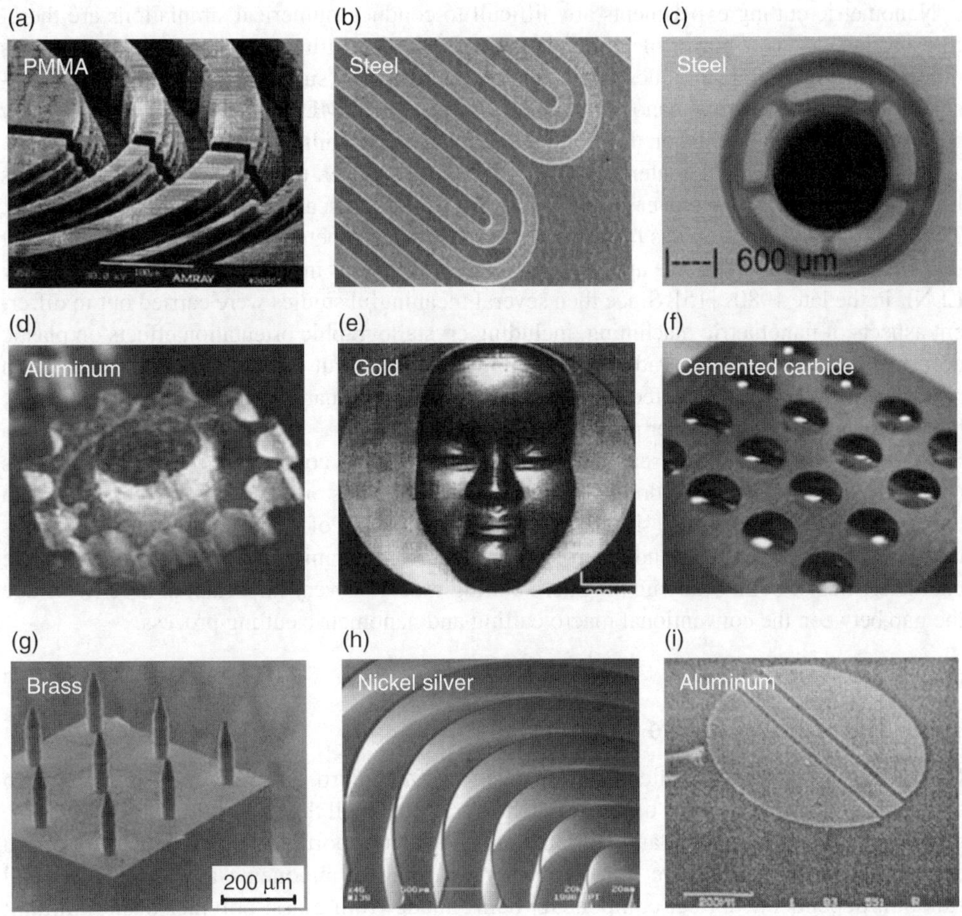

Figure 1.4 Examples of high accuracy micro components and micro structures by micro cutting. Reproduced from [6]: (a) Micro trenches. Reproduced with permissions from [7]; (b) Micro reactor [8]; (c) Micro mould. Reprinted from [9]. Copyright 2001 Elsevier; (d) Micro-gear. Reproduced with permission from [10]. Copyright 2004 ASME; (e) 3D micromachined part – Noh-mark (Fanuc). Images courtesy of FANUC; (f) Micro projection array (Fanuc). Images courtesy of FANUC; (g) Micro needles array. Reprinted from [11]. Copyright 2006 Elsevier; (h) Micro wall. Reproduced with permission from [12]; (i) Target foil for nuclear fusion. Reproduced with permission from [13]. Copyright 2001 EUSPEN

conducted ultra-precision machining experiments under extreme small uncut chip thickness, for example, less than 10nm. This can also be regarded as nanometric cutting. One of the promising applications using nanometric cutting is ductile mode cutting with nanometric level surface roughness and being free from cracks in brittle materials, such as semiconductor materials. But it should be noted that most nanometric cutting experiments were carried out under well controlled conditions in a laboratory environment and not many applications on an industrial scale have been found.

Nanometric cutting experiments are difficult to conduct, numerical simulations are therefore carried out as a powerful tool to study nanometric cutting processes. Among various numerical simulation techniques, molecular dynamics (MD) simulation has played a significant role in investigating nanometric cutting mechanics. MD simulation is an extremely accurate simulation method on the atomic scale and has the ability to fully describe the microstructural evolution of the material being processed. However, the simulation scale of MD is limited by computational power and so far even at the largest scale it can only reach a few μm^3. Therefore, MD simulation has been mainly applied to nanometric cutting where depth of cut is at the nanometric level. The application of MD simulations in machining was pioneered by LLNL in the late 1980s [15]. Since then several meaningful studies were carried out in different aspects of nanometric machining, including crystallographic orientation effects on plastic deformation [16], tool edge radius and minimum depth of cut effects on the chip formation mechanism [17], effects of defect structure in the workpiece material, diamond tool wear [18]; [1], subsurface deformed layer property [19], and so on.

Although the simulation scale of MD cannot directly cover micro cutting processes (typically, a few to a few hundreds of microns), these studies provide valuable base-line data and results for micro cutting simulations. The length scale of micro cutting in nature falls between nanometric cutting and macro cutting, therefore the micro cutting inherently has the characteristics of both. Studying the micro cutting process is very important in order to bridge the gap between the conventional macro cutting and nanometric cutting process.

1.2 Materials in Micro Cutting

One of the advantages of micro cutting over MEMS micro manufacturing is that micro cutting has fewer constraints on material choices. Almost all the material families – metals, polymers, glasses, ceramics and composites have been reported to be processed by micro cutting. As shown in Figure 1.4, materials for micro components are application and function dependent: optical components being made from glass, polymer or aluminium; medical engineering components from polymer or glass; mechanical components from ferrous or non-ferrous metals; and dies/moulds from copper alloys, aluminium or high-hardness steels. Some of the micro components and micro structures require sub-micron accuracy and nanometric surface finishes so diamond machinable materials are used to achieve the accuracy and surface requirement.

Although micro cutting uses the same range of materials as macro cutting, there are a number of material issues in micro cutting which is fundamentally different from macro cutting. These material issues affect micro cutting performance and hence research efforts have been broadened to investigate material behaviours at the micro scale.

Most engineering materials used are polycrystalline materials with typical grain size varying from between approximately 100 nm to 100 μm. When a micro part or feature decreases in relation to this size range, grains are actually equivalent to being either removed or refined. For most metals, mechanical properties are dominated by the presence and mobility of structural dislocations. As equivalent grain size is reduced the maximum spacing between a dislocation and a grain boundary is reduced, the ease of dislocation movement is influenced by any number of obstacles such as grain boundaries, defects and micro part/feature surfaces, and so on material strength is therefore increased. The changed material properties will in turn affect machinability of micro cutting.

On the other hand, the uncut chip thickness in micro cutting is usually in the same order as the material grain size, hence the workpiece material cannot be assumed as homogeneous and isotropic. Experiments on micro cutting of multiphase materials have shown significant varying cutting mechanisms and the associated process response [20], [21], [22].

Various material constitutive models have been employed to model material behaviours in micro cutting. Most of these material models address material behaviours such as strain hardening, strain rate sensitivity and thermal softening. Multiphase FE simulation models for micro cutting were also proposed to address the material size effect mentioned above [20], [22].

1.3 Micro Cutting Processes

Kinematically similar to conventional cutting, typical micro cutting processes include micro turning, micro milling, micro drilling and micro grinding (with shafts particularly). These four micro cutting processes vary in workpiece geometry, machining efficiency and achievable accuracy, although these cutting process mechanics share lots of common characteristics. Table 1.2 summarizes the geometric characteristics of the four micro cutting processes. Chapters 7–10 will discuss these micro cutting processes in detail.

Table 1.2 Geometric characteristics of typical micro cutting operations

	Micro turning	Micro milling	Micro drilling	Micro grinding
Workpiece Shape	Rotational convex shape with large aspect ratio, such as micro shafts, micro pins, etc.	3D shape both convex and concave with high aspect ratios and high geometric complexity	Round holes through or blind	Hard and brittle materials; 3D convex and concave shape using micro grinding tips
	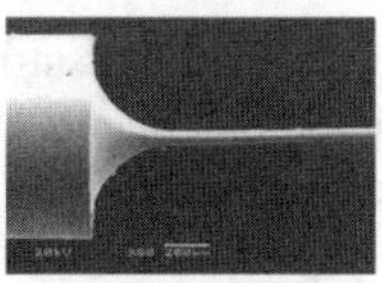	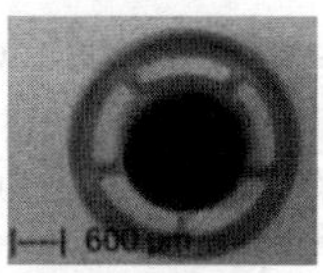	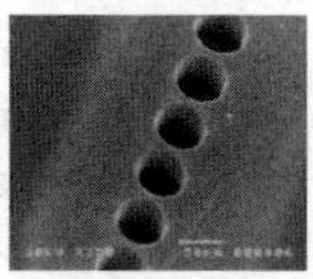	
Typical size	Down to $\phi 5$ μm, though 100 μm above more applicable	50 μm slots are practical applicable	$\phi 50$ μm holes are practical applicable	Micro structures down to 20 μm
Achievable surface roughness	0.1 μm Ra	Optical surface (<10 nm Ra) via diamond milling for non-ferrous materials	0.1 μm Ra	advantageous for brittle materials with optical surface finish (<10 nm Ra)
References	[25] [26]	[9]	[27]	[28] [13]

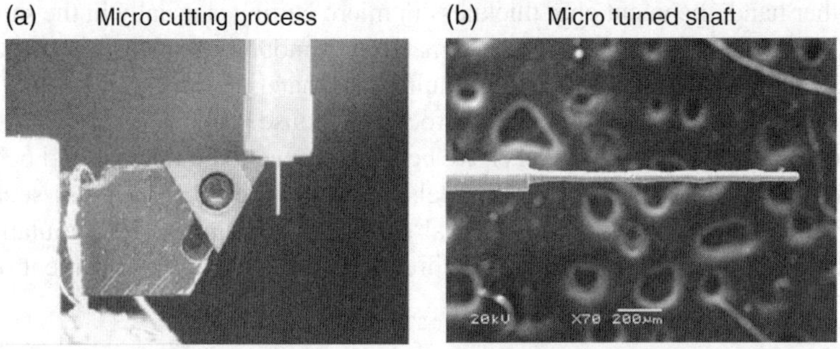

Figure 1.5 An example of micro-turned shaft (Reproduced from [23]). Reproduced with permission from [2]. Copyright 2007 Elsevier

1.3.1 Micro Turning

Micro turning is an effective way to produce micro cylindrical or rotational symmetry components. Figure 1.5 shows examples of a simple micro pin with the diameter of 33μm. A micro part with the high aspect ratio can be achieved using the micro turning [23]. The most serious problem encountered during micro turning is the cutting force which tends to bend the workpiece, and the machining force influences machining accuracy and the limit of machinable size [24]. A detailed analysis on how size effect influences micro part rigidity and deflection is provided in Chapter 7. Micro turning is performed on either a conventional precision machine or a micro turning system.

Diamond turning of the micro structured surface can be regarded as another group of micro cutting. With the aid of fast tool servos (FTS), complex micro structured surfaces can be generated by diamond turning.

1.3.2 Micro Milling

Micro milling is an emerging technology and is the most flexible and versatile micro cutting process. It is able to generate a wide variety of complex micro components and micro structures. In the past decade significant research has been carried out in micro milling modelling and experiments. Most of the micro components shown in Figure 1.3 were machined using micro milling technology.

Micro tooling is crucial to micro milling as it determines the feature size and also the surface roughness. Commercially available micro milling tools have the tool diameter ranging from 25–1,000 μm. Due to the limited rigidity of small diameter tools and difficulty in fabricating a micro tool, most of the micro milling tools have only two flutes, and some very small diameter tools (<100 μm), especially made from natural diamond or CVD, have only single flute or spade type tools. In terms of types of milling operations, micro end milling using either flat end or ball-nosed end mills dominates micro milling applications, and peripheral milling in macro milling is uncommon for micro milling. One of the challenges in micro milling is premature tool chipping and breakage. There are limited choices for micro tool fabrication. Coated micro grain tungsten carbide tools are widely employed, and natural

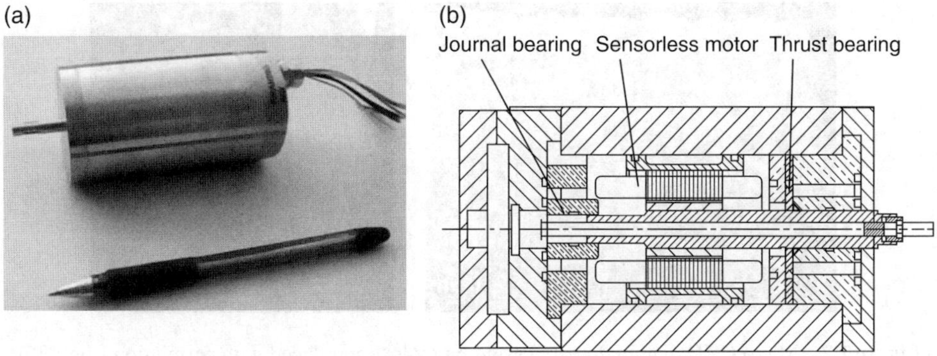

Figure 1.6 The ultrahigh speed aerostatic bearing spindle driven by a sensorless DC motor (a) photograph of the developed miniature spindle; (b) a schematic of the spindle

diamond or CVD micro milling tools are used in some applications requiring very tight tolerance and good surface finishes.

Although micro milling can be performed in a conventional CNC machining centre by retrofitting a high speed spindle, ideally micro milling should be performed in a precision milling machine or micro machine specially designed for micro milling purposes. Chapter 2 presents some industrial precision machine tools and miniature machine tools with micro milling capability. Small diameter micro milling tools require extremely high rotational speeds to achieve even modest machining rates and also a high stiffness spindle to maintain high accuracy in the presence of cutting forces. High machining accuracy also requires low spindle running temperatures to minimize thermal distortion while a fine surface finishing capability can only be achieved with a spindle having low motion errors. So precision high speed spindles with operating speeds of more than 100,000 rpm are commonly used. Figure 1.6 shows an ultra high speed aerostatic bearing spindle with an operating speed range of 20 000 to 200 000 rpm.

1.3.3 Micro Drilling

Drilling is a popular machining method to create a round hole in a part made from many materials. Although it shares many cutting mechanics with other cutting operations, micro drilling has not been researched to the same extent as micro turning and micro milling. This is because micro drilling tools have more complex geometry compared to milling and turning tools. Holes of 50 μm can be practically machined with commercial twist drills. Micro drills of less than 50 μm diameter are also available and normally of the spade type. One of the main applications of high speed micro drilling is printed circuit boards (PCBs) drilling. Micro drills of 50–300 μm in diameter are commonly used in PCB drilling production lines and a hole depth/diameter ratio up to 15 has been achieved [29].

Compared with micro milling, micro drilling is more efficient in creating holes and capable of machining deep holes, although micro drilling cannot machine flat-bottom holes because of the drilling point. Since a micro drill can easily be broken, sensitive torque feedback control

Figure 1.7 Micro CVD diamond-coated grinding tools. Reproduced with permission from [30]

is necessary. But usually a thrust force feedback is employed because of the difficulty in the direct measurement of the torque [5].

Micro drilling has a similar requirement on high speed spindles as micro milling, but speed control is not desirable as with a micro milling spindle. Aerostatic bearing or air turbine spindles with maximum speed more than 100 000 rpm are typically used to improve productivity.

1.3.4 Micro Grinding

Micro grinding has been an effective method to produce high dimensionally accurate parts with superior surface finishes. Due to its low material removal rate, micro grinding is normally used as the final production procedure. Unlike other micro cutting processes, such as micro turning and micro milling where ductile or less hard materials are usually used, micro grinding is capable of machining brittle and hard materials.

Similar to the micro turning operation, micro grinding can be performed using relatively large grinding wheels when the micro features do not require micro grinding tools. But the size and geometry of micro grinding tools determine the limit of the size and geometry of micro parts and micro features. Standard diamond abrasive tools are made by bonding diamond monocrystals, PCD or CVD onto a base body. Micro grinding tools have been fabricated by coating CVD diamond layers onto cemented carbides. Figure 1.7 shows a small CVD diamond abrasive pencil with the diameter of 50 to 100 μm.

1.4 Micro Cutting Framework

This section presents a framework for micro cutting with the aim of highlighting various micro cutting aspects in an integrated environment and how these aspects interact and related to each other. Figure 1.8 shows a representation of the micro cutting framework. Challenges and needs of miniaturization are always the main driving force to push micro cutting science and engineering forward. Existing challenges such as size effects and micro-machinability have raised research issues which are being addressed by the micro machining community. The market need for miniaturized and micro products or components with smaller dimensions/

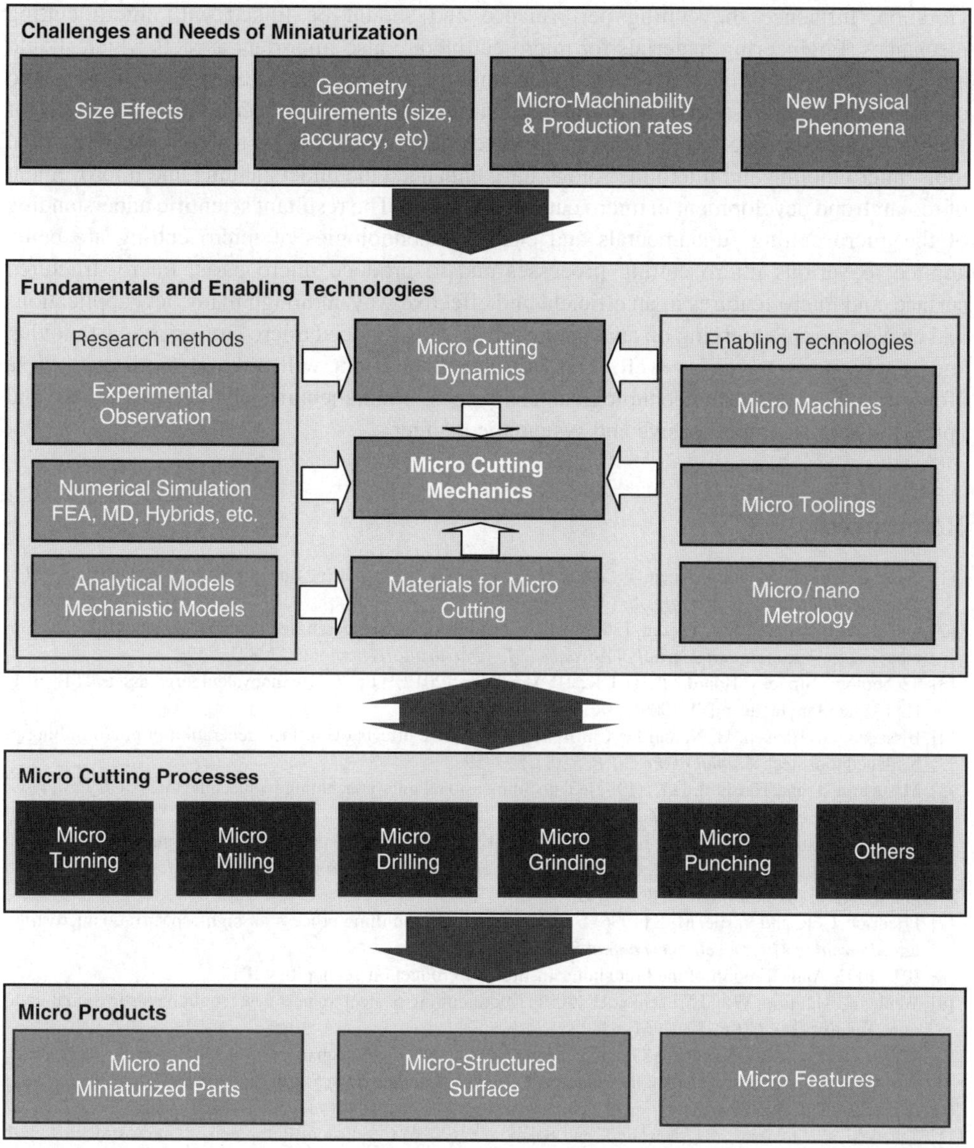

Figure 1.8 A framework for micro cutting

features and tighter tolerance sets the scope of micro cutting and drives the progress of this subject area. Micro-machinability and production rate of micro cutting determine if micro cutting is a feasible and favourable industrial method for a certain micro product.

Micro cutting mechanics are central to the micro cutting fundamentals. Similar to conventional macro cutting mechanics, issues like chip formation, cutting force, cutting temperature, tool wear, burr formation, surface generation, are being investigated, but in the micro domain. Micro cutting dynamics, including tool run-out, tool deflection, micro machining chatter and vibration, influence the cutting performance and should be linked with micro cutting mechanics. Engineering materials for micro cutting are also important aspects which should be taken into account in micro cutting mechanics. The available research methods in macro cutting, especially the analytical and numerical methods, become increasingly attractive for studies of micro cutting. On the other hand, developments on enabling technologies – machine tools, micro tooling and micro metrology have enhanced the understanding and improvement of research and development in micro cutting processes. The resultant scientific understanding of the micro cutting fundamentals and enabling technologies of micro cutting are being applied to various micro cutting processes and to produce micro parts, micro-structured surface, and micro features in an efficient and effective way, although many new applications and challenges are emerging on an almost daily basis, as indeed micro cutting is a fast moving and timely subject area as well. The subsequent chapters will attempt to discuss these aforementioned scientific/technological challenges, fundamentals, engineering issues and applications in a comprehensive and systematic manner.

References

[1] Chae, J., Park, S. S., and Freiheit, T. (2006) Investigation of micro-cutting operations. *International Journal of Machine Tools and Manufacture*, 46 (3–4), 313–332.

[2] Qin, Y., Brockett, A., Ma, Y. et al. (2009) Micro-manufacturing: research, technology outcomes and development issues, *International Journal of Advanced Manufacturing Technology*, 47, 821–837.

[3] Technology Strategy Board (TSB) UK, HVM report. (2012) http://www.innovateuk.org/_assets/TSB_IfM_HighValueManufacturingT12-009%20FINAL.pdf

[4] Bissacco, G., Hansen, H. N. and De Chiffre, L. (2006) Size effects on surface generation in micro milling of hardened tool steel. *Annual of the CIRP*, 55(1), 593–596.

[5] Masuzawa, T. and Tönshoff, H.K. (1997) 'Three-dimensional micromachining by machine tools'. *CIRP Annals – Manufacturing Technology*, 46(2), 621–628.

[6] Huo, D., Cheng, K. and Wardle, F. (2009) Design of a 5-axis ultraprecision micro milling machine – ultraMill: Part 1: Holistic design approach, design considerations, and specifications. *International Journal of Advanced Manufacturing Technology*, 47, 867–877.

[7] Friedrich, C. R. and Vasile, M. J. (1996) Development of micromilling process for high-aspect-ratio microstructure, *Journal of Microelectromechanical Systems*, 5(1), 33–38.

[8] IPT (2003) Annual report of the Fraunhofer Institute of Production Technology IPT.

[9] Weule, H., Hüntrup, V. and Tritschler, H. (2001) Micro-cutting of steel to meet new requirements in miniaturization, *Annals of the CIRP*, 50(1), 61–64.

[10] Liu, X., DeVor, R.E., Kapoor, S.G. and Ehmann, K.F. (2004) The mechanics of machining at the microscale: Assessment of the current state of the science, *Journal of Manufacturing Science and Engineering, Transactions of the ASME*, 126, 666–678.

[11] Takeuchi, Y., Suzukawa, H., Kawai, T. and Sakaida, Y. (2006) Creation of ultraprecision microstructures with high aspect ratio, *Annals of the CIRP*, 56(1), 107–110.

[12] Weck, M., Hennig, J. and Hilbing, R. (2001) Precision cutting processes for manufacturing of optical components, *Proceeding of SPIE*, 4440, 145–151.

[13] Brinksmeier, E., Riemer, O. and Stern, R. (2001) Machining of Precision Parts and Microstructures. Proceedings of the 10th International Conference on Precision Engineering (ICPE), Initiatives of Precision Engineering at the Beginning of a Millennium, Yokohama, Japan: S. 3–11.

[14] Vasile, M. J. and Friedrich, C. R. (1996), The micromilling process for high aspect ratio microstructures, *Microsystem technologies : sensors, actuators, systems integration* 2(3),144.

[15] Belak, J. and Stowers, I. F. (1990) A molecular dynamics model of the orthogonal cutting process. *Proceedings of ASPE Annual Conference*, 100–104.

[16] Komanduri, R., Chandrasekaran, N. and Raff, L.M. (2000) M.D. simulation of nanometric cutting of single crystal aluminum-effect of crystal orientation and direction of cutting, *Wear*, 242 (1–2), 60–88.

[17] Komanduri, R., Chandrasekaran, N. and Raff, L.M. (1998) Effect of tool geometry in nanometric cutting: A molecular dynamics simulation approach, *Wear*, 219 (1), 84–97.

[18] Cheng, K., Luo, X., Ward, R. and Holt, R. (2003) Modelling and simulation of the tool wear in nanometric cutting, *Wear*, 255, 1427–1432.

[19] Zhang, J.J., Sun, T., Yan, Y.D., Liang, Y.C. and Dong, S. (2008) Molecular dynamics simulation of subsurface deformed layers in AFM-based nanometric cutting process, *Applied Surface Science*, 254, 4774–4779.

[20] Abouridouane, M., Klocke, F., Lung, D. and Adams, O. (2012) A new 3D multiphase FE model for micro cutting ferritic–pearlitic carbon steels, *CIRP Annals – Manufacturing Technology*, 61, 71–74.

[21] Zhang, L., Wang, C., Yang, L., Song, Y. and Fu, L. (2012) Characteristics of chip formation in the micro-drilling of multi-material sheets. *International Journal of Machine Tools & Manufacture*, 52:40–49.

[22] Park, S., Kapoor, S. G., DeVor, R. E. (2004) Mechanistic cutting process calibration via microstructure level finite element simulation Model. *Transactions of ASME, Journal of Manufacturing Science and Engineering*, 126(4):706–709.

[23] Rahman, M., Lim, H.S., Neo, K.S., Kumar, A.S., Wong, Y.S. and Li, X.P. (2007) Tool-based nanofinishing and micromachining, *Journal of Materials Processing Technology*,185, 2–16.

[24] Masuzawa, T. (2000) State of the art of micromachining, *Annals of the CIRP*, 49, 473–488.

[25] Lu, Z.N. and Yoneyama, T. (1999) Micro cutting in the micro lathe turning system, *International Journal of Machine Tools and Manufacture*, 39, 1171–1183.

[26] Rahman, M., Asad, A.B.M.A., Masaki, T. et al. (2010) A multiprocess machine tool for compound micromachining, *International Journal of Machine Tools and Manufacture*, 50, 344–356.

[27] Egashira, K. and Mizutani, K. (2002) Micro-drilling of monocrystalline silicon using a cutting tool, *Precision Engineering*, 26, 263–268.

[28] Aurich, J.C., Engmann, J., Schueler, G.M. and Haberland, R. (2009) Micro grinding tool for manufacture of complex structures in brittle materials. *Annual of the CIRP*, 58(1), 311–314.

[29] Watanabe, H., Tsuzaka, H. and Masuda, M. (2008) Microdrilling for printed circuit boards (PCBs) – Influence of radial run-out of microdrills on hole quality, *Precision Engineering*, 32, 329–335.

[30] Gaebler, J. and Pleger, S. (2010) Precision and micro CVD diamond-coated grinding tools, *International Journal of Machine Tools & Manufacture*, 50, 420–424.

2

Micro Cutting Mechanics

Dehong Huo[1] and Kai Cheng[2]
[1] *School of Mechanical and Systems Engineering, Newcastle University*
[2] *School of Engineering and Design, Brunel University*

2.1 Introduction

Micro cutting is kinematically similar to conventional cutting. However, it is not a simple scaling of macro cutting. In micro cutting, when uncut chip thickness becomes comparable to the cutting edge radius of tools or grain size of workpiece materials, a number of critical issues, such as cutting edge radius effect, negative rake angle, tool-workpiece contact at the flank face, minimum chip thickness and micro structure effect, become prominent. These behaviours are categorized as size effects, which can influence underlying cutting mechanics in terms of micro cutting forces and specific cutting energy, chip formation process, surface generation, burr formation and tool wear mechanism. In addition to process and material size effects, micro cutting process performance is strongly based on its enabling technologies, namely, appropriate machine tools, micro tooling and measurement and inspection system.

Research has been carried out in micro cutting mechanics for decades and experimental studies still dominate the micro cutting research. Some analytical and numerical models for micro cutting have been developed based on conventional cutting models and some size effects have been incorporated into these models.

From the perspective of applications, some critical issues, such as excess tool wear, low stiffness of the micro tools, unpredictable tool failure, make micro cutting difficult-to-machine materials particularly challenging. Two scientific approaches are being employed to address these issues, namely, vibration assisted micro cutting and micro-scale laser-assisted milling. Figure 2.1 summarizes the micro cutting mechanics in context in an integrated environment and also provides an outline of the chapter.

Micro-Cutting: Fundamentals and Applications, First Edition. Edited by Kai Cheng and Dehong Huo.

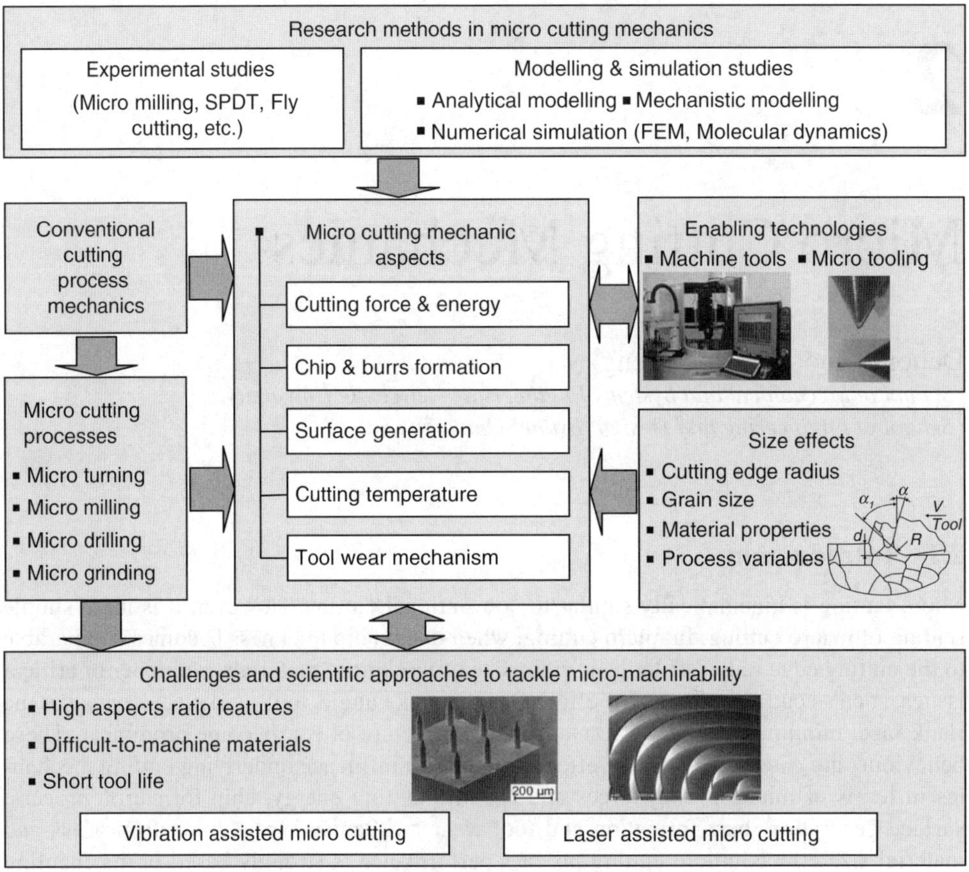

Figure 2.1 Overview of micro cutting mechanic aspects

2.2 Characterization of Micro Cutting

Micro cutting in this chapter refers to mechanical micro machining using a geometrically defined cutter edge, such as micro turning, micro milling, micro drilling and micro grinding. Micro cutting is normally used to machine high accuracy 3D components in various engineering materials with overall sizes or features sizes ranging from a few microns to a few millimetres. Micro cutting is different from conventional cutting in terms of uncut chip thickness.

High-accuracy mechanical miniature components with dimensions ranging from hundred microns to a few millimetres or features ranging from a few to a few hundred microns are increasingly in demand for various industries, such as aerospace, precision engineering, medical engineering, biotechnology, electronics, communications and optics, and so on. Special applications include fuel cells, micro fluidics, moulds for micro optics/lenses and fibre optic elements, micro nozzles, to name a few. Many applications require very tight tolerances and both functional and structural requirements demand the use of various engineering materials, including stainless steel, titanium, brass, aluminium, plastics, ceramics and composites [1]; [2, 3].

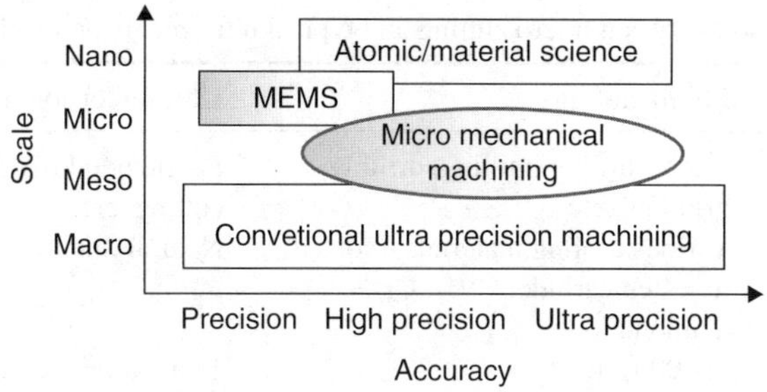

Figure 2.2 Dimensional size for the micro-mechanical machining. Reproduced with permission from [5]

It has long been recognized that traditional MEMS manufacturing techniques, such as chemical etching and LIGA, cannot match future demand-rate for micro components. This is because MEMS-based manufacturing methods are basically planar processes and directed towards semiconductor materials. The relative accuracies of MEMS-based methods are of the order of 10^{-1} to 10^{-2}, whereas the needs of many mechanical miniaturized components require relative accuracies in the order of 10^{-3} to 10^{-5} [4]. On the other hand, many micro component applications also require very high surface quality in terms of surface roughness and surface integrity. Micro cutting can meet the above requirements of producing 3D high accuracy micro components in a wide range of engineering materials. Figure 2.2 compares dimensional size and accuracy with other manufacturing methods.

2.2.1 Micro Cutting and Ultra-Precision Machining

When discussing the micro cutting process, researchers normally compare it with traditional ultra-precision machining. Sometimes distinctions and common characteristics of the two processes can be confusing. This is because:

1. Some micro cutting mechanics research finds its origins in ultra-precision machining studies, such as minimum chip thickness.
2. Micro cutting processes share some common characteristics with traditional ultra-precision machining, such as chip formation and specific cutting forces.
3. There are overlaps between the two processes in terms of the characteristic depth of cut and applications.

However there are many distinctions between the two processes in terms of machining process characters, component size, toolings, applications, and so on. Table 2.1 attempts to compare micro cutting with traditional ultra-precision machining processes in order to gain a better understanding of the scope and main features of micro cutting processes. Generally speaking, ultra-precision machining focuses on achieving highest possible dimensional accuracy and surface finish using diamond tooling regardless of component size, and applications are predominated by electro-optics. Micro cutting, on the other hand, focuses on mechanical micro machining 3D miniature components over various engineering materials using various tools with reasonably high accuracy and surface finish.

Table 2.1 A comparison between micro cutting and typical ultra-precision machining

	Micro cutting	Ultra-precision machining
Processes	Micro turning, milling, drilling, grinding, etc.	Single point diamond turning, fly cutting, etc.
Tooling	Various tooling materials: (coated) tungsten carbide, CVD, CBN, diamond tools	Natural diamond tools
Component size	1–1000 μm	1 mm above, Can be very large
Shape	3D shape with high aspect ratios and geometric complexity	Rotational parts, both spherical and aspheric surface, normally low aspect ratios.
Accuracy	Absolute: <10 μm Relative: 10^{-2}–10^{-5}	Absolute: <1 μm Relative: 10^{-5}–10^{-6}
Surface finish	<100 nm Ra	Typically <20 nm Ra
Machines	Precision machining centres, precision micro machines, ultra-precision turning machines	Ultra-precision turning machines
Applications	Various applications requiring micro components	Electro-optics
Depth of cut (uncut chip thickness)	1–10 μm	0.1 μm – 10 μm

2.2.2 Enabling Technologies for Micro Cutting

Performance of the micro cutting process is strongly dependant on its enabling technologies, namely, appropriate machine tools, micro tools and measurement and inspection system. This section will highlight requirements and state-of-the-art machine tools used for micro cutting, and latest micro tools development.

2.2.2.1 Machine tools

Most of the experimental research for micro cutting processes has been performed in two types of machines, that is, traditional (ultra) precision tuning machines or micro machining centre, and in-house desk-top micro machines or micro factories built by researchers.

The requirements of micro component manufacture over a range of applications are: high dimensional precision, typically better than a few microns; accurate geometrical form, typically better than 100 nm departure from flatness or roundness; and good surface finish, in the range of 10 – 100 nm Ra. These in turn require machine tools to have high static stiffness, low thermal distortion, low motion errors and high damping or dynamic stiffness.

There are a number of industrial ultra-precision turning and milling machines available for precision components manufacture. Most of them are generally aimed at the optical components market and are not well suited to the manufacture of precision micro components due to high investment costs and lack of flexibility. Figure 2.3 shows some examples of industrial ultra-precision machines with micro cutting capability. They fall into two

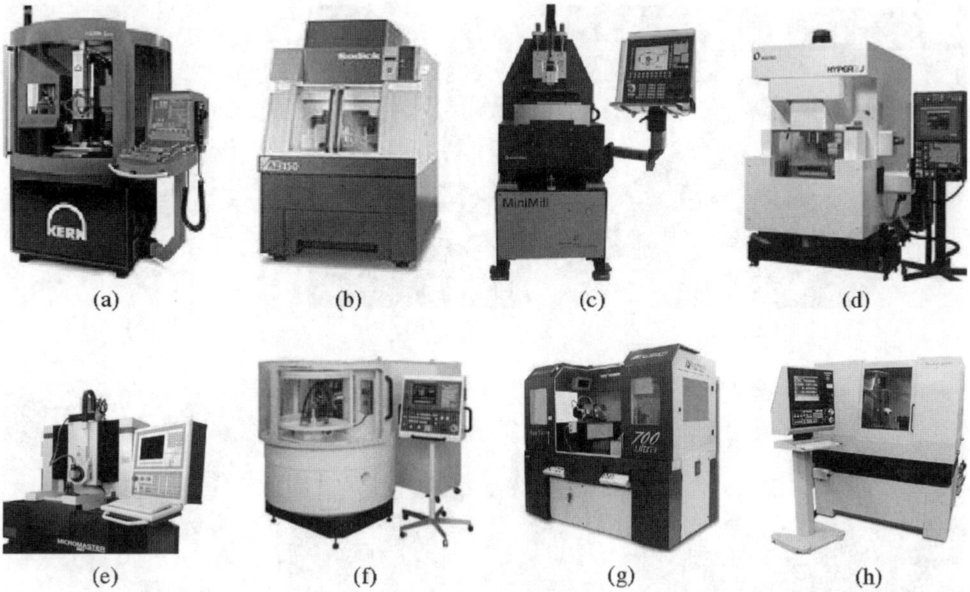

Figure 2.3 Industrial precision machine tools with micro cutting capability: (Reproduced from [2]) (a) Kern micro. Copyright KERN Microtechnik GmbH; (b) Sodick AZ150. Image courtesy of Sodick, Inc.; (c) Fraunhofer IPT Minimill. Reproduced from [6]; (d) Makino Hyper2J. Image courtesy of Makino; (e) Kuglar MicroMaster MM2. Image courtesy of Kugler GmbH; (f) Fanuc ROBOnano. Images courtesy of FANUC; (g) Precitech freeform 700 Ultra. Image courtesy of Ametek Precitech; (h) Moore Nanotech 350FG. Copyright Moore Nanotech

categories. One is conventional ultra-precision machine tools which are designed as diamond turning machine tools with add-on Z-axis, rotary table and a second high speed milling or grinding spindle. Typical examples are Moore Nanotechnoloy Nanotech 350FG and Precitech Freeform 700 ultra as shown in Figure 2.3g and h. Both of them require 5–7 m2 floor space. Their very high cost and low flexibility limit their application to micro components of simple geometries and high added value, such as optical components. Another type of industrial precision micro milling machine tool has emerged in the last decade. A typical example is the Kern micro machine (Figure 2.3a) which meets many applications but still suffers from its machining accuracy for precision micro machining due to the positioning accuracy.

Smaller machine tools are less affected by environmental fluctuations such as changes in temperature, pressure and humidity as compared to their macro-scale counterpart. And the reduced mass of the miniature machine tool reduces the inertia force required to drive the machine tool system, thus consuming less energy and yet providing higher positioning accuracy. Hence, the use of miniature machine tools is also seen as having immense potential in reducing production costs [7].

Numerous research efforts to develop in-house miniaturized machines or micro factories have been undertaken for the manufacture of precision micro components [7]; [8]; [9].

Figure 2.4 Examples of miniature machine tools and micro factories. (Reproduced from [2]) (a) Reproduced with permission from [10]; (b) Reproduced from [11]. Copyright 2005 Spriger; (c) Reproduced from [12]. Copyright 2006 ASME; (d) Reproduced from [13]. Copyright 2008 Springer; (e) Reproduced with permission from [10]; (f) Reproduced with permission from [8]. Copyright 2002 IOP Publishing

Figure 2.4 shows some examples of a miniature machine tool. However, most of them are still at the research stage, and only a few of them have so far found their way into industrial applications, and their application to high accuracy and fine surface quality are still constrained by low static/dynamic stiffness.

2.2.2.2 Micro cutting tools

Micro cutting tooling is another enabling technology for micro cutting, since micro cutting needs to utilize micro tooling to enable micro components and features. Geometry of micro cutting tools is the same or similar to micro components/features to be machined. Cutting tool geometry and material properties have significant influence on chip formation, heat generation, tool wear, surface generation, and so on. In micro cutting the uncut chip thickness and the tool edge dimension are in the same order of magnitude (cutting edge size effects will be discussed in the next section), which requires cutting edges which

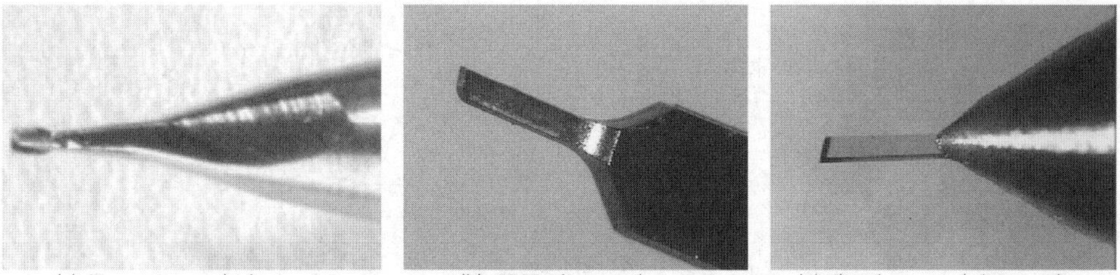

(a) Tungsten carbide (b) CVD diamond (c) Single crystal diamond

Figure 2.5 Micro milling tools (Courtesy of Contour Fine Tooling Ltd). Reproduced with permission from Contour Fine Tooling Ltd.

can withstand high mechanical and thermal stresses, and hence wear resistance, for a prolonged machining time.

Single crystal diamond has been used predominately in ultra-precision machining due to its matchless hardness. Single crystal diamond micro tools have been fabricated to meet the high requirements on accuracy and surface finishes in micro cutting for certain applications. Very little research has been reported on micro cutting performance and tool wear issues using a diamond micro cutter. One the other hand, as diamond has a very high affinity to iron, diamond micro cutting is limited to the machining of non-ferrous materials such as aluminium and copper. Tungsten carbide micro tools are widely used in micro cutting processes. Tungsten carbide has high strength and hardness. In order to increase the wear resistance and hardness, very fine grain size tungsten carbide is fabricated for micro tools and various coated micro tools, such as diamond and titanium coated tools, are becoming popular in micro cutting. Figure 2.5 shows some examples of micro milling tools.

2.3 Micro Cutting Mechanics

Conventional macro cutting mechanics have been researched in the past several decades. Cutting experiments were conducted and various models were developed to understand the mechanics of the cutting process. Most analytical, mechanistic and numerical models have been focused on cutting force prediction and chip formation process. Most conventional cutting models which are based on assumptions such as relative sharp tool edge radius and homogeneous materials cannot be directly applied to the micro cutting process. Micro cutting is not simply scaling down of macro cutting although they are likely kinematically similar. There are a number of critical issues which have arisen from micro cutting which are fundamentally different from conventional machining, that is, material micro structure, micromachined surface generation, subsurface character, and so on. The size effect plays a significant role in explaining these different phenomena and investigating micro mechanics. The following section discusses various size effects and their influences on micro cutting mechanics.

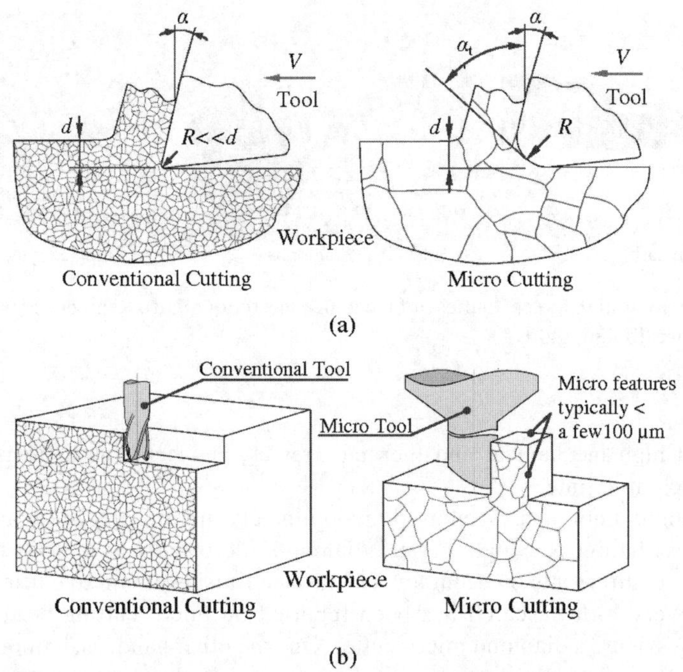

Figure 2.6 (a) Size effects resulting from small ratio of the uncut chip thickness to tool edge radius and small ratio of grain size to tool edge radius. (b) Size effect arising when micro features and grain size are comparable in size

2.3.1 Size Effects

When the cutting parameters, such as depth of cut or the uncut chip thickness, is on the same order of amplitude as the tool parameters such as tool edge radius, the effective rake angle may become highly negative. Size effects resulting from the small ratio of uncut chip thickness to tool edge radius will be a dominant factor for material removal mechanism and chip generation physics in micro cutting. Cutting, ploughing, or slipping phenomenon will occur predominated by this ratio, and eventually influences cutting processes such as surface finishes and surface integrity. Figure 2.6a illustrates the *cutting edge radius size effec*t. In previous ultra-precision machining studies, numerous researchers have investigated this size effect in terms of minimum chip thickness [14] and specific cutting energy [15], ductile mode machining [16], surface generation and burr formation [17]. These researches provided valuable knowledge for investigation of micro cutting.

When the micro structure of workpiece material such as grain size is comparable in size to the tool edge radius and depths of cut, the *micro structure size effect* occurs. Typical grain size of engineering materials varies between 100 nm to 100 μm, while the uncut chip thickness may be a few microns, therefore the chip formation may take place inside the individual grain. The materials in micro cutting cannot be treated as isotropic and homogeneous like in conventional macro cutting. It also differs from conventional cutting of single crystal material since the grain boundary effects cannot be neglected at this size scale. The micro structure size effect will affect the micro cutting processes, including mechanism of chip formation [18], surface

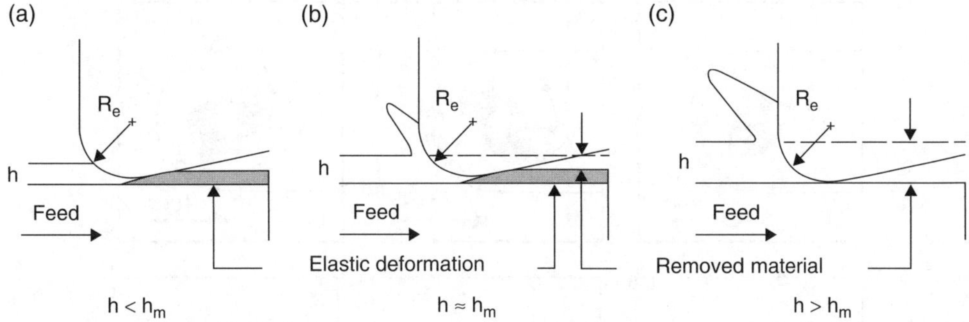

Figure 2.7 Schematic of the effect of the minimum chip thickness. Reproduced with permission from [5]. Copyright 2006 Elsevier

generation [19], and variation of the cutting forces [20]. It is also reported that it is increasingly difficult to control surface finish when micro cutting very small features due to grain size effect [4]. Figure 2.6a illustrates the micro structure size effect associated with micro cutting.

Mechanical property size effect arises where the ratio of a part's minimum feature size to the grain size approach to unity or where the surface area to volume ratio of micro components/features increases. From a material mechanical properties point of view, size effect starts to play a dominant role when one or more of the structural dimensions or features approach down to the scale of a few microns or less. At such a size scale, although there is no significant effect on the elastic properties of the materials since they depend on the bonding nature between the constituent atoms, the onset of plastic deformation strongly depends on dislocation mobility under an induced stress [21]. The ease of their movement can be influenced by any number of obstacles such as grain boundaries, defects and surfaces, and so on. Size effects then begin to govern the plastic behaviour by creating geometrical constraints, which obstruct dislocations to initiate and slide. The changed material properties will in turn affect machinability of micro cutting. Figure 2.6b illustrates the mechanical property size effects which have occurred in micro cutting miniature parts or features.

2.3.2 Chip Formation and Minimum Chip Thickness

In micro cutting, chips may not form when the uncut chip thickness is less than a minimum chip thickness. Instead of a cutting process, rubbing and ploughing phenomena may happen intermittently until the uncut chip thickness exceeds the minimum chip thickness. There are numerous researches being carried out to address chip formation mechanics in micro cutting and quantitatively determine minimum chip thickness for various materials and cutting conditions. Minimum chip thickness research originates in ultra-precision machining research. Ikawa [14] defined the minimum chip thickness in metal cutting as the minimum uncut chip thickness that can be removed stably from a work surface at a cutting edge under the perfect performance of a machine tool.

Figure 2.7 shows the schematic of the effect of the minimum chip thickness [5]. When the uncut chip thickness, h, is less than the minimum chip thickness, h_m, rubbing phenomenon

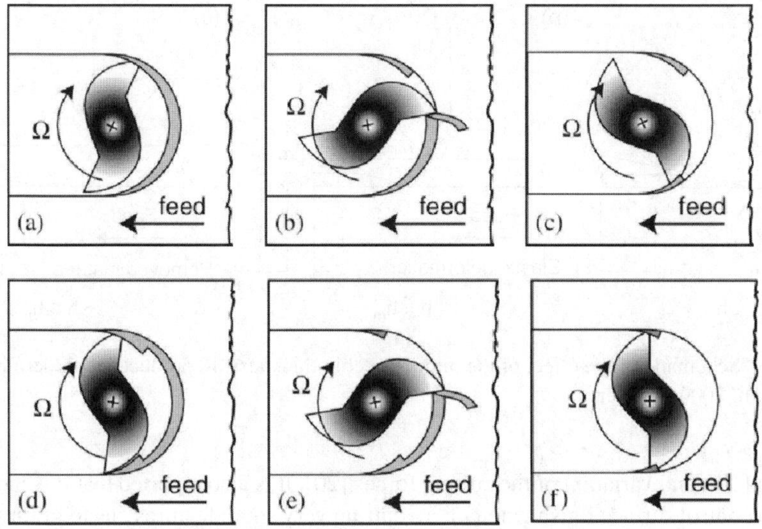

Figure 2.8 Micro milling with minimum chip thickness effect: (a)–(c) first pass, (d)–(f) second pass. Reproduced with permission from [22]. Copyright 2007 Elsevier

occurs. Only elastic deformation occurs and no chip is being formed (Figure 2.7a). As the uncut chip thickness approaches the minimum chip thickness, ploughing phenomenon occurs. Chips are formed with elastic deformation still occurring (Figure 2.7b). When the uncut chip thickness exceeds the minimum chip thickness the elastic deformation decreases and the entire uncut chip is removed (Figure 2.7c).

It is believed that the minimum chip thickness phenomenon is related to the resistivity of materials against plastic deformation, and it is strongly dependant on (a) the ratio of uncut chip thickness to cutting tool edge radius (cutting tool sharpness); (b) workpiece material properties, with more ductile materials exhibiting a higher minimum chip thickness, and (c) friction coefficient between cutting tool and workpiece material.

The ratios of the minimum chip thickness to the edge radius have been reported to be between 5–40% [23], [14], [24], [25], [17]. Shimada conducted molecular dynamic simulation to determine the achievable ultimate accuracy and found that the minimum chip thickness is around 5% of the cutting edge radius for copper and aluminium [23]. Ikawa *et al.* [14] conducted ultra-precision face turning experiments using electroplated copper material and well-defined diamond tools [14], and found out the minimum chip thickness is in the order of 1/10 of the cutting edge radius. Some ultra-precision experiments were conducted using well-defined diamond tools with edge radius of around 10 nm and machining system deflection can be neglected, the ratios of the minimum chip thickness to the edge radius may not be applicable to micro cutting directly. Weule conducted micro milling experiments using an ultra-fine tungsten carbide tool with 5 μm radius and estimated the minimum chip thickness to the edge radius ratio to be 0.293 [24].

Filiz [22] provided a graphical description of the effect of minimum chip thickness in micro slot milling as shown in Figure 2.8. The instantaneous uncut chip thickness during slot milling varies from zero at the initial engagement, to a maximum value (equal to the feed per flute) at approximately the centre of the channel, and back to zero during exit from the cut. A chip is formed or not formed depending on if the instantaneous chip thickness exceeds the minimum chip thickness.

2.3.3 *Specific Cutting Energy and Micro Cutting Force Modelling*

2.3.3.1 Specific cutting energy

Specific cutting energy can be defined as the energy consumed in removing a unit volume of material. Without loss of generality, consider the forces acting on the tool and workpiece during orthogonal cutting. The forces can be separated into two mutually perpendicular components which can be directly measured by a dynamometer:

- Cutting force *Fc*. This force is in the direction of cutting, and same direction as the cutting speed v.
- Thrust force *Ft*. This force is perpendicular to the cutting force.

The specific cutting energy, *E*, is calculated by dividing the cutting force by the chip area:

$$E = \frac{P}{MRR} = \frac{F_c v}{vbd} = \frac{F_c}{bd}\left(\text{in J/mm}^3, \text{N/mm}^2 \text{or MPa}\right) \tag{2.1}$$

Where P is the rate at which energy is consumed in the cutting process in N · mm/s; *MRR* is the material removal rate, in mm^3/s; v is the cutting speed in mm/s; b is the width of cut in mm; d is the uncut chip thickness. The specific cutting energy is a measure of the amount of work (energy) needed to remove a unit amount of material. For a constant width of cut b, cutting force F_c is normally proportional to uncut chip thickness, d. The smaller the uncut chip thickness, proportionally smaller is the cutting force, the specific cutting energy is therefore expected to be constant for a specific material. However both experiments and simulations have observed the phenomenon of the non-linear increase of the specific cutting energy with decreasing the uncut chip thickness. Figure 2.9 shows an example of specific cutting energy at different uncut chip thickness in micro milling. Similar non-linear increase of specific cutting energy has been found in other machining processes such as turning [26] and grinding [27, 28].

In the past few decades, researchers have been conducting experiments and simulations to explain why more energy and cutting forces are needed to remove a unit amount of material in cutting a smaller uncut chip thickness. These explanations can be summarized into two categories:

1. **Material strengthening behaviours and subsurface plastic work**. Material strengthening occurs because of the increase of strain rate and temperature-dependent flow stress of the work material in the primary deformation zone when uncut chip thickness decreases [29]; [30] or lack of defects at micro scale [31]. Shaw attributes the origin of the specific cutting energy size effect to short-range inhomogeneities present in all commercial

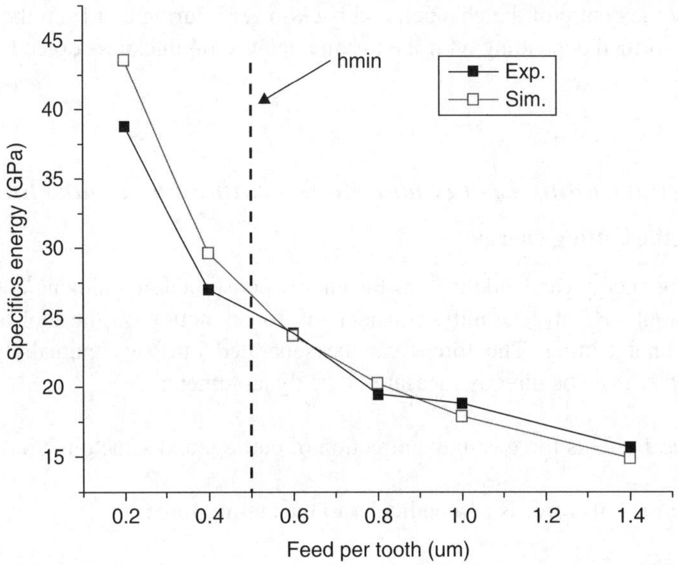

Figure 2.9 Specific cutting energy at different uncut chip thickness in micro milling. Reproduced with permission from [34]. Copyright 2008 Elsevier

engineering materials [28]. Surface integrity in terms of subsurface metallurgical and microstructural changes leads to material strengthening and hence the specific cutting energy size effect [32], [31].

2. **Edge radius size effect**. When the cutting parameters, such as depth of cut or the chip thickness, is in the same order as or even smaller than tool edge radius, the effective rake angle may become negative and contribute to the specific cutting energy. Research on ultra-precision machining reveals that there is a transition from cutting to ploughing due to edge radius size effect and the rake angle has a significant influence on the specific cutting energy at small depths of cut [15, 33]. The specific cutting energy is also related to the minimum chip thickness. When the uncut chip thickness approaches or small then the minimum chip thickness, the specific cutting energy increases exponentially with the decreasing of the uncut chip thickness [34]. Taminiau and Dautzenberg also found in their micro cutting experiments that the specific cutting forces depend on the ratio of the uncut chip thickness to the cutting edge radius when the uncut chip thickness was smaller than the edge radius [35].

2.3.3.2 Micro cutting force modelling

Modelling micro cutting processes, especially predicting micro cutting forces, play an important role in characterizing the micro cutting mechanics. For conventional machining there are several modelling methods available, Ehmann *et al*. summarized these models into three categories: analytical methods, experimental methods and numerical methods [36]. Modelling micro cutting process analytically is still at its early stage, only a few effective models are available. Efforts on analytical micro cutting force modelling are discussed in this section.

There are two major approaches to modelling cutting forces analytically. The first approach calculates cutting forces using the instantaneous uncut chip cross section, examples are Tlusty and MacNeil's cutting force model for end milling [37] and Armarego's unified mechanics of cutting model [38]. The second approach is based on a slip line field model which calculates cutting forces using the shear plane area.

A conventional basic cutting force model based on the instantaneous uncut chip cross section models the cutting process by a proportional relation:

$$F = k_m bh \tag{2.2}$$

Where h is the instantaneous uncut chip thickness variation, b is the width of cut and k_m represents the cutting coefficient. The basic linear cutting model can be refined in Armarego's unified mechanics of cutting model by adding a term proportional to the width of cut through the edge coefficient K_e which represents ploughing at the cutting edge:

$$F = \left(K_m h + K_e\right) b \tag{2.3}$$

Slip line field model is another widely used method to calculate cutting forces. Numerous slip line field models have been proposed in the past six decades [39]. More recently, Waldort proposed a slip line model with emphasis on modelling the ploughing components of the cutting force. According to the model, cutting forces are functions of the shear plane area, the flow stress, the shear angle, the rake angle, the friction factor and the prow angle. Fang proposed a comprehensive slip line model with round edge tool attempting to integrate previous slip line model into a unified analytical model and it is promising in how it addresses cutting edge radius size effect. Detailed descriptions of the two models and their parameters can be found in [40] and [39].

Most of the micro cutting force models are adapted from the two conventional approaches mentioned above but take one or more size effects into consideration in the model. The size effects which have been modelled to predict micro cutting forces include ratio of feedrate to tool radius; cutter edge radius [41], [42]; minimum chip thickness [17]; micro structure effect [10]; and so on.

Ratio of feedrate to tool radius

In order to keep the production rate reasonable, in micro cutting feedrate to tool radius ratio (f_t/r) has to be chosen much higher than conventional cutting. Modelling of uncut chip thickness by considering the cutting feedrate to tool radius ratio becomes important in accurately predicting cutting forces in micro cutting. Bao and Tansel[43] developed an analytical model to predict cutting forces which is based on Tlusty and MacNeil's model. The model is based on geometrical calculation of the chip load by considering the actual trajectory of the tool tip. Zaman et al. advanced Bao and Tansel's micro cutting model by using a theoretical chip area instead of uncut chip thickness [44]. But both models have not taken the tool edge radius size effect into account.

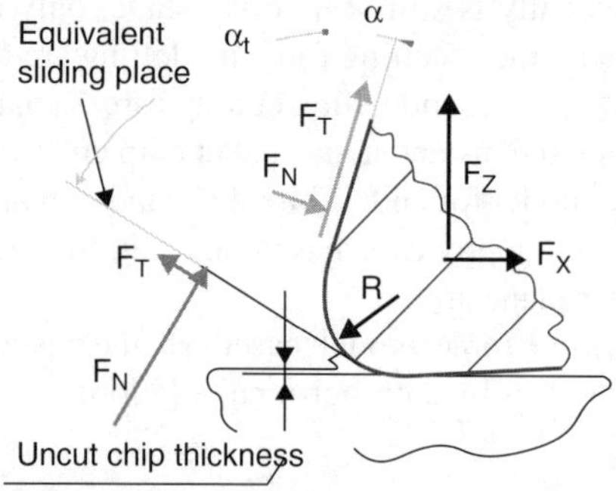

Figure 2.10 Equivalent sliding plane and rake angle. Reproduced from [41]. Copyright 2008 Elsevier

Cutting edge radius effect

The cutting edge radius size effect in a micro cutting force model can be modelled by determining an equivalent sliding plane and calculating an equivalent rake angle. The equivalent rake angle in micro cutting can be defined as the plane tangent to the edge radius at a distance from the machined surface corresponding to the uncut chip thickness as shown in Figure 2.10. There is a limited value of the uncut chip thickness h_{lim}, above which chip sliding occurs on the nominal rake face. The equivalent rake α_t is expressed by:

$$\alpha_t = \arcsin\left(\frac{h}{r_\varepsilon} - 1\right) \quad \text{for } h < h_{\text{lim}}$$

$$\alpha_t = \alpha \quad \text{for } h > h_{\text{lim}} \tag{2.4}$$

Where h_{lim} is given by $h_{lim} = r_\varepsilon(1 + \sin\alpha)$; α is the nominal rake angle; and r_ε is the cutter edge radius. The calculated equivalent rake α_t is then implemented in unified mechanics of cutting model to predict micro cutting forces.

Another size effect issue related to cutting edge radius in modelling micro cutting force is the friction force at the flank surface. When elastic recovery occurs between the flank face and workpiece, a friction force are caused by ploughing or sliding of tool-workpiece contact. This friction force normally accounts for a small amount of total cutting force in macro cutting where the effect of the tool edge radius can be ignored. But in micro cutting this friction force may contribute significantly to the cutting forces and cannot be ignored. The tool-workpiece contact length L_f in the flank face can be obtained by [42]:

$$L_f = \frac{S}{\sin\theta_f} = \frac{K_1 r_\varepsilon H}{E\sin\theta_f} \tag{2.5}$$

Where S is the springback; K_l is a constant, r_ε is the tool edge radius, H and E are hardness and material elastic modulus, and θ_f is the relief angle of the tool.

The contact friction force can be obtained by the tool-workpiece contact length L_f and material yield strength. The force components in cutting and thrust directions can be expressed by:

$$F_{fc} = \frac{CY}{\sqrt{3}} L_f b$$

$$F_{ft} = CYL_f b \tag{2.6}$$

Where C is the constant, Y is the material Yield strength, and b is the width of cut. The contact friction force is then added to the cutting forces calculated from the conventional model.

Minimum chip thickness

Minimum chip thickness effect needs to be incorporated into the micro cutting force model. Vogler *et al.* [17] incorporate the minimum chip thickness and use an equivalent/average rake angle in a slip-line force model. The modelling approach separates the effects due to chip removal and ploughing using a slip line field model and elastic deformation force model. When the uncut chip thickness is greater than the minimum chip thickness a slip line force model is used to predict cutting force. An average rake angle is calculated to quantify the edge radius effect. When the uncut chip thickness is less than the minimum chip thickness, that is, no chip is formed, an elastic deformation force model is employed to predict cutting force. The elastic deformation force is modelled as proportional to the volume of interference between the tool and the workpiece.

2.3.4 Surface Generation and Burr Formation

Machined surface resulting from micro cutting has a significant influence on micro product performance. It is believed that micro components structure integrity, dynamics performances and electrostatic characteristics strongly depend on the achieved micromachined surfaces parameters in terms of surface roughness and surface integrity. Therefore quantitatively characterization of the micromachined surfaces and then optimization of machining parameters are much desired. Much research has been conducted to investigate surface generation and burr formation using both experimental and modelling approaches.

Similar to specific cutting force and energy, in micro cutting surface generation has been found to be strongly influenced by size effects. Normally the smaller uncut chip thickness results in smaller surface roughness value, however, many researchers have observed in some micro cutting research that as the uncut chip thickness is reduced to a certain value, the surface roughness starts to increase. There exists an optimal feedrate which produces the smallest surface roughness and is related to the minimum chip thickness.

Micro burr formation has been studied by Lee and Dornfeld. They have conducted various micro milling experiments to address different micro burr formation types, and size of micro burrs are found to be proportionally bigger than that in conventional milling processes considering the ratio of burr size to chip load. Large tool edge radius to the uncut chip thickness causes rubbing and ploughing instead of cutting generates more burrs [45]. Due to the complexity of its mechanism, no accurate analytical model is available yet to predict micro burr

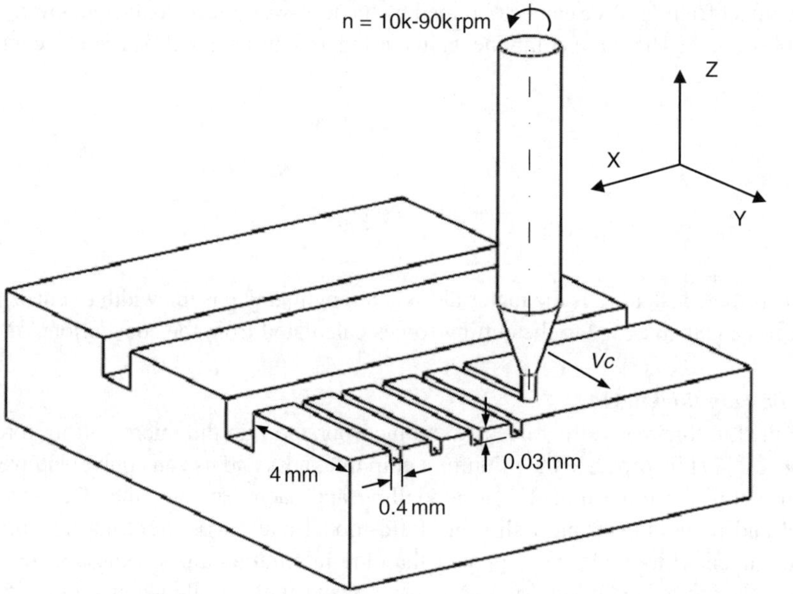

Figure 2.11 Schematic of the micro milling experiment

formation quantitatively. On the other hand, in the micro cutting process, the burr is usually very difficult to remove and, more importantly, burr removal can seriously damage the workpiece. Conventional deburring operations cannot be easily applied to micro burrs due to the small size of parts [46]. It has been found that micro burr height is strongly affected by cutting parameters, such as feedrate and the burr size is related to the amount of tool wear. Excess burr formation can be used as an indicator of tool wear and tool life.

The rest of this section describes some experimental investigations carried out by authors on micro milling of OFHC copper using tungsten carbide, CVD and single crystal diamond micro tools.

The experiments in this work include full immersion slot milling as illustrated in Figure 2.11. For each test, a micro slot of 4 mm long and 0.4 mm wide was milled along the Y direction. OFHC copper was chosen as the workpiece material in this work because it has excellent machinability for both WC and diamond tools, and it is widely used in optics and mould industries. Before experiments, the top surface of the workpiece was prepared using a 2 mm end-mill. A small axial depth of cut of 5 μm at each pass was used for the surface preparation to eliminate surface damage. A machine vision system (InfiniStix™ microscope) is integrated into the machine tool providing 3 μm resolution in Z direction to assist positioning the cutter to the workpiece surface.

The feed per tooth, *Fz*, varied at 0.1 μm, 0.2 μm, 0.5 μm, 1 μm, 2 μm, 3 μm and 4 μm, respectively at a constant axial depth of cut of 30 μm. The wide range of feedrate was chosen to ensure that minimum chip thickness for all three cutters and workpiece combinations fall into this range so that the size effect due to the minimum chip thickness can be investigated. The spindle speed, n is varied, at 10 000, 30 000, 60 000 and 90 000 rpm respectively during these experiments. Effect of axial depth of cut, d, is also studied by varying its value from 10 μm to

Table 2.2 The experimental matrix for micro milling using WC, CVD and single diamond milling tools

No	Depth of Cut, *d* (mm)	Spindle speed, *n* (rpm)	Cutting speed, *Vc* (m/min)	Feedrate, *Fz* (μm/flute)	No	Depth of Cut, *d* (mm)	Spindle speed, *n* (rpm)	Cutting speed, *Vc* (m/min)	Feedrate, *Fz* (μm/flute)
1	0.03	10000	12.56	0.1	21	0.03	60000	75.36	4
2	0.03	10000	12.56	0.2	22	0.03	90000	113.04	0.1
3	0.03	10000	12.56	0.5	23	0.03	90000	113.04	0.2
4	0.03	10000	12.56	1	24	0.03	90000	113.04	0.5
5	0.03	10000	12.56	2	25	0.03	90000	113.04	1
6	0.03	10000	12.56	3	26	0.03	90000	113.04	2
7	0.03	10000	12.56	4	27	0.03	90000	113.04	3
8	0.03	30000	37.68	0.1	28	0.03	90000	113.04	4
9	0.03	30000	37.68	0.2	29	0.01	60000	75.36	0.5
10	0.03	30000	37.68	0.5	30	0.02	60000	75.36	0.5
11	0.03	30000	37.68	1	31	0.03	60000	75.36	0.5
12	0.03	30000	37.68	2	32	0.04	60000	75.36	0.5
13	0.03	30000	37.68	3	33	0.05	60000	75.36	0.5
14	0.03	30000	37.68	4	34	0.06	60000	75.36	0.5
15	0.03	60000	75.36	0.1	35	0.07	60000	75.36	0.5
16	0.03	60000	75.36	0.2	36	0.08	60000	75.36	0.5
17	0.03	60000	75.36	0.5	37	0.09	60000	75.36	0.5
18	0.03	60000	75.36	1	38	0.1	60000	75.36	0.5
19	0.03	60000	75.36	2	39	0.03	10000	12.56	0.1
20	0.03	60000	75.36	3					

100 μm at an interval of 10 μm and fixing spindle speed at 60 000 rpm. Oil-based coolant was mistily sprayed onto the cutting zone during the entire micro milling experiment. The aforementioned experiments were repeated on all three types of micro tools. The full experimental matrix is listed in Table 2.2.

In order to reduce the influence of tool wear on the surface and burr formation, brand new micro tools were used for the experiments. An extra slot (exp #39) was milled at the end of experiments using the same cutting condition as the first slot (see Table 2.2). This validation slot was then compared with the first slot (exp #1). For three types of micro tools, it is observed that there is no significant difference between exp #1 and #39 in terms of surface roughness, slot width and burr height. Therefore the tool wear effect is regarded negligible under this certain cutting distance in this work.

2.3.4.1 Surface roughness

The surface roughness of the bottom surface of the micro milled slots was measured using a white light interferometer (Zygo NewView 5000) along the centreline of the slots. To reduce the measurement uncertainty and assess repeatability, five measurements on a different area were conducted for each slot and an average value of surface roughness (Ra) is used for analysis. There is no standard method of evaluating the burr formation. In this research an average burr height measured across the slot is used to assess the burr formation quantitatively. Five measurements on the burr height were recorded using a surface profiler (Talysurf Serial-2) and an average value was calculated for discussion.

Figure 2.12 shows that average surface roughness, Ra, as a function of feedrate at various cutting speeds for three types of micro milling tools. It can be observed from Figure 2.12b and c that for CVD and single crystal diamond micro tools the tendency is towards higher roughness value with increases in feedrate when feedrate is greater than 0.5 μm/tooth. However, when feedrate is smaller than 0.5 μm/tooth the roughness value decreases with increases in feedrate. The smallest surface roughness for single crystal diamond tools is observed around 11 nm, occurring at 0.2 μm/tooth and a cutting speed of 12.56 m/min; and the smallest surface roughness for CVD diamond tools is observed around 13 nm, occurring at 0.2 μm/tooth and a cutting speed of 37.68 m/min. Good surface finish obtained from micro milling using CVD and single crystal diamond tools indicates that optical surface finish can be achieved in micro milling by carefully selecting cutting conditions.

The tendency of surface roughness results from tungsten carbide tools is towards lower roughness value with increase of feedrate until the feedrate exceeds 3 μm/tooth (see Figure 2.12a), which is different from that of CVD and diamond tools. Under the cutting conditions in this work, surface roughness, Ra, was achieved between 24–50 nm. The smallest Ra value of 24 nm was observed at a feedrate 3 μm/tooth and a cutting speed of 75.36 m/min. It should be noted that the surface roughness results and the smallest Ra value of 24 nm achieved in this work are much better than some micro milling data obtained from miniature micro machine tools in the literature. This indicates the micro milled surface finish depends not only on the cutting tools and cutting conditions but the whole machining system including the motion errors of slideways and spindle runout, and so on.

Generally for all three types of micro tools, spindle speed has no significant effect on the surface finish. However, for CVD and single diamond tools, cutting speeds have a negative effect on the surface, that is, the higher the cutting speed, the higher the surface roughness value, especially for higher cutting speed (see Figure 2.12b and c). By contrast, this trend has not been found in WC micro milling experiments.

When the feed per tooth (i.e. depth of cut) is on the same order as tool edge radius, the effective rake angle may become negative. Size effects resulting from a small ratio of chip thickness to tool edge radius will be a dominant factor for material removal mechanism and chip generation physics in micro machining. Cutting, ploughing or slipping phenomenon will occur predominated by this ratio, and eventually influences machining processes such as surface finishes and burr formation. WC micro tools used in this work have a cutting edge radius in the range of 3–6 μm from SEM measurement results for the same batch of micro tools, whilst CVD Diamond and single crystal diamond micro tools have a cutting edge radius of 0.2 μm as quoted by the manufacturer. The minimum chip thickness can be well used to explain that there exists an optimum feedrate at which micro milling produces the best surface finish as discussed above.

Compared to two-fluted WC micro tools, CVD and single crystal diamond tools with one sharp cutting edge are likely to produce instability because there is always either only one single cutting edge in contact with the workpiece or no contact with the workpiece, which causes cutting force periodical variation, and poor surface finish. For diamond micro milling, surface roughness increases sharply with feedrate when feedrate exceeds 0.5 μm/tooth

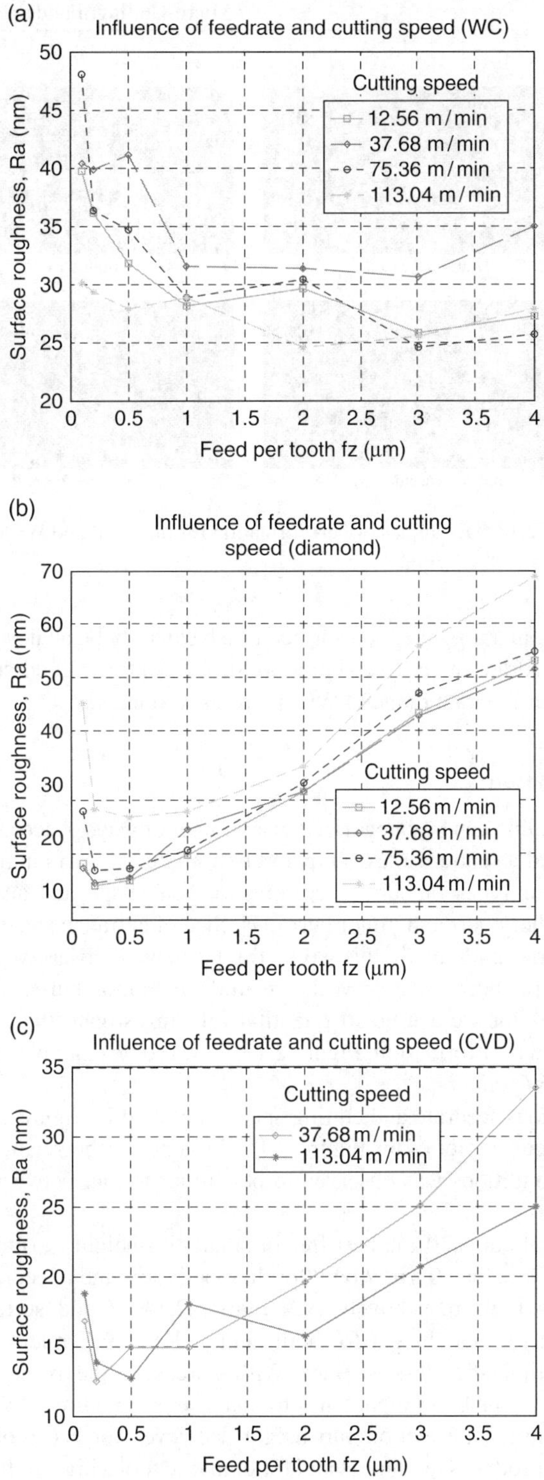

Figure 2.12 Average surface roughness Ra as a function of feedrate and cutting speed. (a)Tungsten carbide tool; (b) single crystal diamond tool; (c) CVD diamond tool

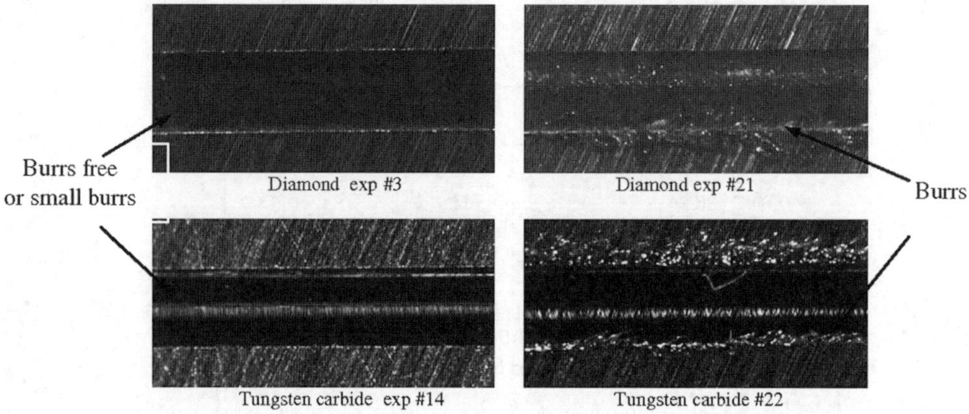

Figure 2.13 Examples of burr formation for diamond and WC cutters

(see Figure 2.12b), indicating that good surface finish can only be achieved in diamond micro milling at limited feedrate range (0.2–1 μm/tooth). By contrast, good surface finish can be achieved from a wider feedrate range in WC micro experiments.

2.3.4.2 Burr formation

Burrs in micro machining can be categorized in a number of ways. According to their position, burrs can be classified as entrance, exit, top and bottom burrs. Burrs in micro machining can be classified into four types of burrs according to their shape and amount: primary burr, needle-like burr, feathery burr and minor burr [47]. Size of entrance and exit burrs were found small and bottom burrs ignorable in this work. Top burrs were observed and measured in this work. Because the top burrs in this work are uniform minor burrs, average burr heights measured across the slot were used to quantitatively investigate the relationship between cutting conditions and burr formation. Figure 2.13 shows some examples of burr formation by using diamond and WC cutters.

Burr height at various feedrate and cutting speed is plotted in Figure 2.14 and its variation with axial depth of cut is plotted in Figure 2.15. In general, the relationship between burr height and cutting conditions was observed to be similar to that between surface roughness and cutting conditions.

When axial depth of cut is 30 μm, burr free or small burr height (<5 μm) cutting was found at a feedrate range of less than 0.5 μm/tooth for diamond tools and a feedrate range of 1–4 μm/tooth for WC micro tools respectively (see Figure 2.14). Good surface finish normally corresponds to small top burr height for both diamond and WC micro tools, which makes cutting parameter optimization less complex. With increase of the axial depth of cut, for both diamond and WC tools, the burr height tends to increase (see Figure 2.15). Cutting speed was found to have no significant effect on burr height. However, for WC tools a medium cutting speed (37.68 m/min) produces least burrs and for diamond tool a high cutting speed (113.04 m/min) helps to reduce burr formation, although more experimental data are needed to further verify this finding.

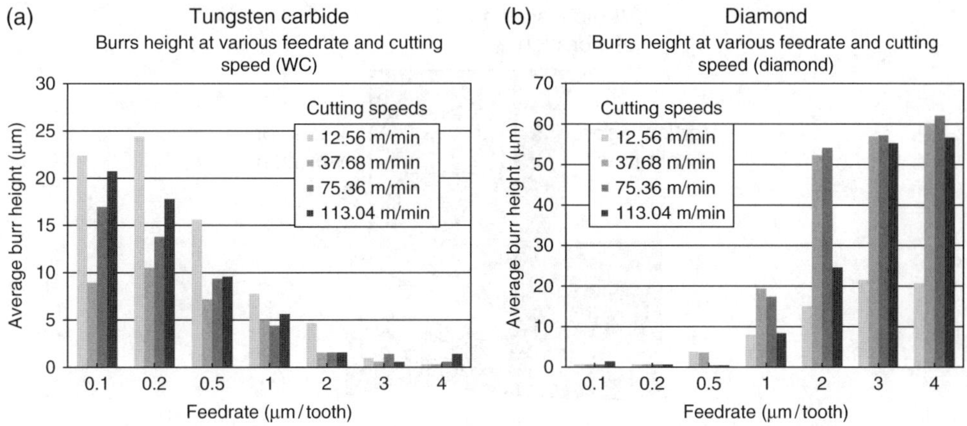

Figure 2.14 Burr heights at various feedrate and cutting speed

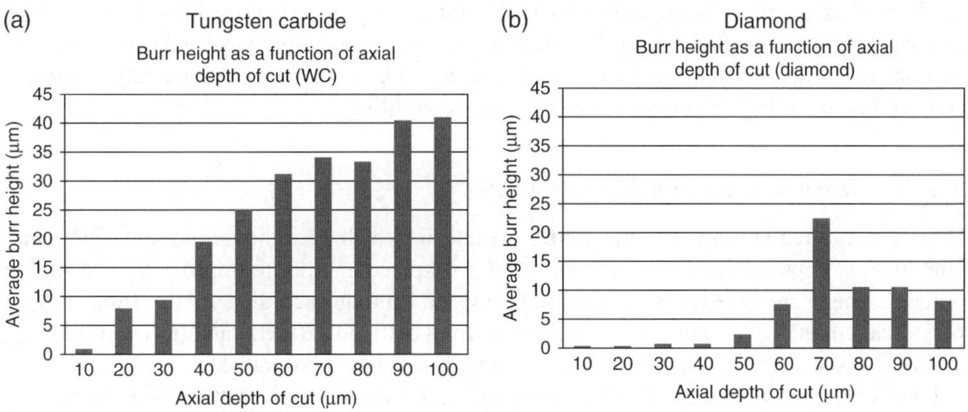

Figure 2.15 Burr heights as a function of axial depth of cut

2.4 Micro Machinability Issues and the Scientific Approaches

Machinability is not a precisely defined term; it is an attempt to account for several factors: tool life, power required for cutting, surface finish obtained, cost of removing material, and so on. The most important factor in most situations is usually tool life and machinability ratings are frequently based on this. Although geometric and material capabilities of micro cutting have been studied by many researchers, industrial application of micro cutting has been hindered by the lack of experience and knowledge on the micro-machinability of materials [22].

Currently most research on micro cutting mechanics employs soft materials for the use of workpiece, such as aluminium, copper, and so on. There are increasing demands on an industrially applicable micro cutting process for hard materials, such as hard steel, titanium alloy, tungsten carbide, ceramic, and so on. In addition, micro components and structures are characterized by high aspect ratios and high geometric complexity. Micro cutting, in particular micro milling process, is capable of machining high aspect ratios and complex geometries

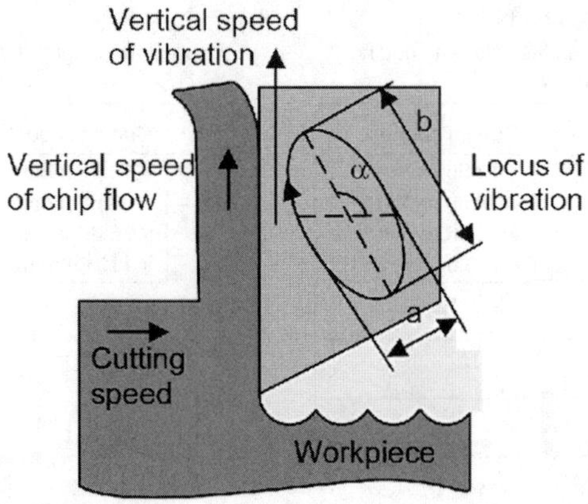

Figure 2.16 Principle of elliptical vibration cutting. Reproduced from [49]. Copyright 1995 Elsevier

with reasonable surface quality, but some issues, such as excess tool wear, low strength and stiffness of the micro tools, unpredictable tool failure, make micro cutting difficult-to-machine materials particularly challenging. Research on micro-machinability has been carried out and two scientific approaches are being employed to address these issues, namely, vibration assisted micro cutting and micro scale laser assisted milling.

2.4.1 *Vibration Assisted Micro Cutting*

Vibration assisted cutting is a machining method in which periodical vibration with small amplitude is imposed to the cutting tool or the workpiece in order to improve the machining process. There are two types of vibration assisted cutting according to the frequency of periodical vibration: (a) ultrasonic vibration assisted cutting (>20 kHz) and (b) low frequency vibration assisted cutting (normally a few hundreds to a few thousands Hz).

Ultrasonic vibration assisted cutting is used to obtain ductile mode cutting of brittle materials and reduce cutting forces in diamond cutting [48]. Ultrasonic vibration assisted cutting is also employed in ultra-precision diamond turning of stainless steel [18] which has been regarded as impractical due to short tool life. [49] proposed ultrasonic elliptical vibration cutting in which synchronized two-dimensional vibrations are applied to the cutting edge in the plane containing the cutting and the chip flow direction in a way that the cutting edge forms an elliptical locus as shown in Figure 2.16. Low frequency vibration assisted cutting has been used in drilling process to prolong tool life and reduce burr sizes [50] [51].

Vibration assisted cutting has been applied to the micro cutting process. In one of Chern's work [52] two-dimensional vibration assisted cutting was utilized in micro slot milling. The experimental results showed that both slot profile accuracy and surface quality could be improved by imposing vibration cutting. Ding et al. proposed a two-dimensional vibration-assisted micro milling model to investigate the kinematics and chip thickness [53].

2.4.2 *Laser Assisted Micro Cutting*

Laser assisted cutting is another promising method to improve machinability of difficult-to-machine materials. Laser assisted cutting is a type of thermally assisted machining technique

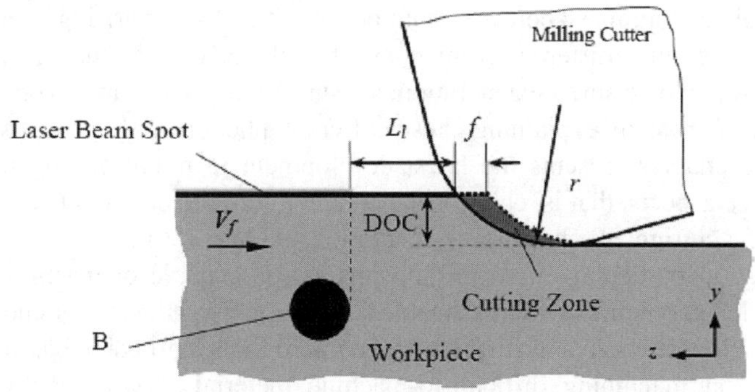

Figure 2.17 Schematic of laser assisted cutting (Reproduced from [60])

in which the workpiece materials are heated and softened locally prior to the cutting tool by external energy source, that is, a focused laser beam, during the cutting process. The reason to heat materials is to enhance machinability, especially for difficult to machine materials. Figure 2.17 shows a schematic of laser assisted cutting.

Laser assisted cutting has been researched extensively in the past two decades at the conventional level. It has been mainly used to process difficult-to-machine materials such as ceramics. For example, [54] studied laser assisted turning of ceramics and found that as the cutting forces were reduced and surface quality was improved, and the tool life had a significant increase compared with that without laser assistance. Yang et al. [55] observed that the cutting forces and specific energy decreased in laser assisted milling of silicon nitride ceramics. Overall laser assisted cutting offers many advantages over traditional cutting process [56], such as:

- reduced power per volume of machined part and higher material removal rate;
- reduced cutting forces and chatter;
- improved machining accuracy and surface finish;
- reduced surface and subsurface damage;
- elongated tool life and avoidance of tool breakage.

Research on laser assisted mechanical machining is still at its nascent stage. Applying this method to micro cutting will offer the opportunity for cutting performance improvement. Preliminary experiments on laser assisted micro cutting have been carried out using both soft materials and hard materials, and promising improvement on tool life, material removal rate, tooling loads and surface quality have been reported [57], [58], [59]. Most of this research is experiment based, and an accurate model needs to be established to understand the fundamental cutting mechanism which governs the laser assisted process.

2.5 Summary

Cutting is a complicated process involving elastic and plastic deformation together with fracture process at high temperature with high strain rate. Research on micro cutting process physics is still at its nascent stage. Micro cutting is not a simple scaling of macro cutting while they are kinematically similar. There are a number of critical issues arising from micro cutting which are fundamentally different from conventional machining, that is, material micro structure, micro-machined surface generation, subsurface character, and so on. The size effect plays a significant role in explaining these different phenomena and investigating micro mechanics. This chapter presents the latest development in micro cutting mechanics with emphasis on three aspects, that is, chip formation and minimum chip thickness; micro cutting force and specific cutting energy; surface generation and burr formation.

Micro cutting, in particular micro milling process, is capable of machining high aspect ratios and complex geometries with reasonable surface quality, However, some issues, such as excess tool wear, low strength and stiffness of the micro tools, unpredictable tool failure, limit its applications on machining difficult-to-machine materials. Two scientific approaches: vibration assisted micro cutting and laser assisted micro cutting to overcome these limitations, are discussed in this chapter.

References

[1] Liu, X., DeVor, R.E., Kapoor, S.G. and Ehmann, K.F. (2004) The mechanics of machining at the microscale: Assessment of the current state of the science, *Journal of Manufacturing Science and Engineering, Transactions of the ASME*, 126 (4), 666–678.

[2] Huo, D., Cheng, K. and Wardle, F. (2009a) Design of a 5-Axis Ultraprecision Micro Milling Machine – UltraMill: Part 1: Holistic Design Approach, Design Considerations, and Specifications. *International Journal of Advanced Manufacturing Technology*, 47, 867–877.

[3] Huo, D., Cheng, K. and Wardle, F. (2009b) Design of a 5-axis ultraprecision micro milling machine – ultramill: Part 2: Integrated dynamic modelling, design optimization and analysis. *International Journal of Advanced Manufacturing Technology*, 47, 879–890.

[4] Ehmann, K., Bourell, D., Culperper, M., Hodgson, T., Kurfess, T., Madou, M., Rajurkar, K and DeVor, R. (2007) Micromanufacturing: *International Assessment of Research and Development*, Springer.

[5] Chae, J., Park, S.S., and Freiheit, T. (2006) Investigation of micro-cutting operations. *International Journal of Machine Tools and Manufacture*, 46 (3–4), 313–332.

[6] Brecher, C., Klar, R. and Wenzel, C. Development of a high precision miniature milling machine. Proceedings of the 3rd International Conference on Multi-Material Micro Manufacture, 4M 2007, 327–330.

[7] Tanaka, M. (2001) Development of desktop machining microfactory. *Riken Review*, 34, 46–49.

[8] Kussul, E., Baidyk, T., Ruiz-Huerta, L., Caballero-Ruiz, A., Velasco, G. and Kasatkina, L. (2002) Development of micromachine tool prototypes for microfactories. *Journal of Micromechanics and Microengineering*, 12(6), 795–812.

[9] Okazaki, Y., Mishima, N. and Ashida, K. (2004) Microfactory – concept, history, and developments. *Journal of Manufacturing Science and Engineering, Transactions of the ASME*, 126(4), 837–844.

[10] Vogler, M.P., Liu, X., Kapoor, S.G., Devor, R.E. and Ehmann, K.F. (2002) Development of Meso-scale Machine Tool (MMT) Systems, *Society of Manufacturing Engineers* MS n MS02-181, 1–9.

[11] Bang, Y.B., Lee, K.M. and Oh, S. (2005) 5-axis micro milling machine for machining micro parts. *International Journal of Advanced Manufacturing Technology*, 25, 888–894.

[12] Lee, S.W., Mayor, R. and Ni, J. (2006) Dynamic analysis of a mesoscale machine tool, *Journal of Manufacturing Science and Engineering, Transactions of the ASME*, 128(1), 194–203.

[13] Li, H., Lai, X., Li, C., Lin, Z., Miao, J. and Ni, J. (2008) Development of meso-scale milling machine tool and its performance analysis, *Frontiers of Mechanical Engineering in China*, 3(1), 59–65.

[14] Ikawa, N., Shimada, S. and Tanaka, H. (1992) Minimum thickness of cut in micromachining, *Nanotechnology*, 3(1), 6–9.

[15] Lucca, D.A. and Seo, Y.W. (1993) Effect of tool edge geometry on energy dissipation in ultraprecision machining, *Annals of the CIRP*, 42(1), 83–88.

[16] Takeuchi, Y., Sawada, K. and Sata, T. (1996) Ultraprecision 3D micromachining of glass, *Annals of the CIRP*, 45(1), 401–404.

[17] Vogler, M.P., DeVor, R.E. and Kapoor, S.G. (2004) On the modelling and analysis of machining performance in micro-endmilling, part I: Surface generation, *Journal of Manufacturing Science and Engineering, Transactions of the ASME*, 126(4), 685–694.
[18] Moriwaki, T., Sugimura, N., Manabe, K. and Iwata, K. (1991) A study on orthogonal micromachining of single crystal copper, *Trans. NAMRI/SME*, 19, 177–183.
[19] To, S., Lee, W.B. and Chan, C.Y. (1997) Ultraprecision diamond turning of aluminium single crystals, *Journal of Materials Processing Technology*, 63(1–3), 157–162.
[20] Lee, W.B., To, S. and Cheung, C.F. (2000) Effect of crystallographic orientation in diamond turning of copper single crystals, *Scripta Materialia*, 42(10), 937–945.
[21] Huo, D., Liang, Y. and Cheng, K. (2007) An investigation of nanoindentation tests on the single crystal copper thin film via an AFM and MD simulation, *Proceedings of the IMechE. Part C: Journal of Mechanical Engineering Science*, 221, 259–266.
[22] Filiz, S., Conley, C.M., Wasserman, M.B., Ozdoganlar, O.B. (2007) An experimental investigation of micro-machinability of copper 101 using tungsten carbide micro-endmills, *International Journal Of Machine Tools and Manufacture*, 47, 1088–1100.
[23] Shimada, S., Ikawa, N., Tanaka, H., Ohmori, G., Uchikoshi, J. and H. Yoshinaga (1993) Feasibility study on ultimate accuracy in microcutting using molecular dynamics simulation. *Annals of the CIRP*, 42, 91–94.
[24] Weule, H., Hüntrup, V. and Tritschler, H. (2001) Micro-cutting of steel to meet new requirements in miniaturization, *Annals of the CIRP*, 50(1), 61–64.
[25] Kim, J.S., Kim, J.H., Kang, M.C. and Seo, Y.W. (2007) A mechanistic model of cutting force in the micro and milling process, *Journal of Materials Processing Technology*, 187, 250–255.
[26] Taniguchi, N. (1994) The state-of-the-art of nanotechnology for processing ultra-precision and ultra-fine products. *Precision Engineering*, 16(1), 5–24.
[27] Shaw, M.C. (1996) *Principles of Abrasive Processing*. Oxford University Press, Oxford.
[28] Shaw, M.C. (2003) The size effect in metal cutting, *Sadhana*, 25 (5), 875–896.
[29] Challen J. M. and Oxley, P. L. B (1984) Slip-Line Fields for Explaining the Mechanics of Polishing and Related Processes. *International Journal of Mechanical Sciences*, 26(6–8), 403–418.
[30] Kopalinski, E. and Oxley, P.L.B. (1984) Size effect in metal removal processes. *Proceedings of 3rd conference on the Mechanical Properties at High Rate of Strain*, Oxford, United Kingdom, 389–396.
[31] Vollersten, F., Biermannb, D., Hansen, H.N., Jawahir, I.S. and Kuzman, K. (2009) Size effects in manufacturing of metallic components, *Annals of the CIRP*, 58(2), 566–587.
[32] Nakayama, K. and Tamura, K. (1968) Size effect in metal cutting force. ASME Transactions-Journal of Engineering for Industry 90(1),119–126.
[33] Lucca, D.A. and Seo, Z.W. (1994) Aspects of surface generation in orthogonal ultra-precision machining. *Annals of the CIRP*, 43(1), 43–46.
[34] Lai, X.M., Li, H.T., Li, C.F. *et al.* (2008) Modelling and analysis of micro scale milling considering size effect, micro cutter edge radius and minimum chip thickness. *International Journal of Machine Tools and Manufacture*, 48, 1–14.
[35] Taminiau, D.A. and Dautzenberg J.H. (1991) Bluntness of the tool and process forces in high-precision cutting. *Annual of the CIRP*, 40, 65–68.
[36] Ehmann, K.F., Kapoor, S.G., DeVor, R.E. and Lazoglu, I. (1997) Machining Process Modelling: A Review, *Journal of Manufacturing Science and Engineering*, 119, 655–663.
[37] Tlusty, J. and MacNeil, P. (1975) Dynamics of cutting forces in end milling, *Annals of the CIRP*, 24, 21–25.
[38] Armarego, E.J.A. (1998) A generic mechanics of cutting approach to predictive technological performance modelling of the wide spectrum of machining operations, *Machining Science and Technology*, 2, 191–211.
[39] Fang, N. (2003) Slip-line modelling of machining with a rounded-edge tool - Part I: new model and theory, *Journal of the Mechanics and Physics of Solids*, 51(4), 715–742.
[40] Waldorf, D.J., DeVor, R.E. and Kapoor, S.G. (1998) A slip-line field for ploughing during orthogonal cutting, *Journal of Manufacturing Science and Engineering*, 120(4), 693–699.
[41] Bissacco, G., Hansen, H.N. and Slunsky, J. (2008) Modelling the cutting edge radius size effect for force prediction in micro milling, *Annals of the CIRP*, 57, 113–116.
[42] Kang, I.S., Kim, J.S., Kim, J.H., Kang, M.C. and Seo, Y.W. (2007) A mechanistic model of cutting force in the micro end milling process, *Journal of Materials Processing Technology*, 187, 250–255.
[43] Bao, W.Y. and Tansel, I.N. (2000) Modelling micro-end-milling operations. Part I: analytical cutting force model, *International Journal of Machine Tools and Manufacture*, 40(15), 2155–2173.

[44] Zaman, M.T., Kumar, A.S., Rahman, M. and Sreeram, S. (2006) A three-dimensional analytical cutting force model for micro end milling operation, *International Journal of Machine Tools and Manufacture*, 46, 353–366.
[45] Lee, K., and Dornfeld, D.A. (2002) An experimental study on burr formation in micro milling aluminum and copper, *Trans. NAMRI/SME*, 30, 1–8.
[46] Lee, K. and Dornfeld, D.A. (2005) Micro-burr formation and minimization through process control. *Precision Engineering*, 29, 246–252.
[47] Chern, G.L., Wu, Y.J.E., Cheng, J.C. and Yao, J.C. (2007) Study on burr formation in micro-machining using micro-tools fabricated by micro-EDM, *Precision Engineering*, 31, 122–129.
[48] Zhou, M., Wang, X.J., Ngoi, B.K.A. and Gan, J.G.K. (2002) Brittle-ductile transition in the diamond cutting of glasses with the aid of ultrasonic vibration, *Journal of Materials Processing Technology*, 121, 243–251.
[49] Moriwaki, T. and Shamoto, E. (1995) Ultrasonic elliptical vibration cutting, *Annals of the CIRP*, 44(1), 31–34.
[50] Adachi, K., Arai, N., Harada, S., Okita, K. and Wakisaka, S. (1987) A study on burr in low frequency vibratory drilling of aluminum, *Bulletin of the JSPE*, 21(4), 258–264.
[51] Chern, G. L. and Lee, H. J. (2006a) Using workpiece vibration cutting for micro-drilling, *International Journal of Advanced Manufacturing Technology*, 27(7–8), 688–692.
[52] Chern, L. and Chang, C. (2006b) Using two dimensional cutting for micro milling, *International Journal of Machine Tools and Manufacture*, 46, 659–692.
[53] Ding, H., Chen, S.J. and Cheng, K. (2010) Dynamic surface generation modelling of two-dimensional vibration-assisted micro-end-milling. *International Journal of Advanced Manufacturing Technology*, 53(9–12), 1075–1079.
[54] Rebro, P.A., Shin, Y.C., and Incropera, F.P. (2004). Design of operating conditions for crack free laser-assisted machining of mullite. *International Journal of Machine Tools and Manufacture*, 44(7–8), 677–694.
[55] Yang, B., Deines, T.W., Geist, C.M. and Lei, S. (2007). An experimental study of laser assisted milling of silicon nitride ceramic. *Transactions of the North American Manufacturing Research Institute of SME*, 35, 473–480.
[56] Yang, B., Shen, X. and Lei, S. (2009) Mechanisms of edge chipping in laser-assisted milling of silicon nitride ceramics. *International Journal of Machine Tools & Manufacture*, 49, 344–350.
[57] Jeon, Y. and Pfefferkorn, F.E. (2008) Effect of laser preheating the workpiece on micro end milling of metals, *Journal of Manufacturing Science and Engineering, Transactions of the ASME*, 130, 1–9.
[58] Singh, R. and Melkote, S.N. (2007) Characterization of a hybrid laser-assisted mechanical micromachining (LAMM) process for a difficult-to-machine material, *International Journal of Machine Tools and Manufacture*, 47, 1139–1150.
[59] Pfefferkorn, F.E., Lei, S.T., Jeon, Y. and Haddad, G. (2009) A metric for defining the energy efficiency of thermally assisted machining, *International Journal Of Machine Tools and Manufacture*, 49(5), 357–365.
[60] Shen, X. (2010) Numerical Modelling and Experimental Investigation of Laserassisted Machining of Silicon Nitride Ceramics, PhD thesis, Kansas State University, Manhattan, Kansas.

Makino. Website: http://www.makino.com/ (Accessed on 25th May 2011)
Moore Nanotechnology System. website: http://www.nanotechsys.com/ (Accessed on 25th May 2011)
Precitech, Inc. website: http://www.precitech.com/ (Accessed on 25th May 2011)
Sodick. Website: http://www.sodick.com/ (Accessed on 25th May 2011)

3

Micro Tooling Design and Manufacturing

Paul T. Mativenga[1], Ampara Aramcharoen[2] and Dehong Huo[3]
[1] *School of Mechanical, Aerospace & Civil Engineering, The University of Manchester, UK*
[2] *Singapore Institute of Manufacturing Technology (SIMTec), Singapore*
[3] *School of Mechanical and Systems Engineering, Newcastle University, UK*

3.1 Tool Size and Machining Scale

The size of the micro tool is a key design factor that determines the smallest feature size that can be machined and the length scale of machining. If the diameter and cutting edge radius of the micro tool are decreased beyond the capability of present micro tool manufacturing technology, then smaller feature sizes and finer detail can be achieved.

Commercially available carbide micro tools are typically in a range from 1000 down to 25 μm in diameter for micro end mills. These tools have edge radius that is a few microns in size. Figure 3.1 shows an image of a commercially available micro end mill. While the flute section diameter is in the micro meter scale the tool shank is made bigger to strengthen the tool. A 3 mm diameter tool shank is quite common for micro cutting tools.

Figure 3.2 shows a closer view of a micro tool and its flute geometry. The cutting tool edge radius r_e is a critical parameter that determines edge sharpness and influences.

The minimum chip thickness that is, the lower limit of micro machining [1–3]. The minimum chip thickness is the smallest un-deformed chip thickness which will result in the micro machining process failing to produce chips. This limit is found to occur at undeformed chip thickness which is significantly smaller than the tool edge radius and a highly negative effective rake angle retards the formation of the chip or its flow on the rake face.

The demands for more accurate micro cutting (closer to nanoscale cutting) can be met by the use of diamond tools which have a smaller cutting edge radius typically in the order of 100 nm and hence promotes smaller minimum chip thickness [4–7]. However the drawbacks of diamond tools are the limited range of workpiece materials that can be machined and the

Micro-Cutting: Fundamentals and Applications, First Edition. Edited by Kai Cheng and Dehong Huo.

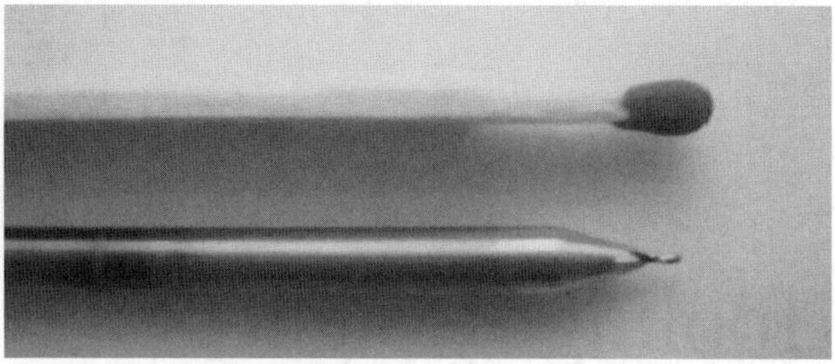

Figure 3.1 Commercial micro tool in relation to matchstick size

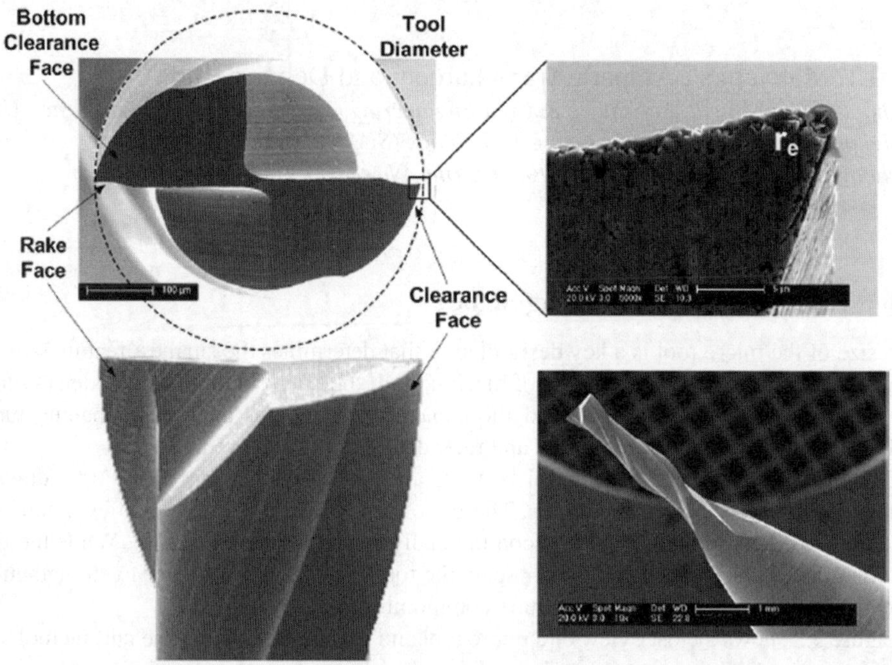

Figure 3.2 Geometry of new micro flat end mill (500 µm in diameter)

high tooling cost (for example in 2009 in the UK, £35 for coated carbide micro tool compared to at least £700 for diamond tipped micro tool). High chemical reaction between the diamond tool and steel workpiece materials can accelerate diffusion wear [8]. Thus, tungsten carbide micro end mills are a compromise in micro milling due to a high toughness, an ability to machine a variety of workpiece materials at relatively lower tooling cost.

3.2 Manufacturing Methods for Solid Shank Micro Tools

The majority of carbide micro cutting tools are now made by a mechanical grinding process starting with an ultra-fine grain carbide rod. However, for micro tools, the grinding process sometimes is not precise enough to fabricate a smaller diameter with more complexity in geometry.

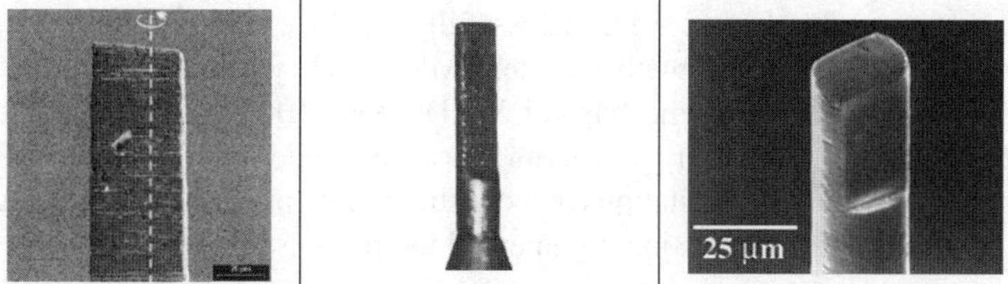

Mechanically ground single cutting edge WC micro end mill	2 flute, 110 µm diameter, tungsten carbide tool made by WEDG	2 flute tungsten carbide micro end mill made by FIB

Figure 3.3 Micro tools produced by mechanical grinding, electrical discharge grinding and focused ion beam machining. **(left)** Reproduced from [15]. Copyright 1999 Elsevier. **(centre)** Reproduced from [11]. Copyright 2007 Elsevier. **(right)** Reproduced from [12]. Copyright 2011 Elsevier

This can be due to the fact that mechanical grinding provides higher cutting force during grinding resulting in tool defection as well as heat affected zones. Schaller *et al.* reported that in grinding 35 to 120 µm diameter micro tools, 50% of the tools broke and they proposed that this was promoted by the brittle tungsten carbide [9]. This high scrap rate in manufacture of micro tools pushes up the cost of carbide micro tools especially when tool diameter is reduced to the lower scale of the micro range. Given the likelihood of micro tool edge fracture and grinding defects it is highly recommended that manufacturers carefully select the suppliers of their micro tool. In this selection it is important especially for carbide tools to select the micro tool with the smallest grain size (ultra fine grain carbide tools have high fracture toughness) and to inspect the cutting edge geometry using high magnification microscopy. Typical grain sizes for carbide micro tools can be in the order of 300 nm to 900 nm. Other micro tool manufacturing processes available include electrical-discharge machining (EDM), wire electrical-discharge grinding (WEDG) and focused ion beam (FIB) processes. Egashira *et al.* [10] fabricated single flute cemented carbide micro carbide tool using micro-EDM. Typical diameter and cutting edge radius of the tools were 20 and 0.5 µm respectively. The smallest diameter achieved was 6 µm. The tools were used to drill micro holes on silicon wafer. Wire electro-discharge grinding (WEDG) enables shorter machining time and low machining force. Chern *et al.* [11] used WEDG to fabricate tungsten carbide micro tools with diameters under 100 µm and a minimum diameter of 31 µm. The tools were successfully used to machine micro slots and thin walls of microscale thickness on 6061-T6 aluminium alloy.

Focused ion beam (FIB) machining is a process which can also be employed to shape or cut off a tiny portion of hard materials. Thus, tungsten carbide micro tools can be fabricated using the FIB method. It offers a precise control over feature size, complex geometries and sharp cutting edge. For FIB, a negligible force acts on the tool during its fabrication [12]. However, the limitation of FIB is that it is time consuming because of the low machining rate. The process is also more complex and more expensive. *Vasile et al.* [13, 14] used FIB to fabricate 25 µm diameter micro tools from cobalt high speed steel tool blanks in order to machine polymethyl methacrylate (PMMA). Adams *et al.* [12] also fabricated micro end mills with 2, 4 and 5 cutting edges using FIB as shown in Figure 3.3. The tools had a diameter and cutting edge radius less than 25 μm and 0.1 μm respectively. The tools made from high speed steel and micro-grain tungsten carbide and were successfully used to machine 6061-T4 aluminium, brass, 4340 steel and polymethyl methacrylate (PMMA). They achieved a surface roughness around 200 nm. Figure 3.3 shows mechanical, WEDG and FIB fabricated research tools.

While EDM, WEDG and FIB processes can fabricate smaller micro cutting tools, there are geometry and production time limitations that constrain these processes from being as popular as mechanical grinding for fabricating commercial micro tools.

3.3 Coatings and Coated Solid Shank Micro Tools

A coating is a thin layer that is applied to an object with the main aim of improving surface properties of a bulk material usually referred to as a substrate. The properties that can be improved are for example wear resistance, corrosion resistance, friction coefficient, thermal barrier capability, appearance, wetability and scratch resistance. Coated cutting tools are important for dry and high speed micro machining operations as well as for cutting high strength and strain hardening materials. In particular, hard coatings provide a higher hardness than the base tooling material and hence improve wear performance of the micro tool. In dry machining, the cutting edges are subject to higher thermal loads and chemical wear compared to wet machining. Therefore, to retain a good performance in dry machining, the cutting tools need elevated hot hardness, chemical stability and low friction coefficient. In addition high toughness is also required for micro milling tools.

Coating materials are deposited on the majority of cutting tools principally by either chemical vapour deposition (CVD) or physical vapour deposition (PVD). Physical vapour deposition generally refers to a variety of methods to deposit thin films in a vacuum by the condensation of a vaporized form of the material onto various surfaces.

The coating method involves purely physical processes such as high temperature vacuum evaporation or plasma sputter bombardment rather than a chemical reaction at the surface to be coated as in chemical vapour deposition. Bonding of the coating should be achieved with high integrity. Bonding of coatings to substrate materials is dependent on several factors: the cleanliness of the surface to be bonded, the energy available for the particle bonding process and the ability to generate a graded bonding layer interface. Adhesion could be improved by substrate treatment before coating for example using micro-blasting [16]. Vapour deposition methods cannot give an acceptable adhesion to the substrate unless the substrate is perfectly cleaned before the deposition process. Cleaning involves the removal of contaminant layers that can sometimes be a few tens of molecules thick.

Vacuum evaporation is a PVD process in which material from a thermally vaporized source reaches the substrate without collision with gas molecules in the space between the source and substrate. The source material is evaporated in a vacuum to allow the vapour particles to travel directly to the target object (substrate), where they condense back to a solid state. In ion plating evaporation is not done into a neutral gas but an 'ionised' gas is used as the scattering medium. The combined high energy (~10 eV) evaporant atoms and continuing substrate bombardment by gas ions promotes excellent film adhesion and denser coating. This technology was developed to deposit wear resistant coatings of oxides, nitrides and carbides. Reactive ion plating was developed because both vacuum evaporation and ion plating are limited to depositing materials as a coating without a change in composition. The principle of reactive ion plating builds on from ion plating and incorporates a reactive gas. The metallic constituent is evaporated into the reactive gas whose reactivity is considerably enhanced by glow discharge. As an example a thin, hard and wear resistant coating of titanium nitride (TiN) can be produced by evaporating titanium into a low pressure glow discharge of a mixture of argon and nitrogen. The reaction that forms TiN can take place in the gas phase or at the surface of the substrate to be coated.

Sputter deposition is a physical vapour deposition (PVD) method of depositing thin films by sputtering (ejecting), material from a target (source), which then deposits onto a substrate. In sputtering the substrate is bombarded with fast heavy particles (usually argon-ion bombardment), and these cause erosion of the target material. The ejected atoms are in the form of high energy electrons which condense on the surrounding surfaces and specimen to be coated. The process occurs in the conditions of a gaseous glow discharge between an anode and cathode and can be enhanced by the choice of a suitable gas and target material. The target is a solid and can be an alloy or compound or an alloy as a mixture of powdered constituencies. The gas ionizing potential is applied as DC or RF.

The gas maybe be inert or reactive and the substrate earthed or biased. Since the high energy of the bombarding ions is sufficient to overcome the binding energy of any substrate atoms and the vapour is generated by ion bombardment and not by heating and the coating temperature is about 2000 °C. The process is very versatile and most widely used for coating cutting tools. Table 3.1 shows a comparison between CVD and PVD deposition techniques.

PVD coatings can be thinner, deposited at low temperature, relatively crack free and have a compressive residual stress and this makes them good for coating micro tools. They can also

Table 3.1 Comparison of CVD and PVD coatings processes for micro carbide tools

	Chemical Vapour Deposition (CVD)	Physical Vapour Deposition (PVD)	Desirable effects
Substrate temperature during deposition	850 to 1000 °C	200 to 500 °C	Low deposition temperature preserves tool substrate edge toughness. This is more critical for micro tools because of their size.
Micro structure and residual stress	Coarse grain /Tensile	Finer grain / Compressive	Compressive residual stress inhibits crack propagation
Coated tool hardness	good	higher	Higher hardness increases with decrease in grain size.
Typical coating thickness	Typically up to 20 μm	Typically less than 5 μm	The ability to deposit thin layers at lower deposition periods makes PVD competitive for micro tools. There is a need to keep the tool edge radius as low as possible.
Comments on adhesion capability	better	good	CVD provides better adhesion due to the higher deposition temperature which promotes diffusion
Geometry capability	Best deposited on honed edge	Can be deposited on sharp cutting edge	Sharp edge deposition capability is very important for micro tools
Typical Process duration	8 to 24 hours	3 to 4 hours	CVD enables higher volumetric deposition rate

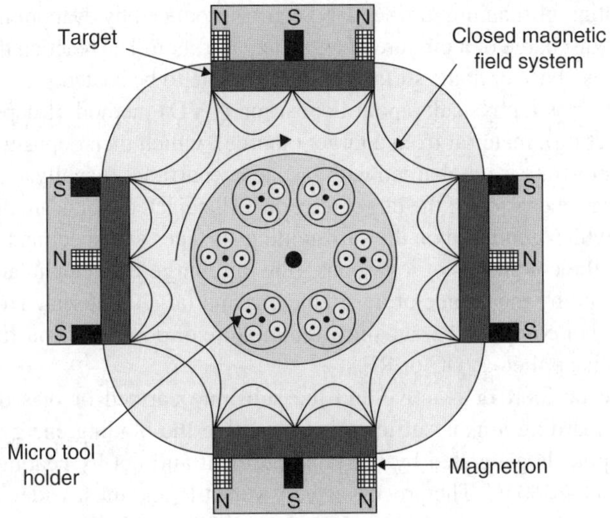

Figure 3.4 Closed Field Unbalanced Magnetron Sputter Ion Plating (CFUBMSIP) developed by Teer Coatings Ltd with micro-tool tool holder platform (adapted from [18])

be deposited as thin multi-layers more accurately around sharp cutting edges for micro tools. Their limitation is a low adhesion of PVD coatings compared to CVD.

Different PVD coating technologies are available commercially such as evaporation, laser ablation and sputtering. An example of a coating process that can be applied for micro tools is Closed Field Unbalanced Magnetron Sputter Ion Plating (CFUBMSIP) shown in Figure 3.4.

3.3.1 Closed Field Unbalanced Magnetron Sputter Ion Plating (CFUBMSIP)

Magnetron sputtering is a powerful and flexible technique to coat a variety of materials and is very competitive compared to other PVD coating techniques [17]. Magnetron sputtering was developed for the Closed Field Unbalanced Magnetron Sputter Ion Plating (CFUBMSIP) technology by Teer Coatings Ltd in 1996 [18]. CFUBMSIP can be defined as the use of closed field unbalanced magnetrons of alternating magnetic polarity as shown in Figure 3.4. As shown in Figure 3.4, four magnetron systems are arrangement such that their polarity is closed to enhance plasma. The link of magnetic field traps and enhances plasma to surround the substrates to be coated. The advantages of this technique are in the ability to deposit denser non columnar coating structures, uniform films, ability to control film thickness, good adhesion and high deposition rate [19].

3.3.2 Coating Layout

The structure of coatings is critical in determining the properties and performance of coated tools. Compared to a single layer, multilayer coatings which consist of a number of different coating materials provide better wear resistance [20, 21]. Nano-scale layer

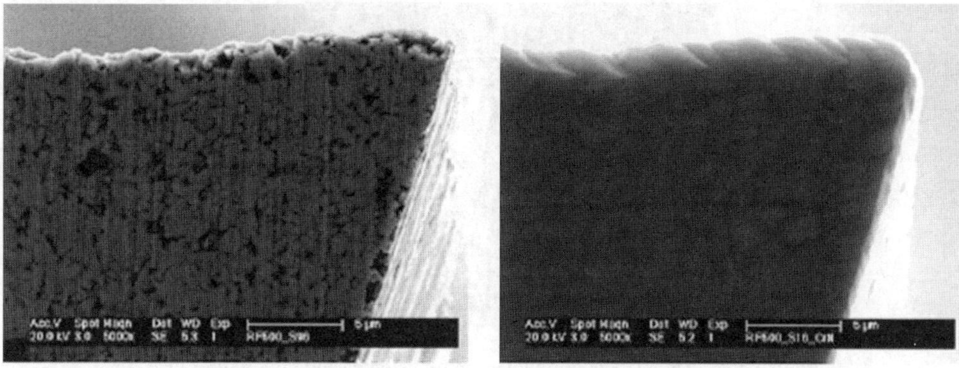

(a) (b)

Figure 3.5 Morphology of (a) uncoated and (b) after CrN coating of ultra fine grain carbide micro end mill

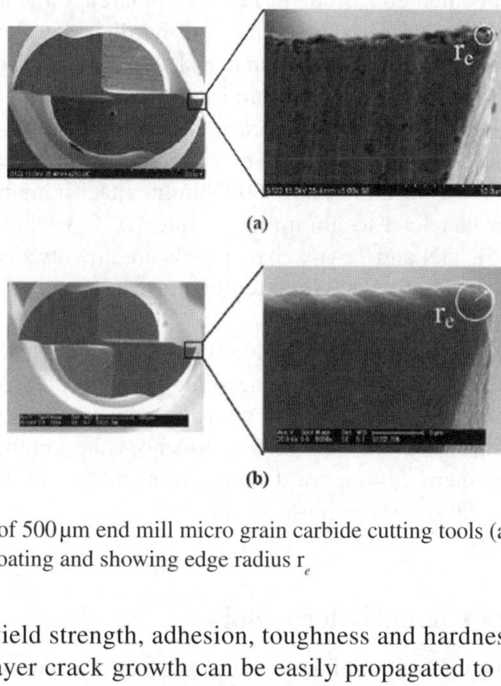

(b)

Figure 3.6 Condition of 500 μm end mill micro grain carbide cutting tools (a) received from tool supplier and (b) after TiN coating and showing edge radius r_e

thickness enhances yield strength, adhesion, toughness and hardness of coating [22]. If a coating was single layer crack growth can be easily propagated to the coating thickness. A thinner thickness of each layer in multilayer coatings results in suppressing crack growth. For multilayer coatings, an adhesive layer is introduced to promote the bonding of the coating to the substrate. Higher hardness and toughness of coatings can be achieved by reducing the coating period. A coating period is the thickness of a uniform deposit that is done continuously before a break in deposition cycle. Typically for PVD coating, hundreds of layers of thickness in nano-scale are applied with a resulting total coating thickness of about 2–5 μm [23]. The thickness of coating is critical for micro tools as it leads to significant increases in the cutting edge radius. Figure 3.5 shows a micro tool cutting edge before and after PVD coating.

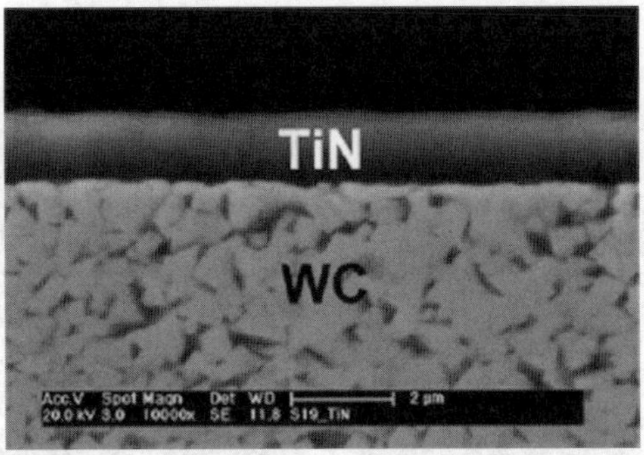

Figure 3.7 Thin TiN coating layer on tungsten carbide micro tool face

Titanium nitride (TiN) was an early coating material applied on a high speed steel tool and later developed for cemented carbide in 1985. Compared to the uncoated carbide tool, TiN coated tool has high wear resistance and high hardness at elevated temperature. It also exhibits good adhesion to the substrate material. Titanium aluminium nitride (TiAlN) was developed from TiN by adding the aluminium (Al). Increasing aluminium led to increasing the operating temperature up from 800 to 900 °C. Higher toughness and hardness, oxidation and corrosion wear resistance are characteristics of TiAlN compared to TiN [21]. At temperatures of more than 750 °C in the machining process, the oxidation reaction of aluminium can lead to aluminium oxide ($A_{12}O_3$) which acts as a lubricant protective layer [24, 25]. TiN and TiAlN coated tools are now used widely in industry for high speed cutting, dry machining and high temperature cutting applications and in micro machining.

A new chromium titanium aluminium nitride (CrTiAlN) coating was developed by Teer Coatings Ltd. It exhibits high hardness, toughness, high thermal stability and good adhesion. It should be noted that in micro milling a low cutting speed is typically used due to the size of the tool diameter. Additionally, the size affect influences the cutting mechanisms. These operating conditions in micro cutting are different from macro machining and hence may demand different functionality for coatings.

3.4 Importance of Coated Micro Tools

The toughness of micro tools for milling applications can be improved by using finer or ultra-fine grain tungsten material. Protective coating materials deposited on micro cutting tools can bring about longer tool life and improvements in cutting performance. Additionally, in industrial macroscale machining, it is well established that appropriate physical vapour deposition (PVD) coatings promote the use of higher cutting speeds and facilitate dry machining or the use of minimum quantity lubrication (MQL).

When micro tools (1 to 999 µm in diameter) are used in milling, the undeformed chip thickness is usually very small and comparable to the cutting edge radius. This condition determines the effective rake angle and hence plays a significant role in the mechanics of micro machining. The differences between macro and microscale machining influence process factors such as cutting interface temperatures, forces and strain fields. The mechanisms through which coatings protect cutting tools are influenced by such process conditions. In addition, the size of micro end mills makes coating deposition challenging especially around the cutting edges. The requirements on the coatings for micromachining tools are not only the desirable properties such as high hardness, high toughness and high chemical/erosive and abrasive wear resistance, but they must also be dense, have a fine micro structure and present a smooth surface to the workpiece, with a reduced coefficient of friction compared to that of the uncoated tool. Critically, the coatings must be free from the droplet-like defects that are inherent in cathodic arc-based coating processes.

In general, the performance of micro cutting tools can be enhanced by shape and geometry designs, substrate material and micro structure modification and the use of thin film material coatings. Coated tools of single layer, multilayer hard coatings and/or solid lubricant coatings have been widely proved to lead to significantly longer tool life and increased productivity. Compared to the use of uncoated tools, coated tools enable a wider process window in terms of cutting feeds and depth of cut that can be used. This ensures that a higher material removal rate and shorter cycle times can be promoted. The improved wear performance can often promote a better surface finish and closer dimensional tolerances. When buying cutting tools, coated fine grain carbide tools are interesting because of lower acquisition cost compared to CBN or sometimes even ceramic tools.

3.5 Diamond Micro Cutting Tools

Diamond cutting tools for micro machining, for example, diamond micro turning, fly cutting and so on, are notable for their ultra-high hardness and significantly smaller tool edge radius. The high hardness is important for reducing wear rate and enables machinability of hard and ceramic materials. The ultra-small sharp edge radius enables smaller undeformed chip thickness to be programmed and hence ultra-precision machining. Diamond is a preferable material for machining of hard materials, materials with inclusions and difficult to cut materials, for glass machining or machining other cutting tool materials such as cemented carbides. Typical processes that use diamond tools are diamond micro turning, and milling and drilling with diamond tooling as well as grinding.

Diamond cutting tool materials for precision micro machining can originate from nature as mono-crystalline diamond [26]. These mono-crystals can also be obtained from high temperature high pressure (HTHP) synthesis. The HTHP process enables growing of bigger diamonds which may be required for heavier cuts and thicker chip dimensions. In selecting diamond cutting tools for micro machining a key requirement is to select the tool with the smallest edge radius and least edge waviness.

Single point diamond tools have been used in industry for many years for producing optics and precision components by diamond turning or fly-cutting. Although the cutting edge radius of single point diamond tools can be sharpened down to 20 nm [27] and edge waviness can be controlled to 50 nm or so, the use of single point diamond tools for micro cutting is mainly for simple convex shape micro parts or micro structures, due to a diamond turning tool clearance angle which limits generation of concave shape with deep or steep cavities.

Micro milling with diamond tools are capable of producing complex concave shape micro structures with a nanometric surface comparable with diamond turning technology. Recently

Figure 3.8 Microlens arrays produced by diamond micro milling tools. Courtesy of Moore Nanotechnology Systems LLC Courtesy of Moore Nanotechnology Systems LLC

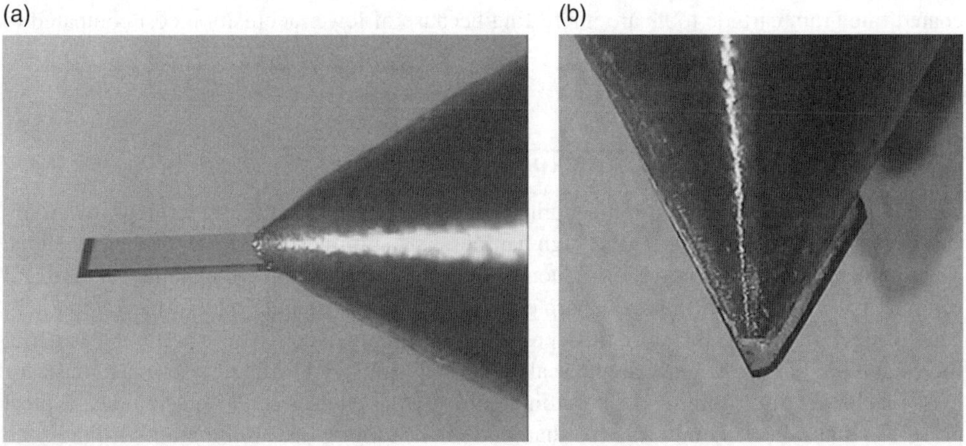

Figure 3.9 Natural diamond (a) cylindrical and (b) ball end micro milling tool [29, courtesy of Contour Fine Tooling Ltd]. Reproduced with permission from Contour Fine Tooling Ltd.

diamond micro milling tools have been used in micro cutting precision micro structures such as micro optics [28]. Figure 3.8 shows micro structures machined by diamond micro milling. All of these micro cutting experiments were carried out in specially designed ultra-precision milling machine tools with aerostatic bearing spindles.

Both cylindrical and ball end natural diamond micro milling tools have been made commercially available. Since diamond turning tools have been made with nanometric accuracy for more than two decades, sharpening natural diamond micro milling to the same accuracy as diamond turning tools is not an issue, but the challenge of fabricating a diamond micro milling tool is to place the edge of the diamond tips precisely with respect to the tool centreline within reasonable tolerances. Figure 3.9 shows natural diamond cylindrical and ball end micro milling tools. Cylindrical milling tools in Figure 3.9a have a diameter of 0.3 mm although smaller

(a) (b)

Figure 3.10 CVD diamond (a) parabolic shape micro milling tool and (b) 500 μm diameter ball end micro milling tool [29, courtesy of Contour Fine Tooling Ltd]. Reproduced with permission from Contour Fine Tooling Ltd.

diameter tools are also commercially available. Ball end mill tools in Figure 3.9b have a standard nose radius of 0.5 mm although smaller radius tools also can be supplied. Due to the difficulty in shaping diamond, unlike tungsten carbide tools, most natural diamond micro milling tools have only one cutting edge.

As discussed in Section 3.3, chemical vapour deposition diamond can be used as a coating material to improve micro tooling performance. CVD diamond can also be used as a tool material to fabricate micro milling tools directly. CVD diamond has similar mechanical and thermal properties as natural diamond, hence CVD diamond tools generally offer similar machining performance as natural diamond tools, but one of the advantages of CVD diamond tools is that they can machine some materials that cannot be processed by natural diamond tooling, such as ferrous metals. Huo *et al.* conducted micro cutting experiments to compare tungsten carbide, CVD diamond and natural diamond tools [30]. Various CVD diamond micro milling tools with complex profiles have been developed. Figure 3.10 shows CVD diamond tools produced by a laser profiling system.

3.6 Micro Cutting Tool Wear

In micro milling, tool wear and tool breakage are factors of major concern when machining hard and difficult to cut materials. Due to the size effect, material spring back, sometimes inefficient cutting due to highly negative rake angles, material phase in-homogeneities and the slender geometry for the solid micro tools promote tool wear. An ideal workpiece material for machining to make high quality micro parts should have low ductility (high hardness), be single phase (eliminate the differential response of material phases to microscale cutting).

A number of workpieces to be machined do not fall into the class of ideal materials for micro manufacture and hence, unpredictable tool life and premature tool breakage before any degree of severe could occur. The classical tool wear progression regime from primary, secondary to tertiary wear can be modified to only primary and secondary wear for some micro tools. The absence of the tertiary wear period presents a challenge for tool condition monitoring in micro machining. Thus micro machining is a knowledge-based application and needs proper selection of tooling, machine tools, cutting conditions and cutting strategies, taking into account size effect theories.

The significantly smaller tool diameters and shaped flute profiles makes the micro tools more fragile, while high stress on micro tools leads to tool deflection and tool run-out. Furthermore, typically small depth of cut and undeformed chip thickness in micro machining increases friction between the cutting tool and workpiece which results in higher tool wear. Wear of micro tools also influences surface finish, burr formation and part accuracy. Tool wear is critical in machining for the reason that it influences productivity in terms of machining quality and economics. In addition, tool wear induces more friction between contact surfaces and hence compromises the surface finish of machined workpiece.

In micro milling, tool wear evaluation is more challenging compared to conventional machining due to difficulty in tool wear measurement. In conventional macroscale milling, tool wear modes are commonly phenomena of progressive material loss on the flank/clearance face (flank wear), rake face (crater wear) or on the cutting edge (nose radius wear) and the most common and critical mode of tool wear is flank wear. The modes of tool wear for micro end mills can be different because of micro size of the tool and the size effect. In addition, tool wear/deterioration for micro tools can be stochastic if the cutting tool edges are not sufficiently strong or the feeds and depth of cut are set to too high values.

For example in micro milling and drilling, the indicators of tool wear for micro tools are, enlargement of the tool edge radius in the form of cutting edge dulling, flank wear, chipping of cutting edges, tool breakage and reduction of tool diameter.

To manage micro machining operations manufacturers need to know when to replace the micro tools during machining before quality is compromised. The international standard of tool life testing in milling ISO 8688-2:1989 [31] recommends a tool life criterion based on the average and localized flank wear to be 0.3 and 0.5 mm respectively. These tool wear and tool life criterion are a considerable percentage of the diameter of the micro tool and in fact larger than some of micro tool. Hence this standard is inappropriate to evaluate the tool life criteria for micro tools. In some cases tool breakage can be used as a criterion especially in micro drilling and milling. While micro carbide cutting tools can have a tool edge radius of a few microns it is found that at tool breakage the edge radius can have grown to 20 to 25 μm when machining hardened material. This growth is also linked to the achievement of minimum chip thickness conditions through tool wear. This is another factor that can contribute to tool life criterion.

In practice the measurement of tool wear for micro tools needs high resolution and high magnification instruments to image the cutting edge radius and tool faces. A good piece of equipment to image micro tools is a scanning electron microscope (SEM). During cutting, the wear of micro tools could be predicted by such process signals as:

- Monitoring cutting force to predict tool wear. An increase in cutting forces can be linked to accelerated tool wear. High force fluctuation due to a continuous shift between ploughing and shearing mechanisms in micro milling can also be linked to sudden tool failure and breakage.
- Burr size can in some cases be an indication of the onset of fracture on the micro tool for drilling process. A rapid increase of burr height can lead to fracture on the micro tool.
- Acoustic emission signals can be used to monitor micro tool condition.

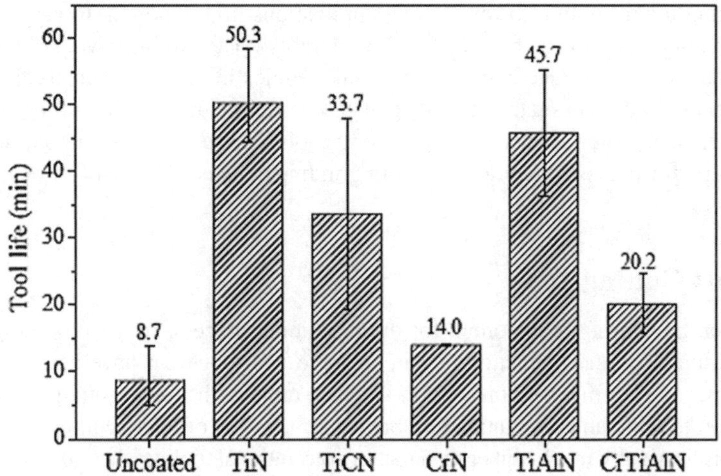

Figure 3.11 Tool life in minutes for 500 μm diameter micro tools in dry slot milling of H13 tool steel of 50HRC hardness (Adapted from [28]). Courtesy of Moore Nanotechnology Systems LLC

Micro machining can be cost effective and the tool life can be very competitive if the process is optimized. For example, in micro machining of 50HRC H13 tool steel, the average time it took for 500 micron diameter micro tools to break in slot milling is shown in Figure 3.11 [32]. The cutting tools had been coated to a 2 μm thick coating layer by CFUBMSIP coating technology. The tool life was 8.7, 14.0, 20.2, 33.7, 45.7 and 50.3 minutes for uncoated, CrN, CrTiAlN, TiCN, TiAlN and TiN coated micro milling tools respectively. This information shows that in micro milling of tool steel, TiN provides better tool life performance compared to TiAlN (a common commercial coating for macro cutting tools). Compared to the uncoated tool (as shown in Figure 3.8), CrN, CrTiAlN, TiCN, TiAlN and TiN significantly improve tool life by 162%, 232%, 388%, 527% and 580% respectively. Thus the use of CFUBMSIP coating technology for micro tools can bring major gains in extending the life of micro tools.

Additionally, coatings such as TiN, TiCN, TiAlN, CrN and CrTiAlN help reduce micro tool cutting edge chipping and edge radius wear as compared to uncoated ultra fine grain carbide end mills. Most coatings lead to reduced burr size compared to uncoated tools. At the beginning of cutting in the (burn-in-period) coatings may not show any improvements in surface finish compared to the uncoated tool. It is common for coatings to have a higher roughness compared to the un-coated tool. This implies the need to produce wear resistant coatings that have an ultra fine surface finish compared to the uncoated tool. The use of CFUBMSIP coating technology for micro tools can bring major gains in extending the life of micro tools. Compared to other coatings or uncoated ultra-fine grain carbide tools, TiN coatings offer the superior performance in micro milling tool steel based on flank wear, chipping, edge radius wear, surface finish, burr size and tool life when milling tool steels [33]. This coating has relatively low hardness coupled with the highest adhesion and fairly good surface finish. This suggests that coating technologies for micro tools should aim at producing tool coatings for the mould and die industry (tool steel) that have an improved adhesion, high toughness and low surface roughness.

In evaluating micro tooling for machining applications it is important to get the micro tool that has high toughness for the base material, preferably coated to improve performance, with relatively smaller positive rake angles or zero rake angles in order to increase edge strength. Manufacturers should focus on the quantity and geometric accuracy of the cutting edge and tool faces as acceptance criteria when buying cutting tools. These issues are important given the variability in performance that could arise from grinding defects and type of carbide grade used.

3.7 Smart Cutting Tools

Tool condition monitoring is important for the machining process. In the process accurate and efficient measurement and prediction of cutting force and tool wear have been a challenge to machining research. For micro cutting, it is essential to monitor the micro cutting process as it suffers from short tool life and pre-mature tool breakage. The cost of micro cutting tools, especially diamond tools, is usually much higher than that of conventional tooling, therefore monitoring tool condition, optimizing micro cutting parameters, and then elongating micro tool life become an important approach to enabling micro cutting more productively and predictably.

The use of a dynamometer to measure cutting forces is a well-known approach and has been widely applied in laboratory environments. But it has several limitations such as high cost, poor performance in harsh industrial machining environments and possible interference with cutting dynamics because of stiffness reduction on the tooling system [34, 35]. On the other hand, because of the construction constraints commercially available dynamometers and force sensors typically specify a bandwidth lower than 5 kHz and even lower in practice. In micro milling operation with tool diameters in the order of 100 microns, in order to achieve even modest machining rates extremely high rotational speeds are required. Micro milling typically operates at a spindle speed in the order of 100 000 rpm and up to 500 000 rpm, therefore the tooth passing frequency for a two flutes micro tool can reach 16 KHz, which is beyond the capability of the measuring range of commercial dynamometers.

Therefore development of an industrial feasible tool condition monitoring device to measure high frequency low amplitude micro cutting force is much needed. Recently a number of novel tool condition monitoring technologies, termed as smart cutting tools in some literatures, have been proposed. Two smart tooling approaches are reviewed in this section; although these approaches were not specifically developed for micro cutting, they provide insight into improving micro tooling technology.

Wang *et al.* proposed a novel smart cutting tool using a separate single-layer piezoelectric film as the sensing unit to detect cutting forces [34]. As shown in Figure 3.12 a single-layer piezoelectric film is placed between the tool insert and tool holder and 6 mm away from the cutting tip. The piezoelectric film acts as a force sensor to detect cutting force during machining. A Kistler dynamometer 9257B was used to measure the cutting force and calibrate the piezoelectric film sensor before using it as a force measurement sensor. Several cutting trails were carried out on a CNC lathe, and the results show that there is a good agreement between dynamometer and piezoelectric film measurement and cutting force during dry machining can be predicted using the smart cutting tool.

Stoney demonstrated a tool condition monitoring method using wirelessly interrogated surface acoustic wave (SAW) sensors which can operate passively and in harsh conditions. A tool holder has been instrumented with the SAW sensor and calibrated by a dynamometer

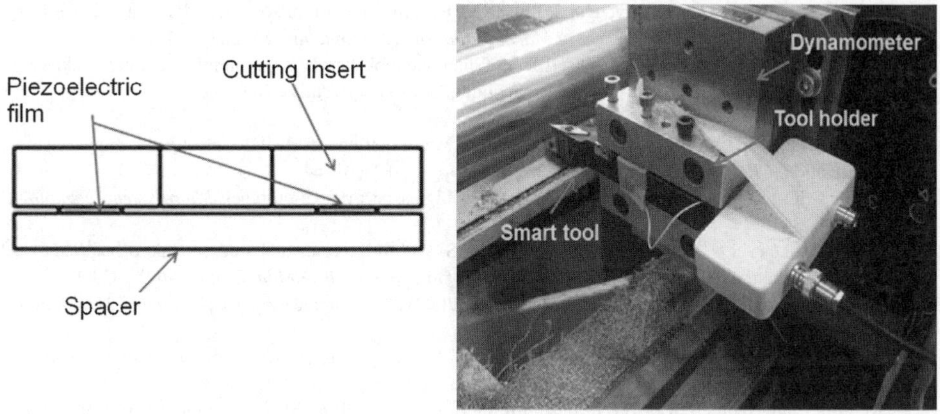

Figure 3.12 Smart cutting tool and Kistler dynamometer on a lathe

Figure 3.13 Tool holder with instrumented SAW for measuring cutting force

as shown in Figure 3.13 where the method was tested in a real machining condition with both constant cut and interrupted cutting operations [36]. Surface acoustic wave (SAW) sensors have higher bandwidth so they are ideal for measuring high frequency low amplitude cutting force and hence overcome many disadvantages of commercial dynamometers. Although this technique can be directly applied to micro turning operation, the use of SAW sensors for micro milling process needs to be further investigated.

References

[1] Aramcharoen, A. and Mativenga, P.T. (2009) Size effect and tool geometry in micro milling of tool steel, *Precision Engineering*, 33, 402–407.

[2] Mian, A.J., Driver, N. and Mativenga, P.T. (2011) Estimation of minimum chip thickness in micro-milling using acoustic emission. *Proceedings of the Institution of Mechanical Engineers, Part B: Journal of Engineering Manufacture*, 225,1535–1551.

[3] Mian, A.J., Driver, N. and Mativenga, P.T. (2011) Identification of factors that dominate size effect in micro-machining, *International Journal of Machine Tools and Manufacture*, 51, 383–394.

[4] Yuan, Z.J., Zhou, M. and Dong, S. (1996) Effect of diamond tool sharpness on minimum cutting thickness and cutting surface integrity in ultra precision machining. *Journal of Materials Processing Technology*, 62(4), 327–330.

[5] Weule, H., Hüntrup, V. and Tritschler, H., (2001) Micro-cutting of steel to meet new requirements in miniaturization. *CIRP Annals – Manufacturing Technology*, 50(1), 61–64.

[6] Friedrich, C.R., (2002) Micromechanical machining of high aspect ratio prototypes. *Microsystem Technologies*, 8(4–5), 343–347.

[7] Ng, C.K., Melkote, S.N., Rahman, M. and Kumar, A.S. (2006) Experimental study of micro- and nano-scale cutting of aluminium 7075-T6. *International Journal of Machine Tools & Manufacture*, 46(9), 929–936.

[8] Alting L., Kimura, F., Hansen, H.N. and Bissacco, G., (2003) Micro engineering, *CIRP Annals – Manufacturing Technology*, 52(2), 635–657.

[9] Schaller, T., Bohn, L., Mayer, J. and Schubert, K. (1999) Microstructure grooves with a width of less than 50 μm cut with ground hard metal micro end mills. *Precision Engineering*, 23(4), 229–235.

[10] Egashira, K. and Mizutani, K. (2002) Micro-drilling of monocrystalline silicon using a cutting tool, *Precision Engineering*, 26 263–268.

[11] Chern, G-L., Wu, Y-J.E., Cheng, J-C. and Yao, J-C. (2007). Study on burr formation in micro-machining using micro-tools fabricated by micro-ECM, *Precision Engineering*, 31, 122–129.

[12] Adams, D.P., Vasile, M.J., Benavides, G. and Campbell, A.N. (2011) Micro milling of metal alloys with focused ion beam-fabricated tools, *Precision Engineering*, 25, 107–113.

[13] Vasile, M.J., Friedrich, C.R., Kikkeri, B. and McElhannon, R. (1996) Micrometer-scale machining: tool fabrication and initial results, *Precision Engineering*, 19, 180–186.

[14] Friedrich, C.R. and Vasile, M.J. (1996) The micro milling process for high aspect ratio microstructure, *Microsystems Technologies*, 2, 144–148.

[15] Schaller, T., Bohn, L., Mayer, J. and Schubert, K. (1999) Microstructure grooves with a width of less than 50 um cut with ground hard metal micro end mills, *Precision Engineering* 23, 229–235.

[16] Bouzakis, K.D., Michailidis, N., Hadjiyiannis, S., Efstahiou, K., Pavlidou, E., Erkens, G. Rambadt, S. and Wirth, I. (2011) Improvement of PVD coated inserts cutting performance, through appropriate mechanical treatments of substrate and coating surface. *Surface & Coatings Technology*, 146–147, 443–450.

[17] Kelly, P.J. and Arnell, R.D. (2000) Magnetron sputtering: a review of recent developments and applications. *Vacuum*, 56(3), 159–172.

[18] Teer, D.G. (1996) Magnetron sputter ion plating, US Patent 5.556.519 UK Patent GB2 258 343 B.

[19] Monaghan, D.P., Teer, D.G., Laing, K.C., Efeoglu, I. and Arnell, R.D. (1993) Deposition of graded alloy nitride films by closed field unbalanced magnetron sputtering. *Surface & Coatings Technology*, 59, 21–25.

[20] Luo, Q., Rainforth, W.M. and Munz, W-D. (1999) TEM observations of wear mechanisms of TiAlCrN and TiAlN/CrN coatings grown by combined steered-arc/unbalanced magnetron deposition *Wear*, 225–229(1), 74–82.

[21] Luo, Q., Rainforth, W.M. and Donohue, L.A., Wadsworth, I. and Munz, W.D. (1999) Tribological investigation of TiAlCrN and TiAlN/CrN coatings grown by combined steered-arc/unbalanced magnetron deposition. *Vacuum*, 53(1–2), 123–126.

[22] Holleck, H. (1986) Material selection for hard coatings. *Journal of Vacuum Science Technology A*, 4(6), 2661–2669.

[23] Weinert, K., Inasaki, I., Suterland, J.W. and Wakabayashi, T. (2004) Dry machining and minimum quantity lubrication. *CIRP Annals – Manufacturing Technology*, 53(2), 511–537.

[24] Dudzinski, D., Devillez, A., Moufki, A., Larrouquere, D., Zerrouki, V. and Vigneau, L. (2004) A review of developments towards dry and high speed machining of Inconel 718 alloy. *International Journal of Machine Tools & Manufacture*, 44(4), 439–456.

[25] Zhou, L., Wang, C.Y. and Qin, Z. (2009) Tool wear characteristics in high-speed milling of graphite using a coated carbide micro end mill. *Proceedings of the Institution of Mechanical Engineers, Part B: Journal of Engineering Manufacturing*, 223, 1–11.

[26] Gaebler, J. and Pleger, S. (2010) Precision and micro CVD diamond-coated grinding tools. *International Journal of Machine Tools & Manufacture*, 50(2010), 420–424.

[27] Miyamoto, I., Ezawa, T. and Nishimura, K. (1990) Ion beam machining of single-point diamond tools for nano-precision turning, *Nanotechnology*, 1, 44–49.

[28] Hurst, P.P. (2012) Moore Nanotechnology Systems' Experience with Ultra Precision milling of Optical Components, 3rd Aachen Precision Days.

[29] Contour Fine Tooling Ltd http://contour-diamonds.com/GB/mircromilling.html

[30] Huo, D. and Cheng, K. (2010) Experimental investigation on micromilling of oxygen-free, high-conductivity copper using tungsten carbide, chemistry vapour deposition, and single-crystal diamond micro tools. *Proceedings of the Institution of Mechanical Engineers, Part B: Journal of Engineering Manufacture*, 224, 995–1003.

[31] International standard: ISO 8688-2, Tool life testing in milling-Part 2: End milling 1989.

[32] Aramcharoen, A. (2009) Milling of Hardened Tool Steel: Process Fundamentals and Extended Life Tool Coatings, Doctor of Philosophy, The University of Manchester.

[33] Aramcharoen, A., Mativenga, P.T., Yang, S., Cooke, K.E. and Teer, D.G. (2008) Evaluation and selection of hard coatings for micro milling of hardened tool steel. *International Journal of Machine Tools and Manufacture.* 48, 1578–1584.

[34] Wang, C., Rakowski, R. and Cheng, K. (2012) Design and analysis of a piezoelectric film embedded smart cutting tool, *Proceedings of the Institution of Mechanical Engineers, Part B: Journal of Engineering Manufacture*, in press.

[35] Stein, J.L. and Huh, K. (2002) Monitoring cutting forces in turning: A model-based approach. *Journal of Manufacturing Science and Engineering, Transaction of ASME*, 124, 27–31.

[36] Stoney, R., Donohoe, B., Geraghty, D. and O'Donnell, G.E. (2012) The development of surface acoustic wave sensors (SAWs) for process monitoring, 5th CIRP Conference on High Performance Cutting 2012, 586–591.

4

Ultraprecision and Micro Machine Tools for Micro Cutting

Christian Brecher and Christian Wenzel
Fraunhofer IPT

4.1 Introduction

There is a continuing drive towards miniaturization of workpieces in the field of micro fluidics, micro mechanics, micro-electronics and micro optics which has left its prototype state and the integration of these components into functional groups of microsystems is state of the art [1].

The mechanical machining by milling, ruling, turning and grinding is of high importance to many such high precision and miniaturized products. The machines being used are designed to achieve form accuracies down into the 100 nm and surface roughness in the Angstrom range. Only the perfect tuning of all relevant components in such a machine system will lead to the highest performance.

Due to the high requirements on precision, automation is only in the early stages. Often such machines are manually operated by highly skilled operators. Depending on the application and the part complexity as well as quantity, automation starts to enter the field of ultra-precision technology and micro machining so to speak for directly machined components. In the large field of the optical and micro-mould making industry, manual tool exchange, manual workpiece alignment and operator vision-based referencing is quite common.

In the case of complex optical surfaces, the program generation is also not standardized and supported by CAD/CAM software for the ultra-precision machining, but is individually programmed and optimized, leading to only a few companies really being able to generate such high quality parts.

In the case of complex optical surfaces, the program generation is also not standardized and supported by CAD/CAM software for the ultra-precision machining, but is individually programmed and optimized, leading to only a few companies really being able to generate such high quality parts.

Examples of directly machined products are CO2 laser mirrors for beam guidance being diamond turned or scanning mirrors for cashiers with mainly flat fly cut surfaces.

Micro-Cutting: Fundamentals and Applications, First Edition. Edited by Kai Cheng and Dehong Huo.

Figure 4.1 Applications of precision and ultra-precision components

Direct machining can also be found in geometrically more complex components such as intraocular lenses with superimposed micro structures or packaging components with high tolerance requirements being used in opto-electronical devices.

The mould making industry for injection and injection compression moulding is used for small part numbers and individually changing requirements. Manual machine adjustment and high efforts for machine set up are state of the art. Such moulds are used for LED illumination optics and automotive sensors, light guides, micro mechanical and micro fluidic components and many others. Figure 4.1 shows some applications of precision and ultra-precision components.

In the following section 4.2, the characteristic machine components for precision and ultra-precision machines will be described and compared. One can find differences in the machine setup depending on the process, the envisaged part geometry, and on the required precision.

A specific description of ultra-precision diamond turning machines will follow in Section 4.3 and will also include auxiliary units such as fast tool servos and aspects of control units.

Section 4.4 covers milling machines starting with an overview on precision machine tools and their setup. The second part of the chapter focuses on ultra-precision technology for precision milling of complex optical surfaces.

4.2 Components of High Precision Machine Tools

The main components defining the accuracy of a machine tool are the materials used for the machine structural elements and the moving components, the used guidance systems, the drive systems for linear and rotory positioning, the measurement systems and, of great influence on the overall performance, the used control systems with the associated amplifiers.

4.2.1 Machine Base Materials

To provide the best attributes for high and ultra-precision machining one can find an overall trend according to the machine base materials. Because of the characteristically low cutting forces (<100 N) and the aim of highest form accuracy and surface quality even over long machining times, the machine base material has to be thermally very stable. This feature is expressed in a low thermal expansion coefficient and a low specific heat capacity. To improve the surface quality, good damping properties of the base material are needed. These attributes are best fulfilled using granite or mineral casting as material.

Mostly granite is used in the field of ultra-precision machine beds because of its very low thermal expansion coefficient of 6.5 μm/mK, 2.5 times less than the one of mineral casting. Additionally, granite with a density of 2.8 kg/dm^3 is beneficial in terms of Eigen frequency and dynamic properties. Figure 4.2 shows a granite base for a large area ultra-precision machine. Figure 4.3 shows the sensitivity to exposure of liquids for granite, CFRP and polymer concrete materials. Air-bearing guideways can be directly machined into the granite material as often being found in coordinate measuring machines or wafer positioning stages.

Machine systems used for chip removal are usually equipped with hydrostatic guideways for improved damping. The granite can also be directly flooded with oil applying the guidance surface on the stone. The oil does enter the stone surface, but saturation takes place not causing accuracy losses in a relevant dimension [2].

With the process steps of sawing, milling, grinding and subsequent manual lapping good geometrical flexibility is given with accuracies below 1 μm/m straightness.

For precision machine systems being sold in greater quantities, mineral casting materials are a good alternative to granite. Unless there is a granite bed, first a negative form of the machine component needs to be produced. This form subsequently will be filled with a mixture of small stones and epoxy resin. Using a reusable form, this procedure can be repeated many times with a high rate of repeatability. The material properties are comparable to the ones of granite, only the heat expansion is about two to three times higher (Table 4.1).

Steel bars as well as pipes can be included during the casting process so that both guide ways or internal channels for cable guidance or cooling can be integrated. High precision surfaces needed as reference in a machine bed for example are often generated subsequently to the casting procedure by a precision casting process using products such as moglice (DIAMANT Metallplastic GmbH) [3] or SKC (SKC Gleittechnik GmbH) [4]. Such materials have a low viscosity and therefore exactly fill up to the precision gauge used as reference for the precision surface. Only a thin layer is applied to achieve accuracies in the one digit micron range per meter [5].

Due to the small modulus of elasticity and the tensile strength, moved components in precision machines such as slides or rotary tables usually are made out of cast iron, steel or aluminium, following design aspects for specific stiffness, feasible light weight design, or thermal requirements. Of interest is that all such materials have nearly the same ratio between the Youngs Modulus and the density. Conventional milling, turning, grinding and lapping operations are conducted to generate the needed geometry and accuracy.

The applied symmetry in the design of the machine tool is crucial for the performance of the system and every detail needs to be considered for precision machine tools. Symmetry accounts for thermal sources and resistances in the structure, the force flux as well as dynamic excitation. Combining different materials, additional aspects in terms of symmetry need to be considered.

Figure 4.2 Granite base of large area ultra-precision machine (Eitzenberger GmbH, Fraunhofer IPT)

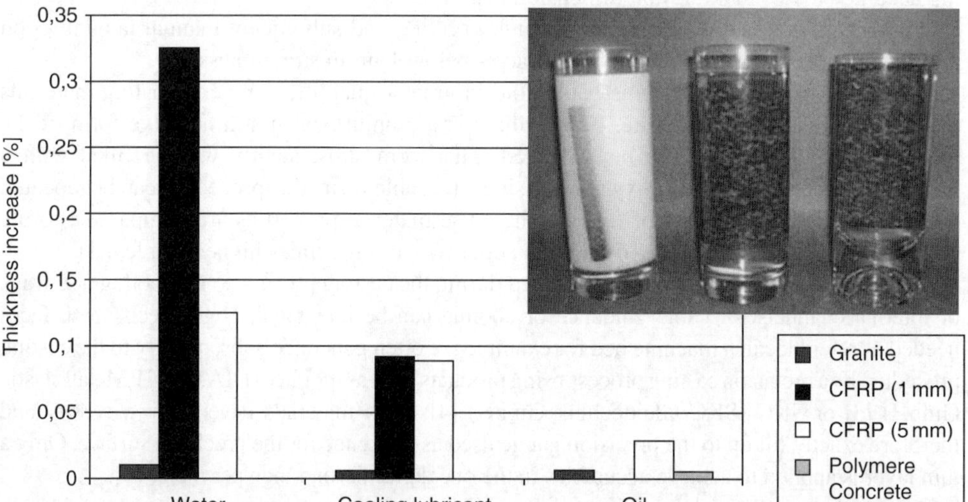

Figure 4.3 Sensitivity to exposure of liquids of granite, CFRP and polymer concrete

4.2.2 Drive Systems

Depending on the process and the needed path feed, different drive technology can be used. To gain the necessary position accuracy the influence of friction has to be minimized. Backlash on the other hand will reduce the position resolution. As a thermal heat source a drive needs to be considered as a disturbing element for the temperature profile in a precision machine system. Such requirements often turn out to be counteracting when detailing a machine design.

Figure 4.4 Customized mechanical design in granite. Image courtesy of JFA

Table 4.1 Material properties for structural elements in precision machine tools

Characteristic	Unit	Steel	Grey Cast Iron	Granite	Mineral Cast	Carbon Fiber Reinforced Plastics
Compression Strength	N/mm²	250–1200	600–1000	70 –300	140–170	n.A.
Tensile Strength	N/mm²	400–1600	150–400	30–35	25–40	400–2400
Modulus of Elasticity	kN/mm²	210	80–120	35–90	30–40	48–360
Heat Conductivity	W/mK	50	50	1,7–2,4	1,3–2,0	1–50
Coefficient of thermal expansion	μm/mK	12	10	6,5–8,5	12 –20	(−1) –0
Density	g/cm³	7,85	7,15	2,9–3,0	2,1–2,4	1,6
Damping (logarithmic decrement)	–	0,002	0,003	0,015	0,02–0,03	n.A.

In many ultra-precision machine tools the use of linear direct drives is state of the art. The theoretical stiffness of these drives is infinite, as long as the force is below its maximum force. The abandonment of mechanical transmission elements supports high dynamics, especially high jerks and KV-factors, which characterizes the ability of high precision at high speeds.

$$K_v = \frac{v}{\delta} \tag{4.1}$$

The KV-factor is the speed amplification factor of the position control where (v) is the path speed and (δ) is the contouring error. Possible KV-factors of linear drives are up to 25 m/(min*mm) because of the lack of mechanical transmission elements. Standard lead screw drive systems reach only up to 3.5 m/ (min*mm) [6].

Two designs of linear direct drives are commercially available, iron core and ironless drives. Different from the ironless motor the iron core motor uses a ferrous base in between the windings. The iron core motor only needs one counter side of magnets leading to a compact design. Ironless motors are built up with two counter facing rows of magnets in between which the spool is moved.

One characteristic shared by iron core drives are strong magnetic forces between the stator and the moving part of the motor in the range of several thousand Newtons. The machine structure and guidance system has to resist the magnetic force. Such magnetic forces are not constant over the moving range of the motor, the so called cogging effect is another limiting factor in terms of precision and continuous movement of iron core direct drives, especially if continuous movement is needed for diamond turning. Advantages of this technology are a slim design and cooling facilities installed in the windings of the drive.

Commonly used for ultra-precision machines are ironless direct drives with three phases, due to their ability to ensure highest positioning accuracy and extremely continuous motion. Intensive investigation on direct drives in combination with air bearings proved possible steps of 10 nm and below depending on the control. In addition, the maximum possible acceleration for ironless motors is higher compared to the iron core ones (Figure 4.5).

In fast tool systems, for example, with a limited stroke the use of ironless motors with one phase only is common. The so called voice coil has characteristically a very linear current force proportion and can be precisely tuned with an adequate controller setup.

For both systems, ironless motors with three phases as well as voice coil drives, heat evolvement in the moving motor part carrying the windings is a critical issue. Commercially available systems only provide limited possibilities for water chilling. Voice coil actuators can reach temperatures up to 100 °C and higher at the operating point.

If high damping properties are needed, for example in grinding machines, another possible drive system is a hydrostatic lead screw. Hydrostatic lead screws are used for wear free transfer of rotation into linear move. The nut is supported by hydrostatic pockets on top of the screw, which apply a pretension in the system, free of backlash and with nearly no friction at low motion. This makes very slow and very small movements possible. Even at maximum load and high speeds in continuous or oscillation move, stiffness and precision remains constant. Hydrostatic lead screws offer on the one hand very high axial damping, due to the squeeze film effect. On the other hand expensive and large size hydrostatic systems have to be integrated in the machine tool and the thermal load depending on the oil viscosity limits the use of hydrostatic lead screws as a standard component. While minimizing the size of machine tools the additional needed space of a lead screw is an additional disadvantage [6].

High precision mechanical lead screws are used in micro milling and conventional precision machining systems. Depending on the design of the lead screw (ball, thread) high precision positioning with steps below 100 nm can be realized. For continuous motion requirements and high dynamic applications however, lead screws do not reach the performance of direct drive units. In ultra-precision machining systems such drives are not state of the art any more.

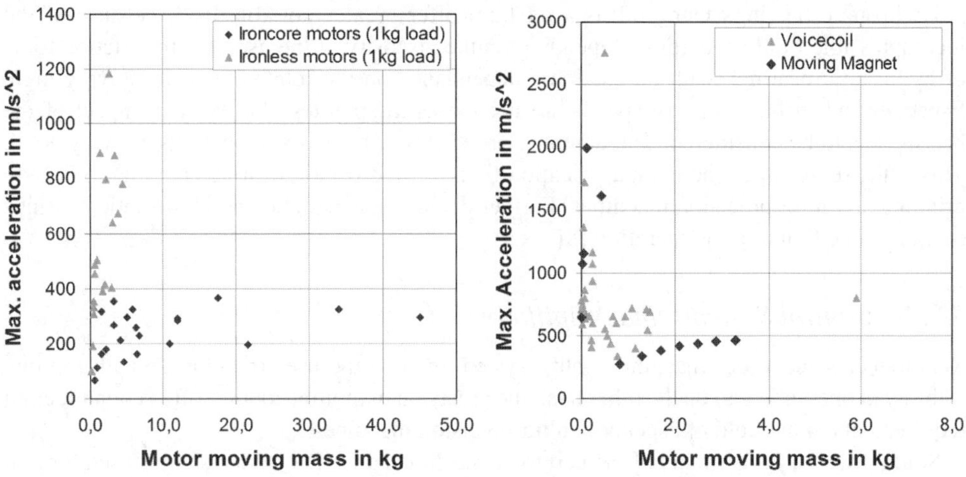

Figure 4.5 Comparison of commercial linear iron core and ironless motors/Voicecoil and Moving Magnet

4.2.3 Guidance Systems

Depending on the drive system, the machining process and the workpiece requests several guidance systems can be used in high precision machines.

Most conventional machine tools are equipped with ball or roller/needle bearings. Using high precision grade guidances, linearity below 1.5 μm/300 mm can be reached with non-recirculating needle bearings (Schneeberger). This allows the integration of these components in high precision machine tools [7]. If optical surfaces are needed, ball bearings can normally not be used because the moving noise will be visible as oscillation on the machined surface. In combination with needle bearings very high stiffness at low construction space can be achieved. This property allows for the integration of ball bearings in compact machine tools with high process forces like milling machines, if no optical surfaces are needed. The characteristic small size is a key factor for compact axis designs to minimize Abbé errors.

Free of direct mechanical contact are fluidic bearings and guideways based on aerostatics or hydrostatics. A thin film of air or oil separates the slide from the guideway and allows for linear or rotary motion in one degree of freedom. The gap for the fluid is in the range of 3 to 10 μm in aerostatic systems and between 7 and 20 μm in hydrostatic ones. In order to preload such a system and to introduce a sufficient stiffness, a wrap-around is needed, usually leading to a fairly large design compared to roller based systems. Alternatively, magnets are used to preload aerostatic bearings and guideways. For hydrostatic guideways the preloading with largely dimensioned vacuum pockets is known.

Fluids, air and oil have different properties, which mainly defines the bearing characteristics. Air as a medium is compressible, limiting the load capacity of the bearings and reducing the damping properties. The system can start to oscillate at stimulation or due to wrong alignment of the aerostatic pockets (Air Hammer). Using aerostatic bearings in combination with low impacts, friction free motion of the slides even at the highest speed can be reached. Precision milling and turning spindles are often equipped with an aerostatic bearing due to high relative speeds between rotor and stator [5].

Hydrostatic bearings using oil as a medium offer greater possible load, because of the incompressible fluid properties. Depending on the viscosity of the used oil, the friction force of hydrostatic bearings with its quadratic dependance on the relative velocity is very high. Especially when high relative speeds are needed as in spindles, the thermal impact of the bearing has to be considered. The damping properties of hydrostatic bearings are very good. This is the reason why the main application field for hydrostatic bearings are high precision grinding and ultra-precision machines. Further details on aerostatic and hydrostatic bearings design can be found in the literature [8].

4.2.4 Control Systems and Amplifiers

Requirements for a tool machine control system include the user friendliness with runtime stability in operation and on the other hand the precision in motion control, often counteracting requirements in the field of especially ultra-precision machines.

Starting with a CAD model of the part to be machined a CAD/CAM system is used for tool path planning providing G-Code for the control system. Geometrically simple parts can be machined directly at the machine with manual G-Code programming or on machine software such as ShopMill (Siemens) or MillPlus(Heidenhain). Some machine control systems are capable of reading polynomial based paths (e.g. Spline, Nurbs), but in a second step they internally break down the polynomial description into discrete points for tool path, velocity and acceleration calculation [6].

In dependently of the approach, the density of the planned set points in G-Code is directly proportional to the deterministic accuracy of the part. Using a low density of points for the parts surface description, the machine control interpolates in between. The operator has little to no influence on this interpolation and therefore cannot control the parts precision. Errors in the range of several microns can easily occur due to unsuited data preparation.

Lowering the set point distance eventually leads to large program files with typically several hundred megabytes of size that are difficult to handle for the machine tool control systems in terms of memory. Either such a data file is internally split and subsequently loaded or is streamed continuously from the hard drive. Both approaches can lead to set point losses during operation if the calculation performance is insufficient.

Next to the supply of set points, the online processing with a high bandwidth is crucial for the performance of the machine tool control. The processing covers the preparative internal calculation routines that are necessary for path control, the axis position and velocity control, the data transmission via bus systems to the amplifier and finally the current control performance of the amplifier itself for the drive system.

State of the art amplifiers for machine tools are digital with clock rates between 4 and 16 kHz. For certain applications, especially in ultra-precision applications, the use of analog amplifiers is preferred to avoid the disturbance of the digital noise [9, 10].

4.3 Diamond Turning Machines and Components

Diamond turning machines are commonly used to manufacture rotational symmetric parts of non-ferrous metal, polymers (especially PMMA) and brittle hard materials such as germanium, silicon and others. Furthermore, non-rotational workpieces are being manufactured on

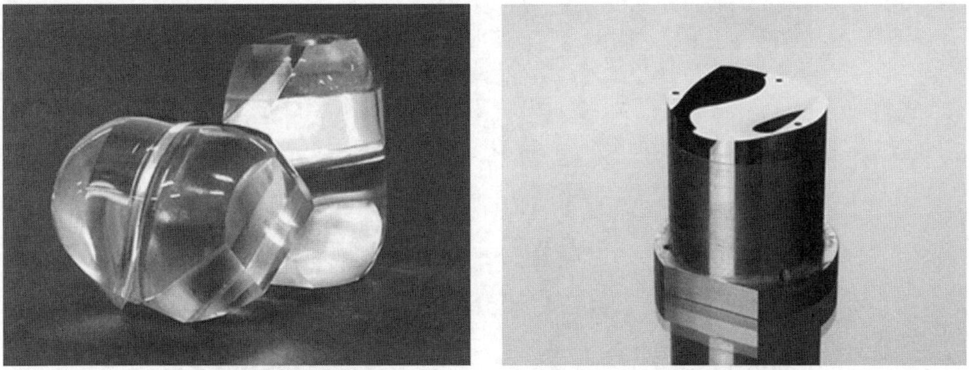

Injection molded freeform lens Mold insert for injection molding

Figure 4.6 Freeform lens for LED illumination and mould insert

ultra-precision lathes using fast tool servo systems or slow-tool functionality as dynamic auxiliary axis in modern ultra-precision machines.

The batch size of diamond turned workpieces varies from individual items to several thousand pieces per batch. The probably most prominent application of high batches being diamond machined are varifocals. Fast tool machining enables the manufacturing of freeform parts for logically varying curvatures on the lens which allow toric lens design and correction of different visual defects of each individual customer.

Small batches are being produced for mould inserts for injection and injection compression moulding of optical polymer parts. Examples are the mass production of contact lenses as well as focusing optics for solar, sensor and LED applications. In particular, focusing optics in LED applications increasingly incorporate complex freeform surfaces which demand specialized manufacturing technologies such as fast tool turning. Familiar examples are LED headlamps in the automotive sector and concerted illumination of street lamps (Figure 4.6).

Other examples for small batch series are metal optics used in laser applications for beam shaping and beam guidance of CO2 laser systems. Some examples are shown in Figure 4.7. Phase modulation mirrors (1) are used for reducing the coherence wavelength of the laser beam. Helix mirrors (2) are used in coaxial lasers and facetted mirrors (3) are used for combined beam shaping and guiding for increased efficiency due to the reduced number of optical elements within the optical path.

Examples of mostly individual items are parts of large telescope mirrors. The single parts of the mirror having spherical radii high above the machinable radii of conventional ultra-precision lathes are being manufactured in off-axis mode using fast tool servo assistance in the manufacturing process.

4.3.1 Typical Machine Setup

Figure 4.8 shows the typical design of a state of the art UP-lathe where the C-axis or spindle is placed upon the first linear axis (x-axis) and the b-axis (mostly optional) is positioned on top of the second linear axis (z-axis). The given design ensures better stability, stiffness and accuracy compared to serial design where more than two axis are stacked.

Figure 4.7 Diamond machined laser mirrors (FHG IPT)

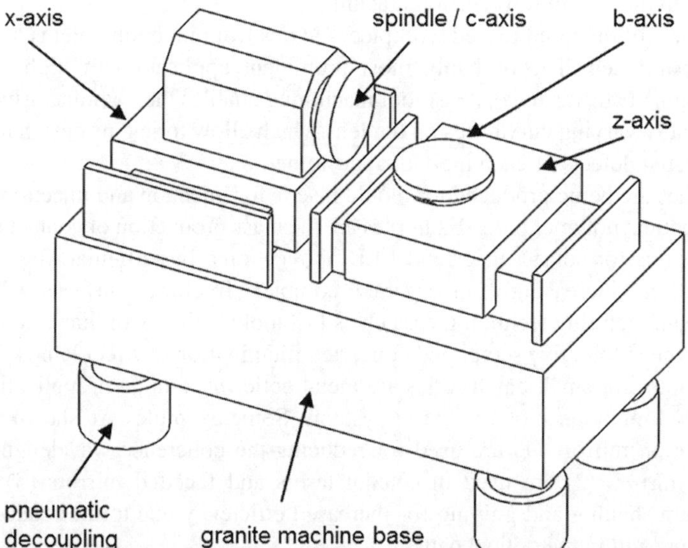

Figure 4.8 Typical 4-axis design of state of the art UP-lathe

The equipment of ultra-precision lathes with mainly hydrostatic bearing systems in all linear slides gives straightness of below 0.1 μm/100 mm moving length. The hydrostatic bearings ensure stick-slip free movement for smallest movements down to the single-digit range of nanometers. Additionally, high stiffness for minimized displacement in the bearing system combined with high damping ensures highest form and surface quality.

Ironless linear motors are state of the art in machine axis to overcome the cogging-effect. Feedback resolution systems in UP lathes are unexceptional optical encoders and incremental scales. Well established manufacturers of encoder systems are Haidenhain, Renishaw and

Sony offering linear scales down to an optical resolution of 138 nm (Sony) and down to 0.034 nm interpolated resolution (Sony).

Due to the high circumferential speed, the spindles usually feature air bearing technology to reduce the heat development in the spindle at high rotational speeds. Nevertheless, high load capacities up to 102 kg (Precitech Heavy duty HD-160 spindle) and highest axial and radial motion accuracy below 12.5 nm (PI 5.5) can be achieved. The damping of the air bearing is again beneficial for the surface quality of the produced workpieces. To allow fast tool servo manufacturing and C-axis functionality many spindles are equipped with high resolution feedback systems.

To finally ensure the desired accuracy, a control system is needed which handles the high accuracies in short control cycles to guarantee accurate movement along the desired tool paths without interpolation errors and loss of set-point data. The most common control system PMAC is supplied by the manufacturer DeltaTau Data Systems Inc. and is used for example in Moore and LT Ultra machines. Precitech machines use a PC based control system by PDMIA running on QNX real time operating systems.

4.3.2 Market Comparison

The market of ultra-precision lathes is rather small compared to the market of conventional machine tools. Three companies represent the available spectrum in the market of ultra-precision lathes, LT Ultra-Precision Technology GmbH, AMETEK Precitech Inc. and Moore Nanotechnology Systems, LLC. Additionally, suppliers for special purpose machines are Schneider GmbH & Co. KG, Satisloh GmbH and DAC International, Inc. Schneider and Satisloh supply machines for the production of ophthalmic lenses whereas DAC supplies specialized lathes for the production of contact lens manufacturing. As the machines from Satisloh and Schneider differ widely from classical UP lathes, the machines are not included in the comparison in Table 4.2. In addition each company supplies machines differing from the machines shown.

The ultra-precision lathe and ultra-precision milling machine Moore Nanotech (USA) 350 FG is designed for manufacturing micro structures and optics. The hydrostatic linear axis is driven by linear direct drives and is positioned in a cross table design. The machine base is made of monolithic mineral casting.

The basic machine setup features two linear axis and one spindle/rotary axis (c-axis) which is placed on top of the x-axis. Additionally, a second rotary axis (b-axis) can be placed on top of the z-axis and a vertical linear axis (y-axis) can be placed between x- and c-axis. Hence 5-axis-machining is possible using a high-frequency milling spindle. The machine is controlled by DeltaTau PMAC and slow-tool machining with a dynamic movement of the z-axis is possible.

AMETEK, Precitech Inc. (USA) ‘Nanoform 700 ultra’ machine incorporates the same basic machine concept as the ‘350 FG’. Also, a x-, z- and c-axis belong to the standard configuration in cross-table design. The optional b-axis can be placed on top of the z-axis. A y-axis is available for a similar machine configuration called ‘Freeform 700a’. The linear axes and b-axis also use linear direct drives and hydrostatic bearings.

Compared to the ‘350 FG’ the machine base is made of natural granite. The machine control is a Precitech specific control system which is based on QNX real time

Table 4.2 Comparison of UP-lathes

	Precitech Nanoform 700	Moore 350FG	LT Ultra MTC 250	DAC ALM Lens Lathe
Axes	X: hydrostatic, 350 mm travel, 4 m/min Z: hydrostatic, 300 mm travel, 4 m/min B (optional): hydrostatic	X: hydrostatic, 350 mm travel, 4.5 m/min Z: hydrostatic, 300 mm travel, 4.5 m/min B (optional): hydrostatic	X: hydrostatic, 270 mm travel, 1.5 m/min Z: hydrostatic, 270 mm travel, 1.5 m/min B (optional): aero- or hydrostatic	X: aerostatic, 305 mm travel, 18 m/min Z: aerostatic, 203 mm travel, 18 m/min
Spindle	Air bearing, 7.000 RPM (3000 RPM as C-axis)	Air bearing, 10.000 RPM (3000 RPM as C-axis)	Air bearing, 4.500 RPM (2000 RPM as C-axis)	Air bearing, 10.00 RPM (2000 RPM as C-axis)
Load capacity spindle	85 kg @ 100 psi	70 kg @ 100 psi/85 kg @ 145 psi	30 kg	Not specified
Resolution	0.032 nm linear, 0.026 arc-sec C-Axis	0.034 nm linear, 0.07 arc-sec C-Axis	0.034 nm linear, 0.026 arc-sec C-Axis	Sub-nanometer, sub-degree
Specified form accuracy	< 0.125 μm (non specified workpiece)	< 0.15 μm (75 mm diameter, 250 mm convex sphere)	< 0.100 μm (100 mm diameter)	+/− 2.5 μm center thickness, +/− 5 μm diameter, +/− 2.5 μm radius of curvature
Specified surface roughness	< 1.0 nm Ra	< 3.0 nm Ra	< 2.0 nm Ra	< 2.0 nm Ra
Control System	UPx controller (QNX)	Delta Tau PC based	Delta Tau PC based	DAC Motion Control System

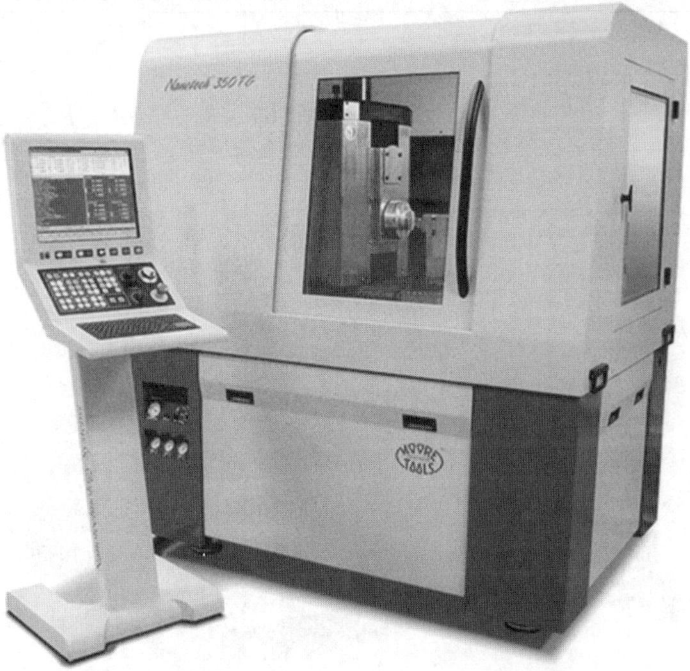

Figure 4.9 Moore Nanotech '350 FG'. (Courtesy of Moore Nanotechnology Systems) Copyright Moore Nanotech

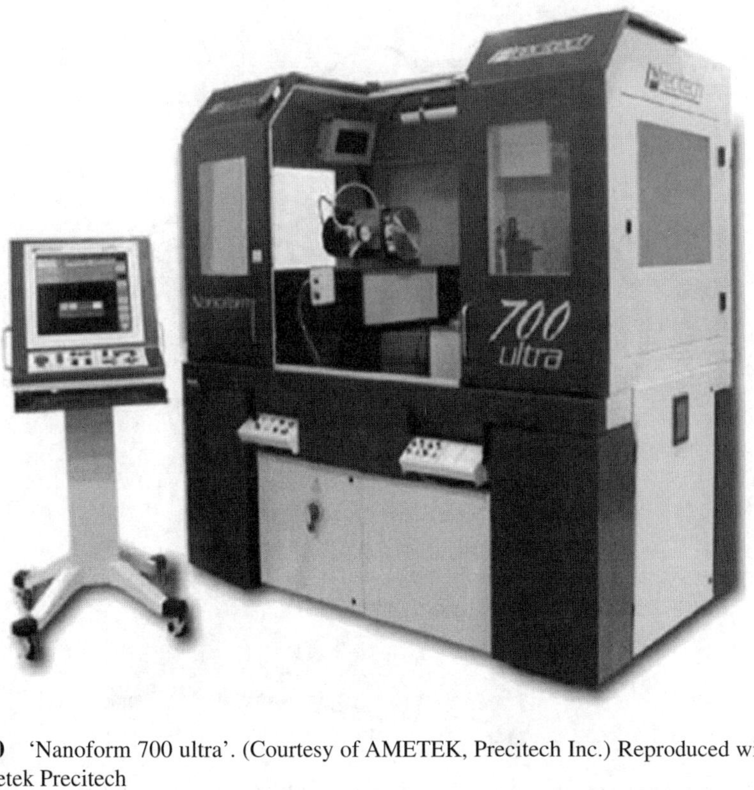

Figure 4.10 'Nanoform 700 ultra'. (Courtesy of AMETEK, Precitech Inc.) Reproduced with permission of Ametek Precitech

Figure 4.11 'MTC 250'. (Courtesy of LT Ultra) Image courtesy of LT Ultra

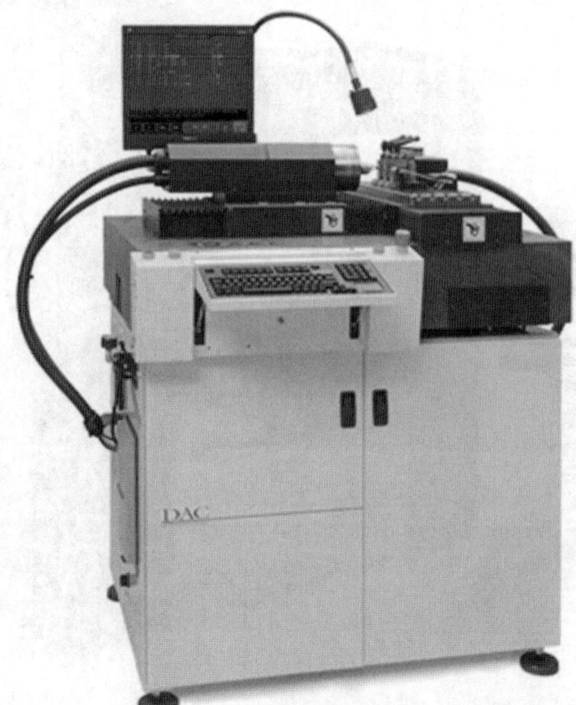

Figure 4.12 "2Axis ALM Lathe - Series IV". (Courtesy of DAC International) Image courtesy of DAC International

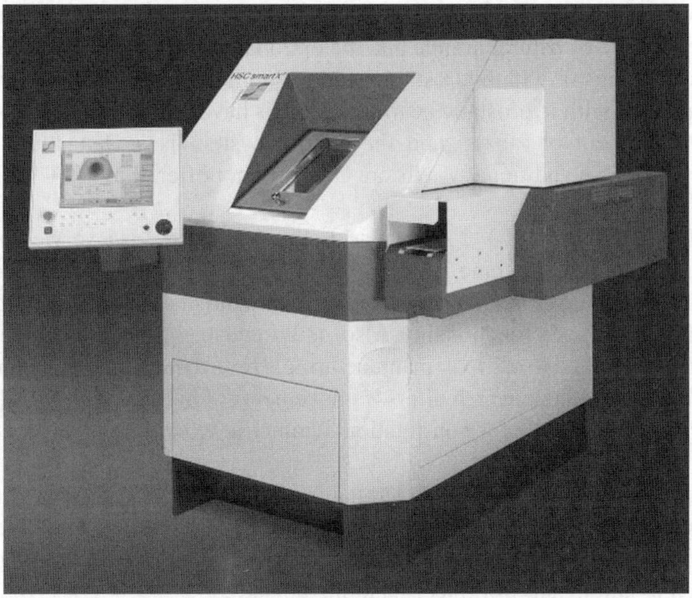

Figure 4.13 'HSC smart X'. (Courtesy of Schneider) Image courtesy of Schneider GmbH

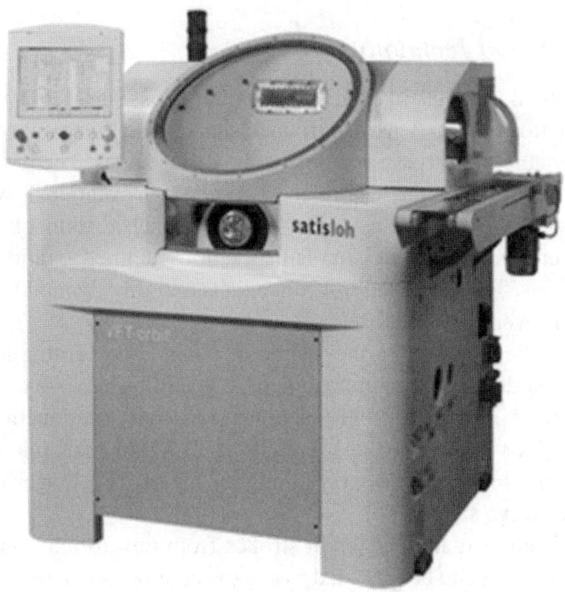

Figure 4.14 'VFT-orbit'. (Courtesy of Satisloh) Image courtsey of Satisloh GmbH

operating systems and freeform machining is also possible using dynamic movements of the z-axis in slow-tool mode.

LT Ultra's (Germany) 'MTC 250' is a smaller machine compared to the '350 FG' and the 'Nanoform 700 ultra'. The larger machine 'MTC 400' has a similar setup. Again, the machine design is a cross-table design coming with x-, y- and c-axis as standard equipment. The machine base is made of natural granite and the optionally available b-axis can be integrated either with air bearings or with hydrostatic bearings onto the z-axis, whereas the linear axes are always equipped with hydrostatic bearings. All axis have direct drives (slow-tool machining/dynamic movement of z-axis) and the control system is a DeltaTau PMAC. Ultra-precision lathes with an additional y-axis are currently not offered by LT Ultra as a standard product.

The DAC International (USA) '2 Axis ALM Lathe – Series IV' is a very compact machine in cross-table design. The machine is designed for small workpieces and especially contact lenses and intraocular lenses manufacturing. All axes are direct driven air bearing axes and the spindle does not offer c-axis functionality. Also, additional axes are not available so the manufacturing is strictly limited to a 2-axis-manufacturing. Nevertheless, an additional 'Oscillating Tool Technology' is available which offers z'-movements of high frequency and short strokes synchronized with the spindle for non-rotation-symmetric workpieces. The control system is developed by DAC international.

The Schneider GmbH & Co. KG (Germany) 'HSC smart X' and the Satisloh GmbH (Germany) 'VFT-orbit' are freeform generators for high-speed cutting of polymer workpieces for the manufacturing of ophthalmic lenses. Both machines include milling and turning stations as well as automated workpiece handling devices.

The machines are designed for economic production of ophthalmic lenses, including varifocals, using fast tool machining.

4.3.3 Fast Tool Servo Technology

Mass customized optical products like varifocals usually possess freeform surfaces to provide different optical functions in one lens. To produce shapes like this a conventional ultra precision machine is only able to process rotational symmetric shapes. The machining of freeform surfaces requires a splitting of the workpiece geometry in rotational symmetric and non-symmetric ratio, as shown in Figure 4.15. As the symmetric ratio of the surface can be machined by a conventional ultra precision lathe the non-symmetric ratio has to be machined using an additional axis. This so called fast tool axis is a highly dynamic axis which provides an additional linear movement.

The fast tool has to be mounted on the z-slide or x-slide of a turning lathe. Depending on the x-position of the slide and the spindle angel φ the additional z*-axis (fast tool) must be positioned. Nevertheless, modern UP-lathes feature slow-tool functionality, offering dynamic movements of the z-axis based on the principle of fast tool machining. This enables the machine to produce freeform surfaces even though it may take a longer manufacturing time compared to fast tool servo systems.

By the use of different actuators different strokes from μm- to mm-range can be realized. Strokes in the low μm range can be provided by piezo actuators and larger strokes are supplied by voice coils or linear direct drives. The possible positioning frequency of the fast tool servo depends strongly on the actuator principle and decreases by increasing strokes of the fast tool and moving mass of the system. Figure 4.16 shows three different fast tool servo systems for different applications. The piezo FTS is designed for short strokes (35 μm) and high bandwidth (1500 Hz). The aerostatic FTS is designed for strokes up to 10 mm and a bandwidth

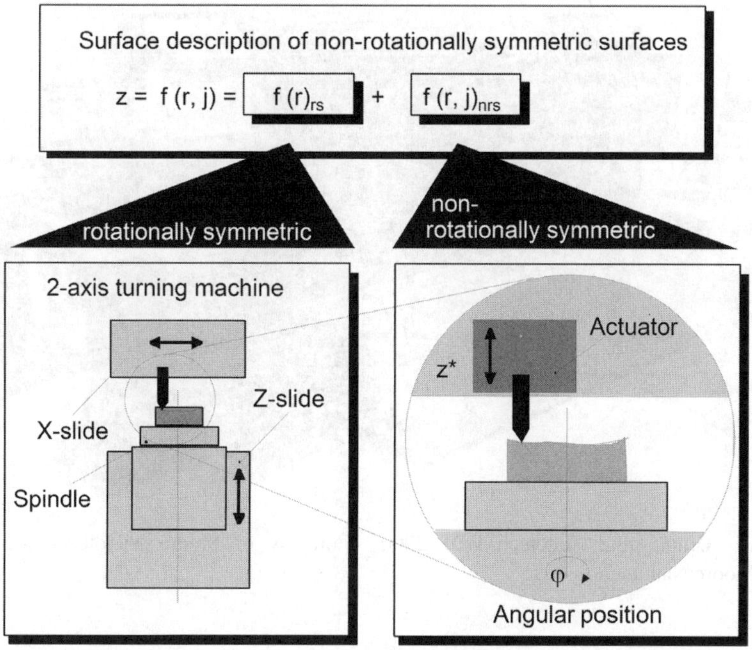

Figure 4.15 Geometry splitting for freeform machining

Figure 4.16 Different fast tool servo systems (Fraunhofer IPT)

up to 600 Hz. The hybrid FTS combines the advantages of long stroke (25 mm) in a voice-coil driven hydrostatic axis and high bandwidth (1500 Hz) with an integrated piezo FTS.

Moreover, drums for roll to roll processes are being diamond turned on special drum roll lathes available from all major suppliers of ultra-precision machinery (Figure 4.17).

4.4 Precision Milling Machines

Due to the modular setup of the above described ultra precision lathes next to turning operations, they are often used for milling of micro lense arrays, freeform shapes, complex optical geometries or ultra-high precision mechanical components.

Figure 4.17 Drum lathe Nanotech HDL-2000. (Courtesy of Moore Nanotechnology Systems) Copyright Moore Nanotech

Different from this recurring machine setup, precision and high precision milling machines are used for micro machining or high precision milling operations as needed in medical applications, mould making or aerospace. Requirements are different for these machines driven by miniaturization with the need of running milling tools with diameters down to 50 μm, and with a precision asking for volumetric accuracy below 5 μm or by stiffness and damping for machining innovative materials such as tungsten carbide hard metals for example.

Figure 4.18 gives an overview of about 63 commercially available machine systems for the 3- or 5-axis precision milling operation. The length of the individual linear guidance systems in the machines is between 100 mm and 1000 mm. The linear position accuracy of the single axis is below 10 μm as specified by the machine manufacturer. Not going into detail, the machine systems differ in their setup as well as in the use of basic components also incorporating roller and needle guideways, roller bearings for spindles and rotational axis as well as ball screws for linear drives [1].

Focusing on the high precision milling machines with a characteristic positioning accuracy of a single linear axis below 1 μm, the field of available machines narrows down drastically (Figure 4.19). For the machining of precision parts the volumetric accuracy is of greater interest, but usually not given as a comparable value by the manufacturers. The performance of a single linear axis is reduced during axis interpolation due to geometrical misalignment, dynamical excitation, deformation due to insufficient stiffness and thermal issues.

For all of these machines common design principles are fluidic guiding principles and drive systems free of direct mechanical design. Especially for 5-axis machines with swivel and rotary axis a cumulated accuracy during interpolation of all axis below 3 μm is not feasible with standard setups.

The company Hemtech (Netherlands) launched a fully integrated 5-axis precision machine in 2006. This machine is conceived for high precision 5-axis hard milling. With the goal of

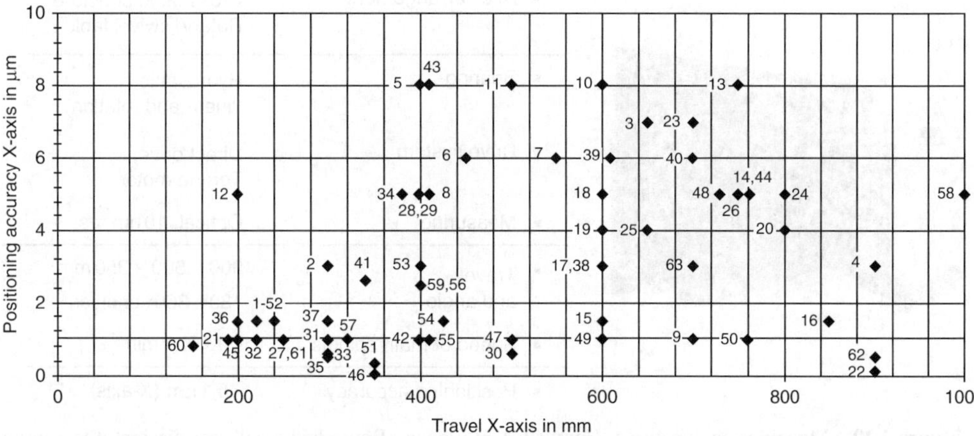

1	Almac CU 1005	17	Fidia HS664 RT	33	Kern twenty five	49	OKK VP 4005AX
2	Almac FC 1105	18	Hedelius RS 60 Single	34	Knuth Twinmaster 380	50	OKK VG 5000
3	Alzmetall GS 650/5-T	19	Hermle C20	35	Kugler Micromaster	51	Precitech Freeform 700
4	Breton XCEEDER 900RT	20	Hermle B300	36	Kugler Microgantry gf1	52	Primacon PFM 24 NGD
5	Bridgeport XR300 5AX	21	Hawema PRÄZISA	37	Kugler Miicrogantry micro	53	Primacon PFM 4024-5D
6	Chiron FZ 08K S MAGNUM	22	Hemtech Nano-Focus	38	MAG CFV 5	54	Röders RXP 500 DSC
7	Chiron FZ 12K S	23	Hüller Hille NBH 500	39	MCM Concept 2g	55	Roku Roku Mega S
8	Compumill HMC 140	24	Hurco VTX U	40	MCM Clock 700	56	Satisloh G1
9	Dixi DHO 50	25	Huron K2X 8 Five	41	Michael Deckel s20e	57	Schütte WU 305
10	DMG DMC 60 U	26	Hwacheon M2 -5AX	42	Michael Deckel s22P	58	Serrtech M3
11	DMG DMU 50 eVo	27	i-mes Premium 3020	43	Micron HSM 400 U	59	StarragHeckert STC 100
12	DMG Ultrasonic 20 linear	28	i-mes Premium 4030 µ	44	Millac Mu-400VA	60	UPM µltramill
13	Emag W04-12	29	i-mes Premium 4820 µ	45	MitsuiSeiki VL30 5X	61	WilleminMacodel W-408s
14	Fadal VMC 3016 FX	30	Kern Pyramid Nano	46	Moore Nanotech 350FG	62	YASDA YBM Vi40
15	Famup LINEARMILL 600 HD	31	Kern Evo	47	Moore Tool 500 CPZ	63	YCM FX350A
16	Fidia G996RT	32	Kern Micro	48	Mori Seiki NMV5000 DCG		

Figure 4.18 Market overview of 5-axis machines (linear positioning accuracy < 10 µm)

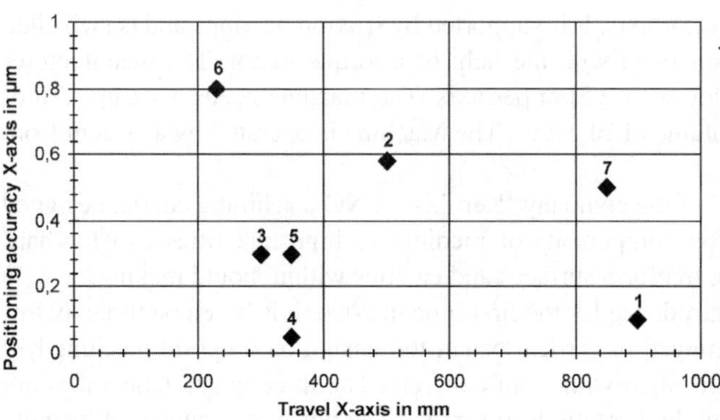

Selection of high-precision 5-axis machines

1 Hemtech NanoFocus
2 Kern Pyramid Nano
3 Kugler MicroMaster
4 Moore Nanotech 350FG
5 Precitech Freeform 700
6 Upm µltramill
7 Yasda YBM Vi40

Figure 4.19 Market overview of 5-axis machines (linear positioning accuracy < 1 µm)

Specification
"Hemtech NanoFocus"

▪ Installation surface	14,4 m^2
▪ Axes arrangement	Cross table structure, Rotary / swivel table
▪ Bearing	Hydrostatic (linear and rotation)
▪ Drive system	Direct drive, Torque motor
▪ Measuring	Optical, 10 nm res.
▪ Travel and angle	900 × 500 × 350 mm^3 130°, 360° endless
▪ Speed of main spindle	< 40.000 min^{-1}
▪ Positioning accuracy	**< 0,1 μm** (X-axis)

Figure 4.20 Machine specifications 'Hemtech NanoFocus'. Reproduced with permission of Hembrug Machine Tools

improved surface quality in the field of mould and die making, subsequent steps such as polishing or grinding are supposed to be reduced to a minimum.

The machine base is built from granite, all linear axes use direct drives. The rotary-axes of the rotary and the swivel axis that are fully integrated into the machine concept possess torque motors. The manufacturer specifies a positioning accuracy of the linear axis of less than 0,1 μm. For this kind of accuracy, the NanoFocus is one of the largest machines of its kind providing a machining volume of 87,5 dm^3. The set-up area of the machine is 14,4 m^2. The machine is originally controlled by a Siemens 840 D system [11].

Comparable with the NanoFocus is the YBM Vi40 from the company Yasda (Japan) as a precision processing centre in bench-top design. It follows specifications for hard milling and the processing of gauges. The linear axes, installed on a cast iron bed are hydrostatically supported and are driven by servo motors in combination with ball screws. The rotary/swivelling unit is set up on the lower linear axis. It is supported by friction bearings and is swivelled by a worm drive, rotation is generated with the help of a torque motor. The manufacturer specifies the positioning accuracy with 0,5 μm per axis. The machine needs a set-up area of 11,7 m^2 and offers a working volume of 91,1 dm^3. The Machine is operated by a system from the company Fanuc [12].

The machine 'Pyramid Nano' of the company 'Kern' is a CNC machining centre, designed for the manufacture of precision components of medium to high quantities and for hard machining of filigree structures, freeform surfaces and cavities within mould making.

The machine was officially introduced for the first time in 2006. It is based on three hydrostatic linear axes, which are arranged in vertical bench-top design on a hybrid machine bed consisting of a steel construction filled with resin-concrete. The three hydrostatic slides are positioned by servo motors using hydrostatic lead screws. The position accuracy of a single axis is in the range of ±0,3 μm as given by the manufacturer.

Five axes functionality can be added with a rotary and swivel unit that is supported by ball bearings with mechanical contact (3 + 2). Due to the fact that the unit is mounted on the

Specification
"Yasda YBM Vi40"

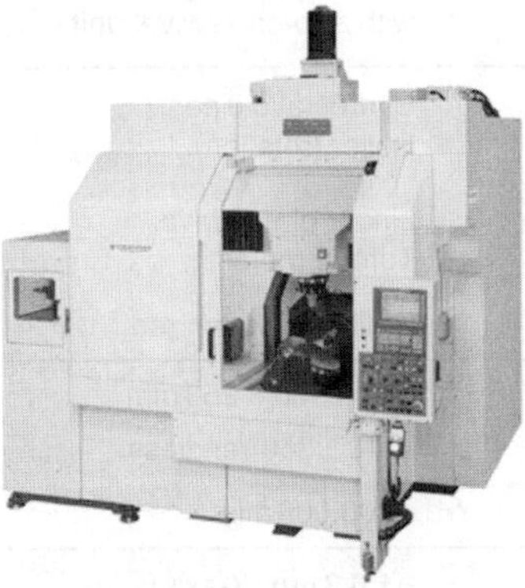

▪ Installation surface	10,2 m²
▪ Axes arrangement	Bench-top design with a rotary / swivel unit
▪ Bearing	Hydrostatic (linear) Friction (rotation)
▪ Drive system	Ball screw, worm drive
▪ Measuring	Optical, 100 nm res.
▪ Travel and angle	850 × 450 × 450 mm³ ±110°, 360° endless
▪ Speed of main spindle	< 24.000 min⁻¹
▪ Positioning accuracy	< **0,5 μm** (X-axis)

Figure 4.21 Machine specifications 'Yasda YBM Vi40'. Reproduced with permission of Yasda Precision Tools K.K.

Specification
"Kern Pyramid Nano"

▪ Installation surface	3,4 m²
▪ Axes arrangement	Bench-top design with a rotary / swivel unit
▪ Bearing	Hydrostatic (linear) Friction (rotation)
▪ Drive system	Direct drive, Torque motor
▪ Measuring	Optical, 20 nm res.
▪ Travel and angle	500 × 500 × 400 mm³ 130°, 360° endless
▪ Speed of main spindle	< 50.000 min⁻¹
▪ Positioning accuracy	<± **0,3 μm** (X-axis)

Figure 4.22 Machine specifications 'Kern Pyramid Nano'. Copyright KERN Microtechnik GmbH

workpiece table, only a reduced part of the total volumetric space of 100 dm³ can be used for 5-axis processing.

The machine is operated by a Heidenhain control as standard setup [13].

The 5-axis machine 'MicroMaster' by Kugler (Germany) is designed for ultra precision processing with forces in the single digit Newton range. All three linear axis are hydrostatically supported and positioned by ironless synchronous linear motors. The machine is arranged in a bench-top design on a granite bed. A friction bearing rotary/

Specification
"Kugler MicroMaster"

▪ Installation surface	11,7 m^2
▪ Axes arrangement	Bench-top design with a rotary/swivel unit
▪ Bearing	Hydrostatic (linear) Roller (rotation)
▪ Drive system	Linear direct drive, Torque motor
▪ Measuring	Optical, 0,25 nm res.
▪ Travel and angle	300 × 300 × 250 mm^3 ±110°, 360° endless
▪ Speed of main spindle	< 200.000 min^{-1}
▪ Positioning accuracy	<± **0,3 μm** (X-axis)

Figure 4.23 Machine specifications 'Kugler MicroMaster'. Copyright Kugler GmbH

Specification
"Upm / Brunel μltramill"

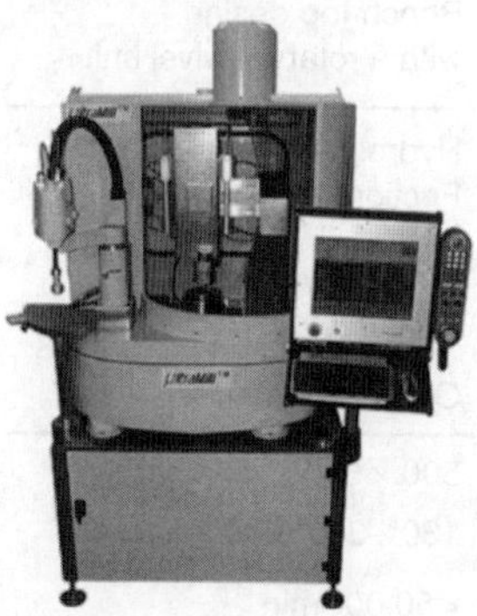

▪ Installation surface	1,9 m^2 (system)
▪ Axes arrangement	Bench-top design with a rotary/swivel unit
▪ Bearing (linear, rotary)	Aerostatic, roller beared / aerostatic
▪ Drive system	Linear direct drive, Torque motor
▪ Measuring	Optical, 5 nm res.
▪ Travel and angle	230 × 225 × 160 mm^3 ±90°, 360° endless
▪ Speed of main spindle	< 200.000 min^{-1}
▪ Positioning accuracy	< **0,8 μm** (X-axis)

Figure 4.24 Machine specifications. (UPM/Brunel University)

swivelling unit is optionally installed on the lower linear axis and is positioned by direct torque drives (3 + 2). The manufacturer specifies the positioning accuracy of a linear axis with ±0,3 μm. The machine covers a traverse space of 22,5 dm^3 and needs an installation surface of 3,4 m^2. The machine is operated by a control system of the company Andron [14].

The 'µltramill', developed by the Brunel University of London in cooperation with the company Upm(GB) is a high precision machine for micro milling, micro drilling and micro grinding. All linear axes are aerostatically supported using linear direct drives for positioning. The linear axes are arranged on a granite bed in a bench-top design. The direct driven rotational axes are carried out as a rotary/swivel unit using friction bearings for the swivel axis and air bearings for the rotational axis with squeeze film dampers. The positioning accuracy of a linear axis is specified to 0,8 µm. The machine provides a working volume of 8,3 dm^3 and needs a set-up area with auxiliaries of approximately 1,9 m^2. A PMAC-control is used for CNC functionality [15].

References

[1] Utsch, P. (2011) *Steigerung der Arbeitsgenauigkeit mehrachsiger Maschinensysteme durch Kompakt- und Integralbauweise am Beispiel einer 5-Achs-Maschine.*
[2] Brecher, C., Hannig, S. and Stimpfl, J. (2009) *Conventional and New Materials for Machine Slides*, lecture on UPT-Meeting October 2009
[3] URL: http://diamant.ph/ (05.05.2011).
[4] URL: http://www.skc-technik.de/ (05.05.2011).
[5] Langenbeck, P. (2009) *Wirtschaftliche Mikrobearbeitung*, München: Carl Hanser Verlag.
[6] Weck, M. (2001) *Werkzeugmaschinen – Mechatronische Systeme, Vorschubantrieb und Prozessdiagnose*, Heidelberg: Springer Verlag.
[7] Schneeberger (2004), *Linearführungen.*
[8] Weck, M. and Brecher, C. (2006) *Werkzeugmaschinen – Konstruktion und Berechnung*, Heidelberg: Springer Verlag.
[9] Brecher, C., Lindemann, D., Merz, M. *et al.* (2010) *Hybrid Control System for Highly Dynamic Axes*, Nürnberg: Proceedings of the PCIM Europe.
[10] Brecher, C., Lindemann, D., Merz, M. *et al.* (2010) *FPGA-Based Control System for Highly Dynamic Axes in Ultra-Precision Machining*, Boston: Proceedings of the ASPE Spring Topical Meeting.
[11] URL: http://hartdrehen.info/ (02.12.2009).
[12] URL: http://www.yasda.co.jp/ (02.12.2009).
[13] URL: http://www.kern-microtechnic.com/ (02.12.2009).
[14] URL: http://www.kugler-precision.com/ (02.12.2009).
[15] URL: http://www.brunel.ac.uk/ (02.12.2009).

5

Engineering Materials for Micro Cutting

Sathyan Subbiah[1] and Shreyes N. Melkote[2]
[1] *Nanyang Technological University, Singapore*
[2] *Georgia Institute of Technology, USA*

5.1 Introduction

Several engineering materials have been processed and new materials are continuously being investigated for processing, using the micro-cutting process to produce micrometer sized parts, features and very fine surface finishes. These materials range from very ductile materials such as copper and aluminium alloys to very brittle materials such as graphite, silicon and germanium and new materials such as bulk metallic glasses. The challenges of micro cutting these materials include: reduced burr formation, obtaining desired surface finishes, reduced tool wear, obtaining precise dimensions, reduced surface and sub-surface damage and economical duration of manufacture. These challenges, while common to macro-cutting, pose unique, perhaps more severe, problems to micro cutting: burr sizes are comparable to feature sizes in micro cutting, the micro-cutting process is more sensitive to tool wear due to the size ranges involved, and residual stresses can change dimensions and tolerances to a larger extent than in macro-cutting. These challenges in micro cutting can be partly addressed by studying the material response at the micrometer scales to the stresses and strains induced by the advancing cutting tool during the material removal process. This chapter explores some of the unique aspects of the science of micro cutting from a materials point of view.

Micro cutting, like its macro counterpart, induces very large deformation in the form of stresses and strains in the material immediately ahead of the cutting tool. The uniqueness of the micro cutting process is that this deformation is induced by a very small area of the tool and is concentrated in a very small volume in the material. Thus, the response of the material can be very unique. One of the early observations of such a unique response in materials to micro cutting was that the cutting forces, and hence the energy consumed in micro cutting, did not scale as expected as smaller and smaller amounts of material were removed. This was

Micro-Cutting: Fundamentals and Applications, First Edition. Edited by Kai Cheng and Dehong Huo.

termed the size effect. Investigations into this size effect led to a complete reassessment of the machining theory and understanding. Aspects of material responses that were either averaged or overlooked were seen to play important roles in micro cutting particularly in explaining the size effects. Hence, this chapter initiates discussions with the size effect; following this the stresses and strains involved in deformation during micro cutting is reviewed. The elastic and plastic behaviour of materials are then discussed leading then on to looking at effects of fracture and then finally concluding with some recent reports of various engineering materials processed via micro cutting.

5.2 'Size' Effects

The need to remove smaller and smaller amounts of materials were driven by three factors: one, to produce very fine surface finishes needed in optics and other areas, two, to produce smaller and smaller features and parts needed to meet modern technology's requirements, and three, to fulfill researcher's curiosity to observe the cutting process at smaller scales. The latter is noted here since observations related to size effect [1] dates back to a time that precedes the more recent micro/nano technological advancements.

Some interesting observations were noted when smaller and smaller amounts of material were removed, particularly in one of the measurable process signatures of machining – the cutting forces. It was observed that the cutting forces do not tend to continue to decrease as the thickness of the material to be removed, so called uncut chip thickness to, was reduced.

The forces were seen to level off (Figure 5.1) below a certain range of critical uncut chip thickness, meaning that the force needed to cut was larger than anticipated at lower uncut chip thickness. The implication of this is more pronounced in the specific cutting energy, u, defined as the energy needed to remove a unit volume of material. It was seen that the specific cutting energy increases without bounds (Figure 5.1) as the uncut chip thickness was reduced; one would expect the energy per unit volume, indicating unit effort, to remain constant. In the case of orthogonal cutting where the cutting edge is perpendicular to the velocity of cut, the specific cutting energy can be defined as:

$$u = \frac{F_c V}{b t_o V} = \frac{F_c}{b t_o} \tag{5.1}$$

where F_c is the cutting force, V the cutting velocity, b the width of cut. The observed fact that the force, F_c, does not decrease proportionally to t_o causes the specific cutting energy to increase dramatically. This observation is termed a size effect in micro cutting and has been observed in many engineering materials (brass, aluminium alloys, steel, Germanium, PMMA, CaF_2), under various machining conditions (V ranging from 0.1 to 200 m/min; t_o ranging from 10 nm to 300 μm), in various cutting processes (grinding, turning, milling) and using tools of varying edge conditions ([1–4]).

There are several varied explanations for this size effect in micro cutting (Figure 5.2). The explanations can be broadly classified into those that are related to material properties and those that are independent of material properties. The material related factors uniquely influencing the micro-cutting process is the main focus of this chapter and is explored in the following sections. The non-material related factors are briefly explored here before closing this section.

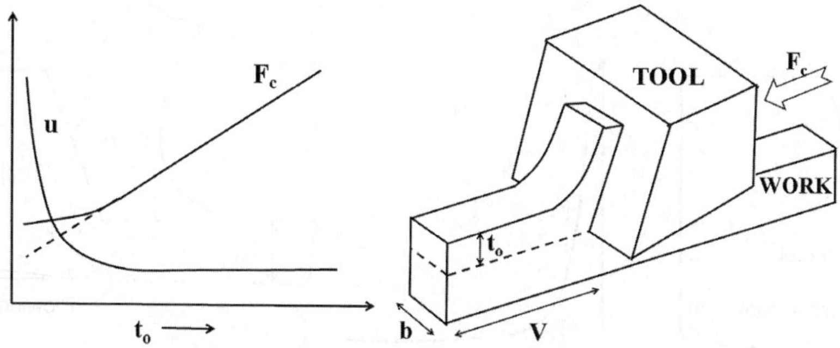

Figure 5.1 Size effect in orthogonal cutting

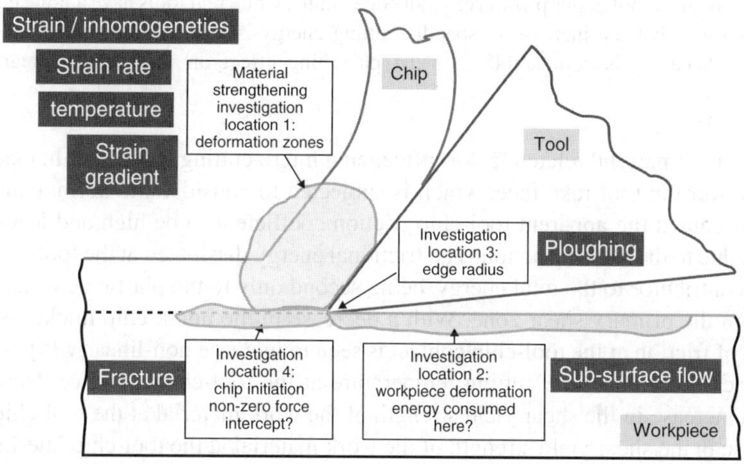

Figure 5.2 Overall explanations of size effect in micro cutting

Two main non-material related factors that influence micro cutting are the edge condition of the tool and frictional effects. The effect of cutting edge radius on the machining process at small values is known to give rise to an important size effect. Cutting tools are never perfectly sharp and hence the cutting edge formed at the intersection of the tool rake and flank faces is never an ideal straight line or curve (the rake face is the face over which the chip flows, while the flank face is the relief face of the tool closest to the machined surface). There is always a small radius present at the cutting edge. Studies have shown that the force and energy trends change in the region where the uncut chip thickness becomes comparable to the cutting edge radius, even when the tool is extremely sharp with edge radius in the few tens of nanometers [5]. A tool with a radiused cutting edge is essentially a blunt tool and this causes some pushing of the material immediately ahead of the cutting edge into the chip and into the machined work surface and possibly to the sides. This is commonly referred to as 'ploughing (Figure 5.3). A blunt tool also results in a large negative effective rake angle, which tends to push the work material [6]. This leads to higher forces and energy needing to be expended in the cutting zone and also in turn affects the integrity of the machined surface.

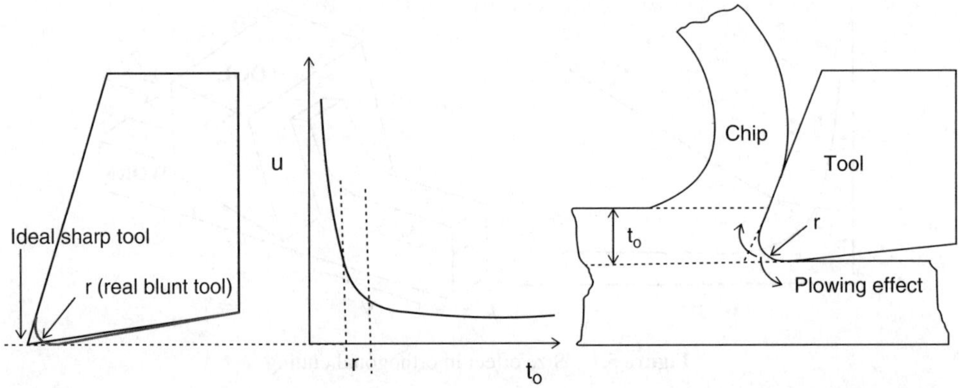

Figure 5.3 An ideal tool is sharp with zero tool edge radius, while real tools have a finite edge radius, r. Experiments show that the increase in specific cutting energy is seen to start in the region where t_o approaches r. When t_o is comparable to r the ploughing effect on the surface integrity is more significant.

The other non-material related factor influencing micro cutting is friction. In machining the chip slides over the tool rake face, which is subjected to considerable normal and frictional forces. This causes the apparent tool-chip friction coefficient to be high and leads to energy dissipation due to the rubbing action. The frictional energy dissipated at the tool-chip interface is a major contributor to the total energy, being second only to the plastic deformation energy dissipated in the primary shear zone. With a decrease in the uncut chip thickness, the mean coefficient of friction at the tool-chip contact is seen to increase non-linearly [5]. As the uncut chip thickness decreases, the cutting temperature at the tool-chip interface decreases. This leads to an increase in the shear yield strength of the work material at the tool-chip interface. The increase in the shear yield strength of the work material at the tool-chip interface tends to increase the friction coefficient. A similar trend in friction coefficient can also be seen in the forging and extrusion processes with decrease in the specimen size [7].

The material related factors affecting the micro-cutting process can be understood better by studying the stresses and strains associated with the deformations caused by the tool-work interaction in the micro-cutting process. The following section discusses these aspects before specifically focusing on the material related size effects in the subsequent sections.

5.3 Strain and Stress in Cutting

The stresses and strains of concern are mainly in the area of deformation directly ahead of the cutting tool as the chip is being formed, particularly in the primary shear plane and in the secondary shear zone (Figure 5.4). Like its meso/macro counterpart deformations remain large even in micro cutting with strain values in the range of 1–5 and time rates of strain ranging between 10^2–$10^6\,s^{-1}$. A good understanding of the material behaviour under these conditions is necessary to explain the various characteristics of the micro cutting process.

Both analytical methods and numerical methods are used to understand the stresses and strains in machining. Analytical efforts and associated understanding of stresses and strain are mainly discussed in this section, while numerical models are explored in the following chapter.

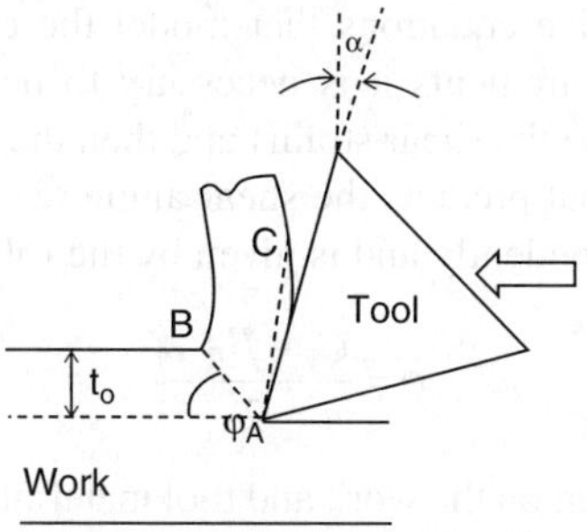

Figure 5.4 Deformation in micro cutting, like its macro counterpart, are mainly in the primary shear zone AB and secondary shear zone AC

While turning, milling and grinding are complex micro cutting processes, let's use a simpler problem to understand stresses and strains. Consider the case of steady state (as opposed to incipient) chip formation in orthogonal cutting where the cutting edge is perpendicular to the cutting speed (Figure 5.4) – the tool cutting edge is perpendicular to the page and the tool is moving horizontally from right to left.

In terms of the measurable cutting and thrust force components, the shear and normal stresses in the primary shear plane is given as,

$$\tau = \frac{(F_c \cos\phi - F_T \sin\phi)\sin\phi}{bt_o}$$
$$\sigma = \frac{(F_c \sin\phi - F_T \cos\phi)\sin\phi}{bt_o} \tag{5.2}$$

where F_c is the cutting force, F_T the thrust force, b the width of cut, t_o the uncut chip thickness and φ the angle of the shear plane. The amount of shear the material undergoes as it crosses the shear plane is given as,

$$\gamma = \frac{\cos\alpha}{\sin\phi\cos(\phi-\alpha)} \tag{5.3}$$

where α is the rake angle of the cutting tool. The time rate of this shear strain can be calculated as,

$$\dot{\gamma} = \frac{\cos\alpha}{\sin\phi\cos(\phi-\alpha)}\frac{V}{\Delta s} \tag{5.4}$$

where V is the cutting speed and Δs, usually very small and measured experimentally, is the thickness of the primary shear plane. For more details on the derivation of these basic equations the reader is referred to books on metal cutting (see [8]). The use of these equations to determine strain, strain rate and stresses involve experimentally measuring various parameters such as the cutting and thrust forces, chip thickness (to determine the shear angle) and the primary shear plane thickness.

In order to develop predictive equations that model the cutting process without relying heavily on experimental measurements it is necessary to be able to first predict the shear angle, as we can then determine the shear strains and then the stresses experienced. The most effective analytical equation that predicts the shear angle reasonably was developed by both Merchant and Piispanen independently and is given by the following relationship:

$$\phi = \frac{C_1 - \beta + \alpha}{2} \tag{5.5}$$

where C_1 is a constant dependent on the work and tool material combination and β, the friction angle. The key assumptions behind this relationship are: (a) shear (τ) and normal (σ) stresses acting on the primary shear plane are linearly related (in fact, $C_1 = \cot^{-1}(K)$ where K relates σ and τ), and (b) the shear angle adjusts itself to minimize the force and hence energy expended in the process. Typical values of C_l range from 70–79 for steel-HSS/carbide work-tool combinations, 83–84 for aluminium-HSS/carbide, and 47–49 for copper-HSS/carbide. For a given work material, the values of C tend to be 10–30% higher when the tool material is diamond.

Having now the ability to predict the shear angle and hence the shear strain, the next task is to relate the stresses and strains. Not much work has been reported in understanding stress-strain behaviours in materials subject to strains as large as that encountered in micro cutting. The most popular relation between stress and strain is: $\sigma = K\,\varepsilon^n$. This is usually valid for relatively small strains, close to unity. For strains larger than this the stress-strain curve is usually linear and varies as: $\sigma = A + B\,\varepsilon$. This has been experimentally verified by a few researchers (e.g. [9] in the case of uniaxial and plane strain tension/compression, and even under a combination of high compressive and shear stresses [10]. These relations indicate that stress increases monotonically with strain. However, in specially designed experiments, when the normal compressive stresses acting in combination with shear is lower [11] a different trend was observed – shear stress was seen to decrease with increasing strain. This trend has been attributed to the formation of micro-cracks; when compressive stresses are very large such cracks are not formed, but under moderate stress values such micro-cracks prevent the shear stress from increasing monotonically. Such a scenario may very well apply to machining where some evidence of micro-cracks in the shear plane has been shown [6].

Now, the material undergoing micro-cutting is subject to not only just high strains, but a combination of high strain, large strain rates and temperatures (leading to complex thermo-mechanical coupling). Analytical methods cannot capture these effects easily, and an alternative is numerical modelling. One of the challenges involved in numerical modelling, and understanding the cutting process has been identifying suitable material constitutive relations that link the stresses developed in the material subject to the high strain, strain rates and temperatures. An empirical material model that captures the effect of high strains, strain-rates and temperatures has been developed by [12]; this was originally developed in the context of modelling material behaviour under ballistic impact conditions. This model has now been widely adopted in the numerical modelling, especially using the finite element method, of various machining processes. The model is of the following form:

$$\sigma = (A + B\varepsilon^n)(1 + \ln \dot{\varepsilon}^*)(1 - T^{*m}) \tag{5.6}$$

where A, B, C_2, m and n are constants determined experimentally, σ is the stress, ε is the plastic strain, $\dot{\varepsilon}^*$ the ratio of rate of strain to a reference strain-rate, and T^* is the homologous

Table 5.1 Johnson-Cook constitutive equation model constants [12]

	A (MPa)	*B* (MPa)	*n*	C_2	*m*
OFHC Copper	90	292	0.31	0.025	1.09
C2600 Brass	112	505	0.42	0.009	1.68
1006 Steel	350	275	0.36	0.022	1.00
4340 Steel	792	510	0.26	0.014	1.03
S-7 tool steel	1539	477	0.18	0.012	1.00
Al-2024	265	426	0.34	0.015	1.00
Al-7039	337	343	0.41	0.010	1.00
Pure Ni (Nickel 200)	163	648	0.33	0.006	1.44

temperature. The constants are determined using data obtained from a combination of tests such as torsion tests at different rates of strain, dynamic Hopkinson bar tensile tests and static tensile tests. The values of these constants for a few engineering materials are provided in Table 5.1.

Cylinder impact computations using this model, carried out at strain rates close to $10^5 s^{-1}$ and strains exceeding 2 have shown good agreement with experimental results. These strains and strain-rates are partly in the range of what is normally observed in micro-machining and hence these material models can be used for computational modeling and preduction of trends in micro cutting.

There have also been some attempts to use the machining process itself to develop suitable material models. In this so called indirect method, first proposed and developed by [13], and subsequently enhanced by several researchers, the material constitutive law constants, and other parameters (e.g. friction) are adjusted until the model predicted cutting forces match that measured experimentally. This approach is based on average values of stresses, strains, strain-rates and temperature in the primary shear zone; in reality these parameters are always varying across the primary shear zone. The approach also assumes that the tool is sharp, and plane strain steady state conditions exist. The procedure consists of assuming initial values for three parameters: C_3, the constant that determines strain rate from the shear plane velocity and length of the shear plane (L), δ the ratio of the secondary shear zone thickness to the thickness of the chip and the shear angle φ. These values are adjusted until the shear stress in the secondary shear zone matches the frictional stress (determined from a suitable friction model), the normal stress on the primary shear plane matches that determined at the free surface using a slip line field analysis, and the cutting forces are determined to be the minimum for a given set of conditions. Based on these iterations, the stresses, strains, strain rate, temperatures and hence the forces are determined. The material model constants can now be adjusted until the model calculated forces match the experimentally measured values, essentially a form of regression fit to the data. Figure 5.5 illustrates the method proposed by [13] on developing the material model from machining tests.

More recently, [14] have developed another analytical method to obtain material constitutive laws from machining process simulations that account for distributions in stress, strain, strain rates and temperatures in the primary zone (in the Oxley approach, only average values are considered). A new analytical model is developed to accurately describe the distributions of stress, strain, strain rate and temperatures in the primary zone. This model, referred to as the

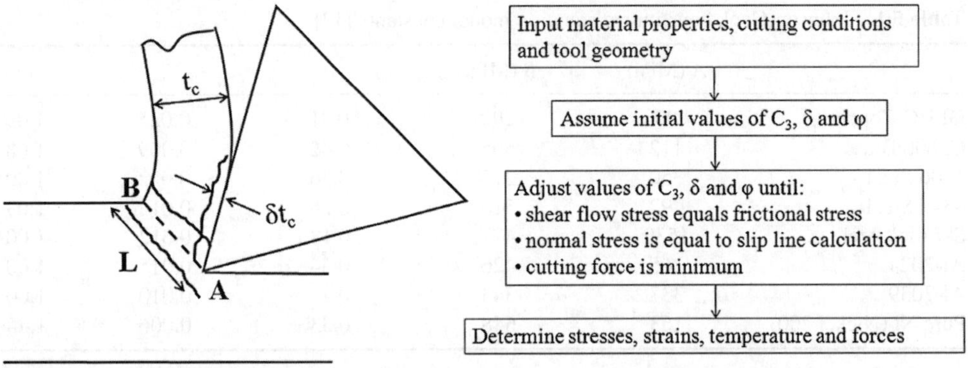

Figure 5.5 Illustrating the method proposed by [13] to develop material model from machining tests. (Adapted from [15]

distributed primary zone deformation (DPZD) model, is based on conceptual considerations, FE simulations and mechanics of the orthogonal metal cutting. FE simulations are used to characterize the plastic deformation in the primary zone.

This section considered the basics of the stresses and strains seen in the machining process. Some material models used to model and understand the process were also introduced. The following section considers the plastic behaviour of materials particularly at the micro-scale highlighting differences in behaviour, which affects the micro-cutting process, compared to that at the macro scale.

5.4 Elastic and Plastic Behaviours at the Micro-scale

Micro cutting just like its macro counterpart involves large deformation of material, albeit within a smaller volume, and hence the elastic and plastic behaviours of the material play a critical role. Particularly any size effects seen in material behaviour can be expected to affect the micro-cutting process.

Size effects in materials are well known where in the materials' response to plastic deformation is different at the micro scales from that of meso/macro scales. The elastic behaviour or temporary reversible deformation of materials, for which the bond length is the only 'size' of concern, is largely unaffected by the scales under discussion here. It is the plastic or permanent deformation behaviour that displays the effect of sizes. Material related size effects, particularly in metals, have been studied extensively by researchers in the fields of materials science and engineering (see reviews by [16] and [17]). Size effects are seen when either the deformed structure is very small or only a small volume of the material is subjected to plastic deformation. It is now well established that such size effects can be observed under two conditions: one, when size parameters (e.g. grain size) and characteristic length scales (e.g. dislocation loop diameter) become comparable and more recently, when a dimensional constraint is introduced by reduced physical dimensions (e.g. thin films, MEMS devices) in which case, for example, the deformation mechanism may start to interact with free surfaces and interfaces. The former is called intrinsic size effects and the latter extrinsic size effect. Size effects are

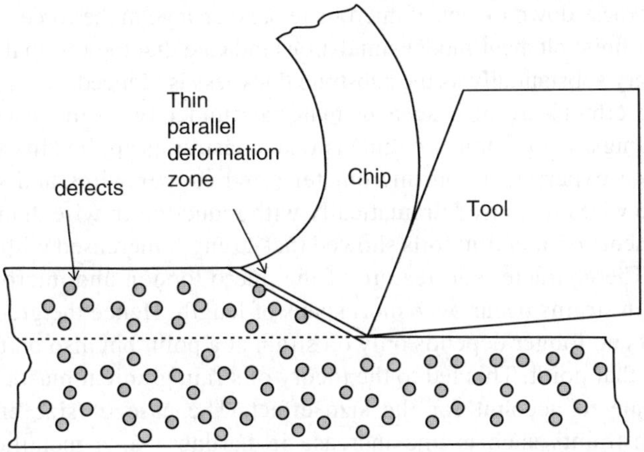

Figure 5.6 In micro cutting, deformation in a small volume and the chance of encountering a defect is less

largely seen to increase the strength of the material; in other words, the material resists plastic deformation much more strongly than expected. The idea that under small scale deformation the material is stronger could possibly explain the size effects seen in micro cutting; higher cutting forces than expected could be explained thus. Many micro-cutting researchers have reported such explanations.

Consider the case of extrinsic size effects in materials in one dimension. As early as in the 1950s, researchers [18] have reported a tensile test conducted on very thin whiskers of pure metals, a few micrometers in diameter and a few millimeters long; such tests showed that the strength of these materials approaches the theoretical strength calculated from bond breakage calculations. The only way to explain such high resistance to plastic deformation is that in such small volumes of material defects are very less. Plastic deformation in materials usually starts in areas where defects (e.g. dislocations) are present and hence the presence of defects lowers the material strength. Defects are also the reason why brittle materials such as silicon and glass are seen to have strengths lower than their theoretical values [19]; in this case the defects are in the form of voids or micro cracks. The role played by defects in reducing material strengths at meso/macro scales and in increasing material strengths at micro scales, can be used to explain the size effect in micro cutting also. A consequence of the reduced material removal (smaller t0) in micro cutting is that deformation happens in a much smaller volume of material and at smaller volumes of deformation the chances of encountering a defect are less (Figure 5.6) and this can cause an apparent increase in the material strength thus affecting the forces and the specific cutting energy [20]. As explained earlier, most of the deformation in machining occurs in an idealized narrow parallel zone ahead of the tool; this zone becomes even thinner at smaller uncut chip thickness values. Hence, chances of encountering a defect in this small zone are reduced.

Extrinsic size effects with one dimensional constraint are also reported in thin films of materials – both free standing and adhering to a substrate. Size effects are seen when the thickness of the films is either in the order of the grain sizes or dislocation loops. Not much work has been reported in micro cutting of thin films or substrates. Recently [21] have reported some experiments on machining thin substrates held adhesively to a substrate. As they reduce the substrate thickness down to tens of micrometers, no change in the forces or chip thickness are observed, but finite element model simulations indicate that the residual stresses induced in the material vary substantially as the substrate thickness is reduced.

Extrinsic size effects are also seen in materials under two dimensional constraints; torsion and bending are common two dimensional constraints applied in experiments. [22] conducted torsion experiments on micrometer sized copper wires and showed that the strength of these wires increased dramatically with reduction in wire diameter. Similarly, bending experiments on metallic foils showed that strength increased with decrease in foil thickness [23]. The characteristic feature of the micro torsion and micro bending is that large variations in strains occur over short spans of length. Hence the gradient of strain is high. Thus, stress no longer depends only on strain at a point but also on the strain values near and around that point. This led to the theory of strain gradient plasticity by introducing a length scale to account for the size effect. The case of size effect with three dimensional constraints such as the increase in hardness of a metallic material with decrease in indentation depth can also be attributed to the dependence of the flow stress of the metal on the spatial gradient of strain. This concept of strain gradient strengthening has been applied by researchers to explain observations in micro cutting. Dinesh et al. [24] linked the size effect observed in micro/nano indentation to that in machining. They suggested that the size-effect in machining can be explained by the theory of strain-gradient plasticity since strain gradients in machining are very intense. Building upon this work [25] presented an analytical model for orthogonal cutting that incorporates a material constitutive law with strain gradient effects. However, their model only considers strain gradient produced in the primary shear zone. More recently, strain gradient incorporated into the flow stress equations has been implemented in a finite element model to simulate machining by [4]. This work incorporates strain gradient in all aspects of deformation in the chip and not just the primary deformation zone. They have shown (Figure 5.7) that a strain gradient plasticity based model of orthogonal micro cutting is able to capture the size effect in specific cutting energy for the aluminum alloy Al5083-H116 examined. Strain gradient strengthening contributes significantly to the size-effect at low cutting speed (< 10 m/min) and small uncut chip thickness (< 10 μm).

Material size effects due to intrinsic reasons have not been utilized to observe nor explain any corresponding effects in micro cutting. Consider grain size effects – it is well known that finer grain size materials display higher strengths than the coarser counterparts; this effect is called the Hall-Petch effect. For example, experiments have not been reported on materials with different grain sizes to show that higher force and hence higher energies are needed to cut fine grained materials.

Apart from the size effects seen in materials that have led to a redefinition of classical plasticity, factors that influence material strength within the realm of classical mechanics have also been used to explain the size effects seen in micro cutting. The stress response of a material is usually dependent on strain at a point, strain rate (time) and temperature. If any of these three parameters alter in micro cutting they may favour increase in material strength. Larsen [27] explained the scaling phenomenon in machining in terms of the strain-rate sensitivity of the workpiece material. Their reasoning is based on empirical data drawn from experiments on plain carbon steel, over a range of cutting speeds (17 to 817 ft/min) and uncut chip thickness (0.005 in to 0.0108 in), which suggest that the maximum shear strain rate, $\dot{\gamma}$, within the primary shear zone is inversely proportional to the uncut chip thickness and is given as:

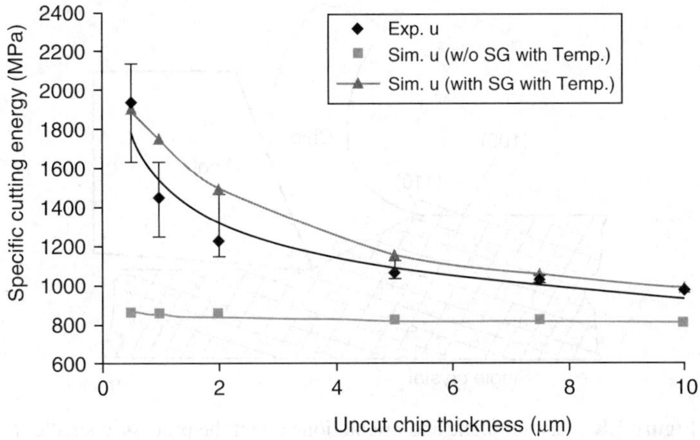

Figure 5.7 Importance of strain gradient in explaining size-effect. Reproduced with permission from [26]. Copyright 2006 ASME

$$\dot{\gamma} = C_4 \frac{V_S}{t_o} \tag{5.7}$$

where V_s is the shear velocity, C_4 a material-dependent constant. Therefore, a decrease in the uncut chip thickness will leave the strain occurring in the shear zone unchanged but the strain rate will increase inversely with the uncut chip thickness t_o. For most metals, an increase in the strain rate causes an increase in the flow stress with the strain-rate sensitivity of flow stress increasing rapidly in the range applicable to machining processes. This could therefore explain the increase in specific cutting energy with reduction in uncut chip thickness. Kopalinsky [28] took into account the effect of temperature in later work and attributed the size effect in the specific cutting force to a decrease in the shear plane angle due to decrease in the tool-chip interface temperature. This, they contended, leads to an increase in the shear strength of the work material. Furthermore, they acknowledge that the temperature effect does not explain the size effect observed at uncut chip thickness less than 50 μm, which is possibly because of the increasing sensitivity of flow stress to strain rate within this range.

Other well known material related effects such as crystallographic orientations and phases/inclusions are known to affect micro cutting. It is well known that crystals in metals are anisotropic due to the nature of atomic arrangements and that crystalline metals plastically yield by slipping on closely packed planes with different critical shear stress of slipping in different planes and directions. In a polycrystalline work material with many grains the anisotropic effect is masked since grains are distributed in a random fashion. If either there is preferred orientation of grains or if the number of grains is very less compared to the volume of deformation then the anisotropic effects will have to be considered in micro cutting. At smaller scales of material removal, the averaging effect of various crystalline grain orientations normally seen in a polycrystalline material is no longer present. The moving cutting edge now encounters individual grains and crystallographic planes in the grains (Figure 5.8) and since the material property changes with the crystal orientations this affects the forces, energy and surface finish [29]. The effects of grain level anisotropy on the machining response has been experimentally observed and studied. The chips and surfaces generated in

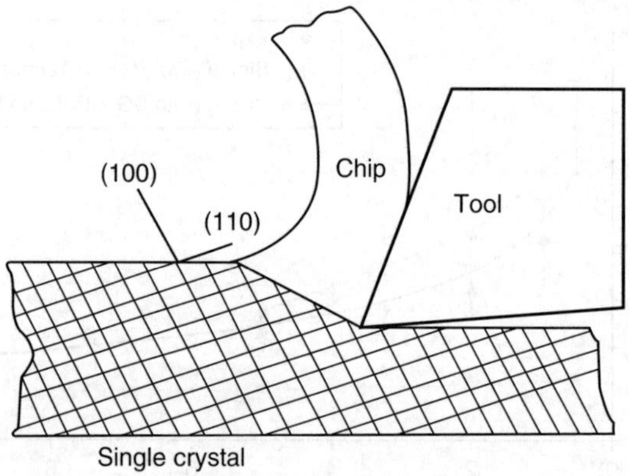

Figure 5.8 Crystal plane and orientations affect the process at smaller t_o

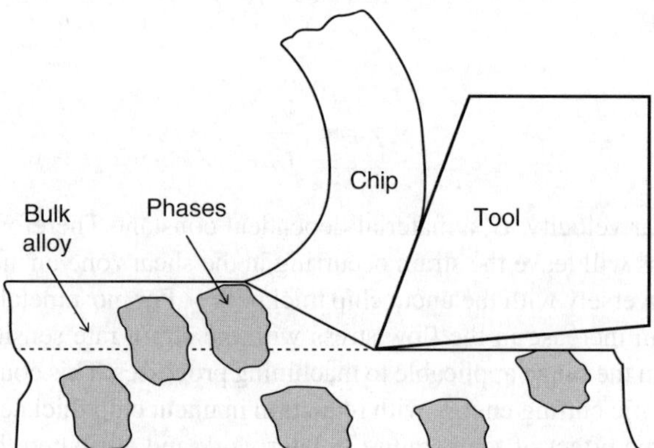

Figure 5.9 Cutting through a heterogeneous material with phases at small t_o

machining of single crystal copper show that the chip thickness and shear angle vary considerably with change in crystallographic orientations [30]. Surface finish is also affected when the cutting direction relative to the crystal orientation is changed. In addition, orthogonal micromachining of single-crystal aluminium have shown that the machining forces vary with crystallographic orientation [31]. This study also reports that the uncut chip thickness has an orientation-dependent effect on the machining forces. In addition, the chips formed in single crystal aluminium have different morphology and displayed shear-front lamella, the periodicity of which varies with crystallographic orientations and uncut chip thickness.

Engineering alloys also usually have several phases present in them and the phases have different material properties from the bulk alloy. In conventional scale machining, the effects of these phases average out. At smaller scales of machining the tool cutting edge now individually encounters these phases as it moves from cutting one phase to another (Figure 5.9). This causes variations in forces, energy and surface integrity. These variations have been studied in detail [32]. For example, in micro milling of cast iron it is shown that frequencies in the cutting force signal more highly than those attributable to the process kinematics can be explained by explicitly considering the interaction between the tool and the multiple phases present in the material. Experiments performed on two compositions of ductile iron phases – pure ferrite and pearlite - show that the nature of variation in the ductile iron cutting force signals can be attributed to the mixture of the phases. In addition, simulation studies show that the frequency of variation is related to the spacing of the secondary (ferrite) phase and the variation magnitude is determined by the size of the secondary (ferrite) phase particle.

The following section considers the effect of fracture on micro-cutting process characteristics.

5.5 Fracture

The question of whether chip formation in metal cutting involves fracture leading to material separation has been a controversial one in literature. There are two schools of thought. One school of thought regards machining as a special case of wedge indentation process and that the chip is generated without any fracture. The other school of thought supports the idea of material separation and fracture leading to the generation of new surfaces. This idea of fracture and associate energy consumed has been used to explain the size effect in micro cutting and is discussed in this section here. Among the proponents of fracture in machining is Atkins who suggests that the machining process adjusts itself to simultaneous actions of cracking and plastic flow with friction playing a superimposed role [33, 34, 35]. Some researchers have speculated on the presence of fracture leading to chip formation. In a report on discontinuous chip formation, [36] have acknowledged that even in the perfectly continuous chip some fracture of the material is involved leading to the development of new surfaces. It is noted, albeit by a few ([36] [35], that for the card number 7 (in the classic Piispanen card model shown in Figure 5.10) to slide up the rake face, it has to be first separated from the base workpiece at region X-Y and hence material separation has to occur here.

From the direction of the required fracture (in the direction of cut) it would seem that it must result from tensile stress acting at the tool point [36]. The argument for this is made by [36] as follows: No matter how carefully a tool may be ground there will always be a finite radius of curvature at its point. If the metal cut consisted of a number of plates then the shear surface might be expected to be as shown in Figure 5.11.

The rake angle associated with plate 1 is very small and hence the corresponding shear angle is also expected to be small. The chip from plate 1 would be bent toward the tool by plate 2. The shear angle would increase for subsequent plates in as much as the rake angles associated with them increase. This variation in shear angle would occur up to point A beyond which the rake angle and hence the shear angle should remain constant. As a consequence, chips 1, 2, 3 and 4 would be of different lengths. In practice, the metal cut does not consist of separate plates and the chips cannot be of different lengths. Hence, it is argued that the chip in the vicinity of the tool face must deform plastically in tension. The region subjected to tension will extend up to point A. The tensile field of stress that is developed as a result of the curvature of the tool point can thus produce the crack necessary in the development of the new surface. The material beyond point A, it is argued, is subjected to large compressive stresses and the crack will be quenched upon reaching this region in case of continuous cutting. Hence, it is argued [36] that the possibility of observing cracks in front of a tool increases as the radius of curvature at the

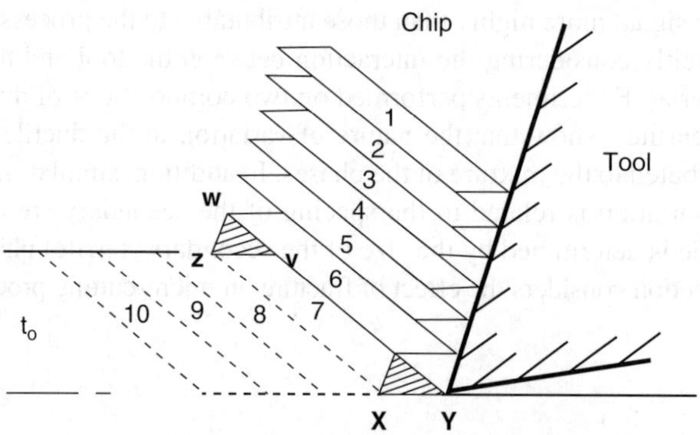

Figure 5.10 Card model of machining. Reproduced from [41]. Copyright 1948 American Institute of Physics

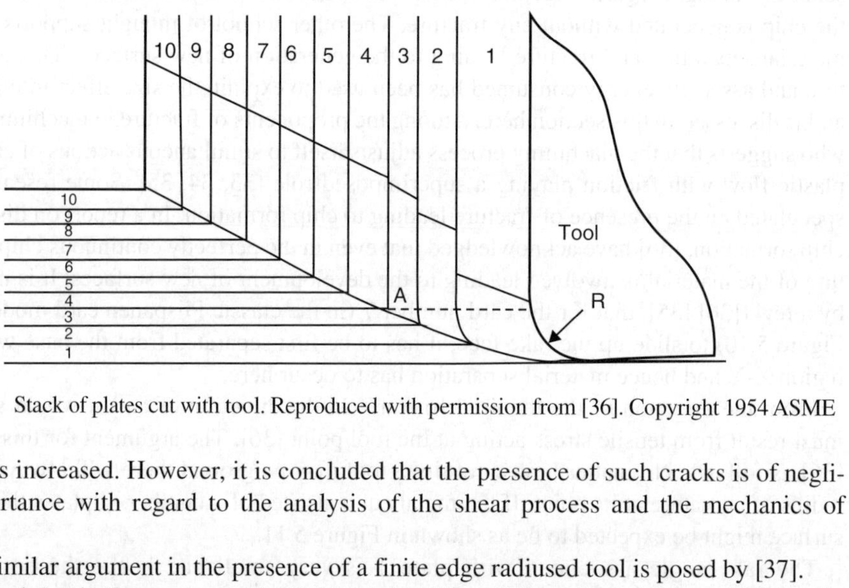

Figure 5.11 Stack of plates cut with tool. Reproduced with permission from [36]. Copyright 1954 ASME

tool point is increased. However, it is concluded that the presence of such cracks is of negligible importance with regard to the analysis of the shear process and the mechanics of cutting.

A very similar argument in the presence of a finite edge radiused tool is posed by [37].

It is argued that in the presence of a finite edge radius, all the material below point D (see Figure 5.12) must pass below the tool and be compressed. All material above D eventually moves parallel to the rake face. Thus, if a vertical element of material is considered to approach D in the direction of cutting, that part of the element above the horizontal through D will move upwards while that part below the horizontal through D will move downwards and hence the element will be subject to a tensile stress. It is not speculated however that fracture can occur because of this tension, nor do [37] include fracture in their analysis.

That the machined surface is a newly generated surface is generally accepted [38] [20]. Generation of new surfaces means that at some point ahead of the cutting edge material separation has to occur. Material separation implies that fracture has to be involved and energy has to be consumed in this process. In Merchant's analysis, it is specifically stated that the work required to separate the chip from the metal at the very cutting edge, thus producing two new surfaces, has been neglected [38]. Lately, this assumption has been increasingly questioned [34, 35]. The traditional argument to neglect this energy is as follows [38, 20]. When a new

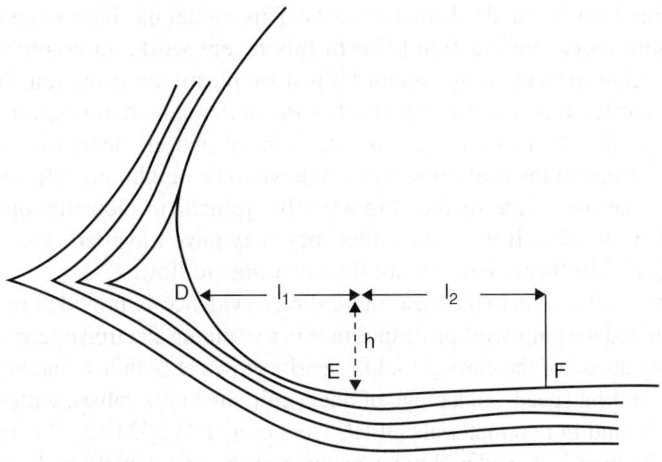

Figure 5.12 Presence of tension ahead of the cutting tool (Adapted from [37])

surface is generated in a solid substance, sufficient energy must be supplied to separate the ions at the interface. A certain energy is thus associated with the formation of a new surface called the surface energy of the substance (analogous to surface tension of liquids). This value of surface energy for most metals is very small, about 1 x 10^{-3} N/mm, and hence the energy associated is considered negligible. Merchant goes on to say that only when the dimension becomes of the order of 0.25 μm or less would the surface energy begin to become significant. He also goes on to add that this estimate of energy required to separate the chip from the metal at the very cutting edge would be increased by considering the work expended in the local plastic flow of the metal adjacent to the newly created surfaces on the chip and work. However, Merchant speculated that even after including this plastic work it would still be negligibly small compared to the total work. It is to be noted that upon inclusion of this plastic work, the energy needed for material separation may become important at values much larger than what Merchant speculated earlier as 0.25 μm. In a related work, [39] have argued that deformation during chip formation by the action of the tool is similar to piercing a hole using a punch. There is some material displacement or pushing before separation occurs. They acknowledge though that material separation does occur. Thus, they argue that there is always a constant force involved in the pushing action, regardless of the uncut chip thickness, that gives rise to the scaling of the specific cutting energy.

In more recent work [34] has argued for the applicability of modern ductile fracture mechanics in modelling the cutting process. He cites the work associated with the chip separation criterion in Finite Element simulations of machining to be orders of magnitude different from that associated with surface energy or surface tension, and comparable to fracture toughness values. The work done in cutting is considered to be expended in (a) plasticity along the shear plane, (b) friction at the tool-chip interface and (c) formation of new surfaces. From this, it is seen that if shear and friction can be minimized, then for a unit width of cut the fracture energy term naturally gives rise to a constant component of the cutting force viz., R, the fracture toughness. However, when Atkins applied this theory to data in the literature, the values of R needed to get a good fit of data were different from that obtained using standard fracture tests. This model has been extended to oblique cutting in a more recent publication [35]. In this recent work, an argument is made for why chip formation in machining cannot be just by plastic deformation. In plastic flow, elements of material that are neighbours before permanent deformation are the same neighbours after flow. In machining, this would mean that elements just above, and just below the parting line at the tool tip would still have to be neighbours afterward [35]. That is, elements on the underside of the chip are still 'joined' to elements on the machined surface, however far away from one another they may have travelled. This is implausible and suggests that chip formation is more than just plastic flow.

In addition to such indirect evidence, more direct evidence is provided recently by [40]. They have examined the chip-workpiece interface in a scanning electron microscope. Evidence of ductile tearing ahead of the cutting tool is clearly shown at both low and high rake angles, at high and low cutting speeds, in a range of uncut chip thickness values, with small and large edge radius tools, and in two materials OFHC Copper and Al-2024T3. This material separation by ductile fracture is considered to be a source of the force levelling discussed earlier in the section on size effect. Also, as discussed earlier in the context of stresses and strains, presence of microcracks in machining a low carbon steel, ahead of the cutting tool, has indeed been shown by [6].

A detailed approach to model the cutting process that captures ductile fracture leading to material separation has been developed by them using the finite element method. The model is implemented in a commercial software ABAQUS using the explicit formulation. Material separation is modelled via element failure. The model is then validated using the measured cutting and thrust forces and used to study the energy consumed in cutting. As the thickness of layer removed is reduced, the energy consumed in material separation becomes important. Simulations show that the stress state ahead of the tool is indeed favourable for ductile fracture to occur. In addition, simulations performed at varying rake angles show the trend towards a constant force component similar to that observed in experiments. The numerical model is also used to study the difference in energy consumption between a sharp tool and a tool with a non-zero edge radius. By presenting indirect and direct evidences, they have conclusively shown the presence of ductile fracture leading to chip formation in micro cutting of ductile metals. It also shows the importance of including this phenomenon in modelling the micro-cutting process and in explaining the scaling of the specific cutting energy. More recently, [41] has refuted the argument that ductile fracture energy needs to be accounted for in modelling machining forces and energy. Childs has shown via finite element simulations, for work materials made of steel, that the cutting and thrust force (in good agreement with experimental results) confirm the previously accepted view that the size effect is quantitatively explained by only two factors: one is the ploughing forces caused by the non-zero edge radius of cutting tools, two, the increase of shear flow stress with reducing uncut chip thickness that results from a reducing temperature and increasing strain-rate in the primary shear zone.

Thus in summary, while some researchers have argued that there is no fracture involved in machining, some have argued and shown evidence of fracture, while others acknowledge that fracture occurs in machining but there is no need to include its effects in modelling of machining processes.

That fracture occurs during micro cutting of brittle materials such as silicon and germanium is beyond any doubt. In fact, uncontrolled fracture and crack propagation, leading to poor

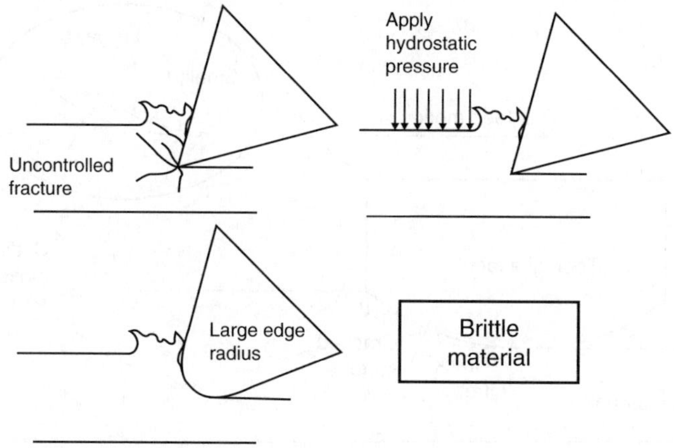

Figure 5.13 Controlled fracture is necessary to machine brittle materials

surface finish and sub-surface damage, is a problem to be resolved in micro cutting these materials. Some techniques adopted to solve the problem include performing machining under hydrostatic pressure conditions and more popularly by controlling the uncut chip thickness vis-à-vis the tool edge radius (Figure 5.13).

The notion of fracture free deformation in brittle materials seems similar to that seen in wear/scratch tests and in indentation studies. Although machining is not quite simply an indentation or a scratching process some parallels can be drawn. Studies indicate that when the volume of material undergoing the deformation is low then plastic flow can occur even in a brittle material and fracture can be avoided. The concept seems to be that the energy for plastic deformation scales with volume while the energy for fracture depends on the surface area. Hence, when scales are smaller, plastic deformation is energetically favoured compared to fracture. [42] predicted that a critical load on the diamond indenter is required for fracture initiation to occur. The threshold indentation load P, was determined as

$$P = \frac{\Omega K_c^4}{H^3} \tag{5.8}$$

H, being the hardness, K, the fracture toughness, and Ω a normalizing parameter that depends on E/H, where E is the elastic modulus. Using such analysis, the critical depth hc, of penetration by the indenter for crack to initiate can be determined as,

$$h_c = \psi\left(\frac{E}{H}\right)\frac{K_c^2}{H^2} \tag{5.9}$$

where Ψ is a dimensionless constant dependent on the indentation geometry. Given that the ratio E/H is similar for many ceramic-based brittle materials, the fracture toughness and hardness emerge as key material properties in Equations (5.6) and (5.7). The term (Kc/H)2 is referred to as the brittleness index. It is hypothesized by Blake and Scattergood (1990) that a critical depth parameter (hc) will divide fracture from ductility in ultra-precision machining. The factor Ψ will depend in a complex fashion upon machining

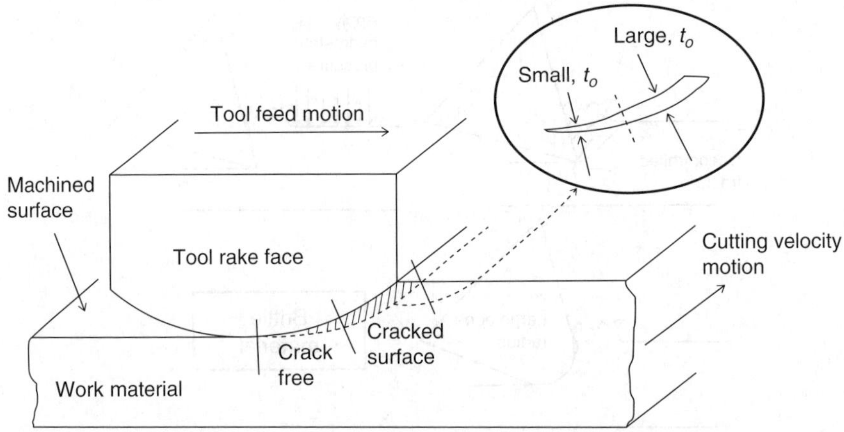

Figure 5.14 At small t_o crack-free surfaces can be obtained in machining a brittle material

parameters and tool geometry. [43] then extended this idea to machining where a critical t_o (Figure 5.14) was identified; when machining is performed below this critical t_o crack-free machined surfaces were obtained on brittle materials such as Silicon and Germanium [43]. In order to obtain consistent and uniform crack-free surfaces, machining has to be performed on special equipment such as an ultra-precision lathe that maintains good position control with high rigidity and low vibrations. As shown in Figure 5.14, the material removal mechanism changes from a ductile crack-free mode to a brittle mode as the uncut chip thickness value reaches a critical value, t_oc. The ability to create crack-free surfaces in brittle materials below a certain t_oc is commonly referred to as ductile regime machining. The critical uncut chip thickness depends on the material properties; for example it is higher for silicon than germanium. Parameters such as the tool rake and clearance angles also affect t_oc, while other parameters such as the surface cutting speed have a negligible effect. More recently, [44] have shown that the cracks are absent when the shear stress in the chip formation region is greater than the critical shear stress for chip formation and the stress intensity factor is less than the fracture toughness of the material. The point of transition is said to take place when the fracture toughness is equal to the stress intensity factor.

In some very fundamental experiments, [45] report results on plunge cutting on an inclined surface in monocrystalline brittle materials such as silicon and germanium. The inclined cut permits the study of varying depths of cut along the tool path (Figure 5.15).

The inclined angle plunge cut experiments show clearly the presence of a critical depth above which brittle material removal was seen and below which ductile mode of removal was observed. A small zone of elastic deformation involving rubbing or friction was also observed. Similar observations were noticed for different crystal orientations of both germanium and silicon. This technique is now widely used by researchers to determine the ductile mode cutting conditions for several materials.

Having now described the various material related effects important in the study and understanding of micro cutting we will now conclude this chapter by reviewing some literature reports on micro cutting various materials, ductile, brittle and others.

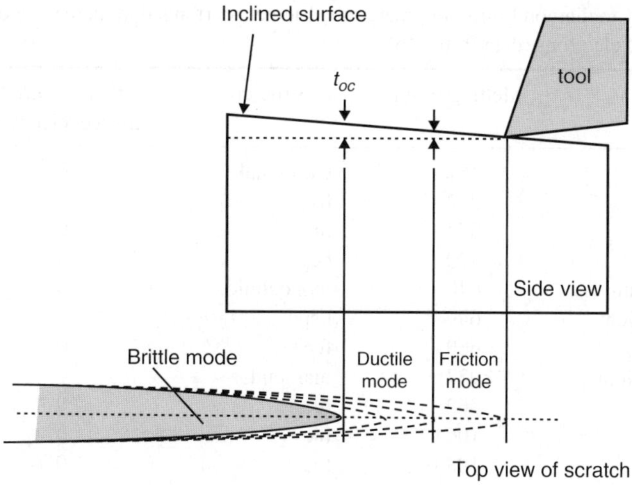

Figure 5.15 Schematic of an inclined angle plunge cut (Adapted from [45])

5.6 Metals, Brittle Materials and Others

As mentioned at the beginning of this chapter, several engineering materials including pure elements have been processed, and new materials are continuously being investigated for processing, using the micro-cutting process. These materials range from pure materials, very ductile materials such as copper and aluminium alloys, to very brittle materials such as graphite, silicon and germanium and new materials such as bulk metallic glasses. This section highlights some reported works in the area of micro cutting these various materials.

5.6.1 Pure Materials

An interesting study comparing the machinability of various pure materials by single point diamond turning was reported by [46]. They summarized the data as shown in Table 5.2, which include data from literature and some of their own tests on some rare earth elements. The data in the table is sorted by the number of unpaired d-shell electrons. Many materials cannot be processed by diamond turning particularly due to the large amount of tool wear in a short machining time. Experimental observations indicate that more often than not the nature of the wear is chemical. From the table it is clear that there seems to be a correlation between the amount of diamond wear seen and the number of unpaired d-shell electrons in the material – the more the number of electrons the more unlikely is the ability to diamond turn these materials. [46] speculate that the chemical wear seen in diamond tools is related to the formation of carbon-metal complexes in the presence of unpaired d-shell electrons. They argue that this hypothesis better explains the tool wear seen experimentally than all other explanations such as hardness, melting point and so on. This study underscores the importance of understanding the structure of materials in relation to the micro-cutting process.

Table 5.2 Ability to diamond-turn pure materials. Elements arranged in increasing order of number of unpaired d-shell electrons (data from [46]

		Melting Point	Structure	# of unpaired d-sheel electrons	Diamond turnable?
In	Indium	157	tetragonal	0	Y
Sn	Tin	232	fcc	0	Y
Pb	Lead	373	fcc	0	Y
Zn	Zinc	420	hcp	0	Y
Pu	Plutonium	640	monoclinic	0	Y
Mg	Magnesium	649	hcp	0	Y
AI	Aluminum	660	fcc	0	Y
Ge	Germanium	937	diamond	0	Y
Ag	Silver	962	fcc	0	Y
Au	Gold	1064	fcc	0	Y
Cu	Copper	1083	fcc	0	Y
Be	Beryllium	1277	hcp	0	Y
Si	Silicon	1410	diamond	0	Y*
Yb	Ytterbium	819	fcc	0	Y
Nd	Neodymium	1021	hcp	0	Y
Sm	Samarium	1077	rhombohedral	0	Y
Pd	Palladium	1552	fcp	0	Y
U	Uranium	1132	orthorhombic	1	N
Ce	Cerium	799	fcc	1	N
Ni	Nickel	1453	fcc	2	N
Ti	Titanium	1660	hcp	2	N
Rh	Rhodium	1966	fcc	2	N
Co	Cobalt	1495	hcp	3	N
V	Vanadium	1890	bcc	3	N
Ru	Ruthenium	2310	hcp	3	N
Ta	Tantalum	2996	bcc	3	N
Fe	Iron	1535	bcc	4	N
Nb	Niobium	2468	bcc	4	N
W	Tungsten	3410	bcc	4	N
Mn	Manganese	1244	bcc	5	N
Cr	Chromium	1857	bcc	5	N
Mo	Molybdenum	2617	bcc	5	N
Re	Rhenium	3180	hcp	5	N

5.6.2 *Ductile Metals*

Ductile metals, in general, when processed using the micro-cutting process, provide continuous chips and good surface finish due to absence of fracture and damage in the surface and sub-surfaces. However, they suffer from some disadvantages such as formation of burrs, build up of material on the tool edge, and in their hardened state can cause large tool wear and higher cutting forces and hence tool deflection.

Some of the early work on micro cutting was done on ductile metals such as aluminium alloys mainly driven by the need to create metallic mirrors, for optics applications, with extremely fine surface finishes. Using a single crystal diamond tool with a sharp cutting edge the turning process was utilized to generate these surfaces. [47]report results from diamond turning a Al-Mg alloy (used in magnetic disks) using a diamond tool with different nose radii. They observed that even with tool wear fine surface finishes were obtainable. The surface roughness was seen to increase with feed rates and nose radius, showing some correlation to thrust forces. The surfaces obtained were seen to have compressive residual stress with sub-surface deformation of only a few sub-micrometers. The earlier discussed work on size effects by [26] and [5] were on a different aluminium alloy, Al-7075. In the experiments reported by [5], continuous chips were observed over the entire range of undeformed chip thicknesses (10–2000 nm) investigated. The machined surface at an undeformed chip thickness smaller than the edge radius of the tool (10–60 nm) was observed to be rougher than at larger undeformed chip thickness for the two cutting speeds tested (10 m/min and 150 m/min). [48] report micro cutting pure aluminum and Al-2017 alloy. They have carried out surface characterizations using interferometry microscopy and grazing incidence X-ray diffraction. Lower roughness values were obtained on the pure aluminium work material while higher roughness was seen in the 2017 alloy and was attributed mainly to the presence of precipitates (CuAl2) in the alloy. Optical observations show a very high concentration of precipitates near the surface induced by machining. Scanning electron microscopy (SEM) observations have qualitatively confirmed that the density of the precipitates was, in the surface, higher than in the bulk. Thus, precipitates were shown to play a key role in the surface integrity obtained.[49] Lucca and Seo studied micro cutting of electrodeposited copper and a Cu-Te alloy. Copper was also the subject of interest for Wang *et al.* [50] who studied the effect of machining parameters on the surface roughness. Moronuki *et al.* [51] showed the effect of crystalline grain structure on the cutting forces in micro cutting aluminum, while Komanduri *et al.* [52] studied crystallographic orientation effects on micro cutting using single crystal aluminum.

5.6.3 Brittle Materials – Glass, Silicon, Germanium, Tungsten Carbide

To date several brittle materials have been successfully processed using single point micro cutting including silicon, glass, germanium, tungsten carbide and others. The challenge in machining these materials remains the same as explained before: prevent fracture formation and limit deformation to a small enough volume to induce only plastic flow. In what appears to be the first reported study [8], Japanese researchers Fujita and Shibata, and independently, Takahashi and co-workers reported in a Japanese language journal, single-point groove cutting and fly cutting, respectively, of glass. They report the use of negative rake angles and depths of cut below the edge radius of the tool to achieve crack free machining. Very early studies on the use of single-point cutting of silicon and germanium were by Blake [43]. Blake used a sharp diamond tool with a very large nose radius and a very small feed in a face turning operation. In this situation, the uncut chip thickness varies along the portion of the tool edge that is engaged in cutting. By rapidly removing the tool out of cut Blake was able to observe defects (pits) along the cutting edge direction. There appeared to be a critical uncut chip thickness below which no pits were detected. This value was found to increase with rake angle and decrease in clearance angle. Fang and Venkatesh [53] report single point micro cutting of silicon. They used inclined angle cuts to determine critical depth of cut for ductile mode machining. They also emphasize the importance of having a higher thrust force (normal to the work surface) component in relation to the cutting force to achieve fracture free surfaces. The nose radius of the tool and a negative rake angle provide such higher thrust forces. Liu and Li [54] have micro cut tungsten carbide (WC) in a ductile manner avoiding fracture; they have used experimental and theoretical means to determine the critical depth of cut. Beltrao *et al.* [55] have reported successful use of ductile mode cutting of commercial PZT (piezoelectric transition) ceramics to obtain good surface finishes. While single point turning and fly-cutting are the most popular methods for investigating ductile mode cutting of brittle materials, a few researchers are now shifting focus to micro milling of brittle materials such as glass while employing the concept of ductile mode cutting. Arif *et al.*, [56] have shown that it is possible to micro mill slots in glass by controlling the feed per tooth, effectively cutting in the ductile mode. By reducing the radial depth of cut ductile mode cutting can be achieved at a higher feed per tooth. Tool wear and chipping can cause rapid shift from ductile mode to the brittle mode of cutting.

5.6.4 Other Materials – Amorphous Alloys, Graphene and Embedded Polymers

Several other newer materials have been reported to be processed using the micro-cutting process. These include bulk metallic glass, graphite in an attempt to produce graphene and a novel process to create nanostructures using embedded polymers.

Bulk metallic glass (BMG) materials are amorphous alloys that exhibit no long range order and possess unique mechanical, thermal and magnetic properties. The presence of any long range order amidst the amorphous structure can deteriorate the properties of the material. These materials are often produced by a casting process involving rapid quenching to prevent crystallization. Machining processes may be needed for finishing operations. A notable feature of BMG is high fracture roughness and low thermal conductivities.

Ueda and Manabe [57] studied the chip formation mechanism of an amorphous metal by micro cutting in-situ inside a scanning electron microscope. The chip morphology was seen to exhibit typical lamellar structure attributed to the formation of the localized shear band. The concentrated shear band widths were observed to be very thin in the order of a few tens of nanometers. The width was seen to scale linearly with the uncut chip thickness. No evidence of crack of fracture was observable in the chips.

Bakkal et al. [58] investigated the effects of machining induced deformation on the properties of Zr–Ti–Al–Cu–Ni based BMG. Ductile chip formation was observed, as expected owing to high fracture roughness. At high cutting speeds intense light emissions were observed caused by oxidation of Zr; the chips were also seen to get crystallized. The low thermal conductivities are attributed to the serrated lamellar chip surfaces observed (Figure 5.16). Chip morphology showed the presence of shear bands, void formation and viscous flow.

Efforts have been made to cut layered materials such as graphite in order to separate the layers and produce useful materials such as graphene using the micro/nano cutting process. Jayasena and Subbiah [59] demonstrated a novel mechanical cleavage technique to produce thin layer graphene from bulk graphite using an ultrasharp diamond wedge assisted by ultrasonic oscillations, as illustrated in Figure 5.17. AFM measurements of the layers produced indicate that the proposed mechanical cleaving method is capable of producing thin layer graphene with a thickness of tens of nanometers. TEM studies reveal

Figure 5.16 Machining BMG materials. Intense light emissions and lamellar chips were observed at different cutting speeds. Reproduced with permission from [58]. Copyright 2004 Elsevier

that there is a considerable amount of attention required to understand the edge formation with ultrasonic oscillation usage since structures that seem to resemble nanohorns were observed. Application of ultrasonic vibrations along the tool edge is seen to significantly reduce the ID/IG ratios seen in a Raman spectrum. Hence, the applied oscillations may

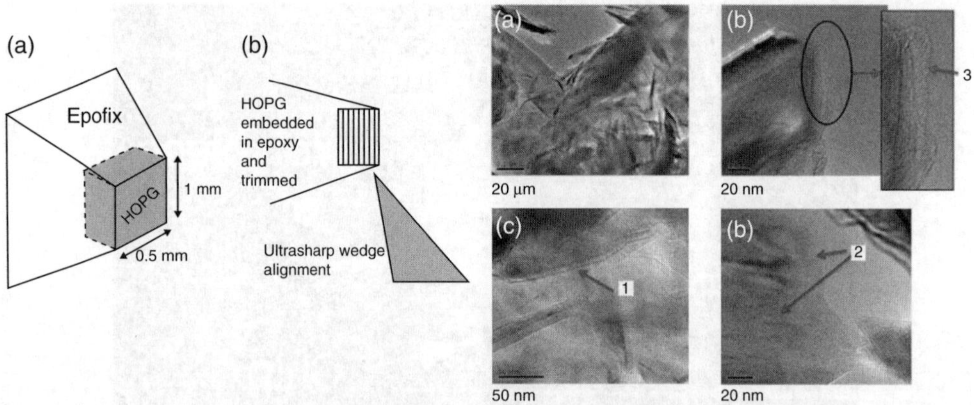

Figure 5.17 LEFT: HOPG mounted in epofix and trimmed to pyramid shape. (b) Setup showing wedge alignment with HOPG layers. RIGHT: TEM images with ultrasonic oscillation (a) FLG, (b) Edge of graphene sheet, (c) and (d) folded FLG. Reproduced from [59]. Copyright 2011 Springer

have the potential to reduce the defects in cleaved layers. The application of ultrasonic vibration also reduces the crystallite size.

A process called nanoskiving, essentially a micro/nano cutting process has been used to produce some interesting nanostructures by a group in Harvard University. Nanoskiving is shown to have potential to fabricate complex nanostructures that are difficult or impossible to achieve by other methods of nanofabrication. In a recent paper Lipomi et al. [60], the group reports the use of this technique on several materials such as metals (aluminum, titanium, nickel, copper, palladium, gold, lead), brittle materials (silicon dioxide, alumina) and several semiconducting and conducting polymers. It is seen that films of soft compliant metals stay intact upon sectioning while hard stiff materials fragment upon sectioning. Fragmentation is lower when the cutting direction is perpendicular to the exposed edge. Several defects are seen to be generated including scoring from cutting edge damages, delamination and compression of the embedded nanostructure/films. Cutting speed is seen to not affect the defect formation. As an example of applying this technique, the work carried out by Xu *et al.* [61] is shown in Figure 5.18. They describe the use of the technique of nanoskiving to fabricate complex metallic nanostructures by sectioning (using orthogonal cutting process) a polymer material containing small, embedded metal structures. The method begins with the deposition of thin metallic films on an epoxy substrate using a suitable technique. After embedding the thin metallic film in an epoxy matrix, the orthogonal cutting process with a sharp cutting tool generates sections (basically chips with thickness as small as 50 nm) of epoxy containing metallic nanostructures. The cross-sectional dimensions of the metal wires embedded in the resulting thin epoxy sections are controlled by the thickness of the evaporated metal film (which can be as small as 20 nm) and the thickness of the sections cut; they used a microtome system with a standard 45° diamond knife (edge sharpness < 2 nm) and routinely generated slabs 50 nm thick.

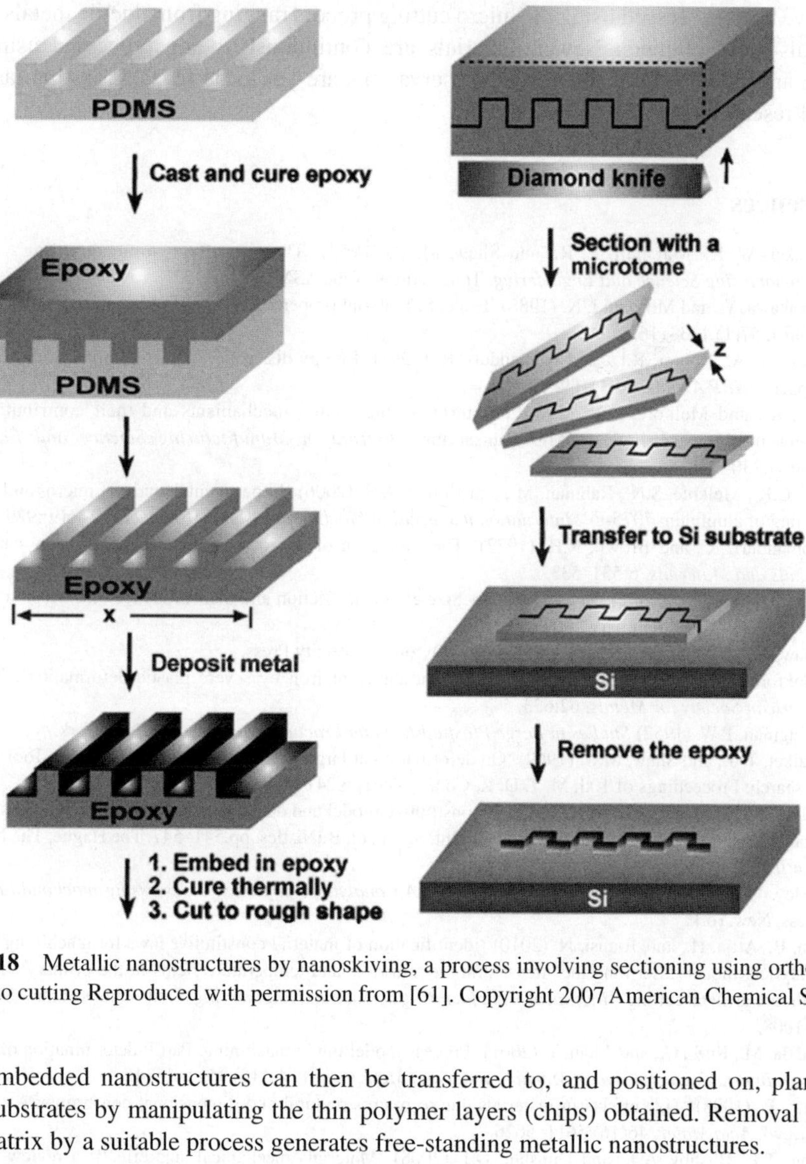

Figure 5.18 Metallic nanostructures by nanoskiving, a process involving sectioning using orthogonal micro/nano cutting Reproduced with permission from [61]. Copyright 2007 American Chemical Society

The embedded nanostructures can then be transferred to, and positioned on, planar or curved substrates by manipulating the thin polymer layers (chips) obtained. Removal of the epoxy matrix by a suitable process generates free-standing metallic nanostructures.

5.7 Summary

This chapter provided a summary of the various material related factors that have to be taken into consideration while understanding the micro-cutting process. The material aspects of size effects in the micro cutting range from increasing in material strength at small scales, influence of phases and crystallographic planes, and the importance of fracture in promoting ductile and brittle material micro cutting. The chapter also provided an overview of the various materials that have been processed using the micro cutting process ranging from ductile metals to brittle materials and polymers. Newer materials are continuously being processed using micro cutting and perhaps more interesting observations are yet to be revealed in this fascinating field of research.

References

[1] Backer, W. R., Marshall, E. R., and Shaw, M. C. (1952) 'The size effect in metal cutting'. *Journal of Manufacturing Science and Engineering*, Transactions of the ASME, 74:61–72.

[2] Furukawa, Y., and Moronuki, N. (1988) 'Effect of material properties on ultra precise cutting processes'. *CIRP Annals*, 37(1):113–116.

[3] Lucca, D.A., Rhorer, R.L., and Komanduri, R. (1991) 'Energy dissipation in the ultraprecision machining of copper'. *CIRP Annals*, 40(1): 69–72.

[4] Liu, K., and Melkote, S.N. (2006) 'Material strengthening mechanisms and their contribution to size effect in micro-cutting'. ASME Transactions *Journal of Manufacturing Science and Engineering*, 128(3):730–738.

[5] Ng, C.K., Melkote, S.N., Rahman, M., and Kumar, A.S. (2006) 'Experimental study of micro- and nano-scale cutting of aluminum 7075-t6'. *International Journal of Machine Tools and Manufacture*, 46(9):929–936.

[6] Komanduri, R., and Brown, R.H. (1972) 'The formation of microcracks in machining a low carbon steel'. *Metals and Materials*, 6:531–533.

[7] Tiesler, Nicolas A. (2002) 'Microforming - Size effects in friction and their influence on extrusion processes', *Wire*, 52(1): 34-38, February 2002.

[8] Shaw, M.C. (2005) *Metal Cutting Principles*, Oxford University Press.

[9] Lankford, G. and Cohen, M. (1969) 'Strain hardening of iron by severe plastic deformation'. *Transaction American Society for Metals*, 62:623.

[10] Bridgman, P.W. (1952) *Studies in Large Plastic Flow and Fracture*, McGraw Hill, New York.

[11] Walker, T. J., and Shaw, M.C. (1969) 'On deformation at large strains'. Advances in Machine Tool Design and Research: Proceedings of 10th M. T. D. R. Conference, pp.241–252.

[12] Johnson, G.R. and Cook, W.H. (1983) 'A constitutive model and data for metals subjected to large strains, high strain rates and high temperatures,' Proc. 7th Int. Symp. on BuNistics, pp.541–547. The Hague, The Netherlands (April 1983).

[13] Oxley, P.L.B. (1989) *Mechanics of Machining, An analytical approach to assessing machinability*, Halsted Press, New York.

[14] Shi, B., Attia, H., and Tounsi, N. (2010) 'Identification of material constitutive laws for machining—Part I: an analytical model describing the stress, strain, strain rate, and temperature fields in the primary shear zone in orthogonal metal cutting'. *Journal of Manufacturing Science and Engineering*, Transactions of the ASME, 132(5): 051008.

[15] Shatla, M., Kerk, C., and Altan, T. (2001) 'Process modelling in machining. Part I: determination of flow stress data'. *International Journal of Machine Tools and Manufacture*, 41(10):1511–1534.

[16] Arzt, E. (1998) 'Size effects in materials due to microstructural and dimensional constraints: A comparative review'. *Acta Mater*. 46(16):5611–5626.

[17] Zhu, T.T., Bushby, A.J., and Dunstan, D.J. (2008) 'Materials mechanical size effects: a review'. *Materials Technology*, 23(4):193–209.

[18] Brenner, S.S. (1956) 'Tensile strength of whiskers'. *Journal of Applied Physics*. 1956, 27:1484–1491.

[19] Griffith, A.A. (1920) 'The phenomena of rupture and flow in solids'. *Philosophical Transaction A*, 221:163–198.

[20] Shaw, M.C. (1950) 'A quantized theory of strain hardening as applied to cutting of metals'. Journal of Applied Physics, 21: 599–606.

[21] Saptaji, K. and Subbiah, S. (2011) 'Orthogonal cutting study of the micro-cutting thin workpiece'. Proceedings of the ASME 2011 International Manufacturing Science and Engineering Conference MSEC2011, June 13–17, 2011, Corvallis, Oregon, USA (paper #: MSEC2011-50256)

[22] Fleck, N.A., Muller, G.M., Ashby, M.F. and Hutchinson, J.W. (1994) 'Strain gradient plasticity: Theory and experiment'. *Acta Metallurgica et Materialia*, 42(2): 475–487.

[23] Solken, J.S. and Evans, A.G. (1998) 'Microbend test method for measuring the plasticity length scale', *Acta Materialia*, 46(14): 5109–5115.

[24] Dinesh, D., Swaminathan, S., Chandrasekhar, S. and Farris, T.N. (November 11–16 2001) An intrinsic size-effect in machining due to the strain gradient. Proc. ASME IMECE, New York, 197–204.

[25] Joshi, S.S. and Melkote, S.N. (2004) 'An explanation for the size-effect in machining using strain gradient plasticity'. *Journal of Manufacturing Science and Engineering, Transactions of the ASME*, 126(4): 679–684.

[26] Kai, L. and Melkote, S.N. (2006) 'Material strengthening mechanisms and their contribution to size effect in micro-cutting', Transactions of the ASME. *Journal of Manufacturing Science and Engineering*, 128(3): 730–738.

[27] Larsen-Basse, J., and Oxley, P.L.B. (1973) 'Effect of strain-rate sensitivity on scale phenomenon in chip formation'. Proceedings 13th Int. Machine Tool Design & Research Conference, Univ. of Birmingham, 209–216.

[28] Kopalinsky, E.M., Oxley, P.L.B. (1984) 'Size effects in metal removal process'. 3rd Conf. Mech. Prop. High Rates of Strain, Oxford, 389–396.

[29] Cohen, P.H., Black, J.T., Horne, J.G., Shih, A.A. (1981) 'Orthogonal machining of single crystals', *Manufacturing Engineering Transactions*, 389–396.

[30] Lee, W.B., To, S., and Cheung, C.F. (2000) 'Effect of crystallographic orientation in diamond turning of copper single crystals'. *Scripta Materialia*, 42(10): 937–945.

[31] Lawson, B.L., Kota, N., and Ozdoganlar, O.B. (2008) 'Effects of crystallography anisotropy on orthogonal micromachining of single-crystal aluminium'. *Journal of Manufacturing Science and Engineering*, Transactions of the ASME, 130(3): 03116-1-11.

[32] Vogler, M.P., DeVor, R.E., and Kapoor, S.G. (2003) 'Microstructure level force prediction model for micro milling of multi phase materials'. *Journal of Manufacturing Science and Engineering*, Transactions of the ASME, 125(2): 202-209.

[33] Atkins, A. G. (1974) 'Fracture toughness and cutting'. *International Journal of Production Research*, 12(2):263–274.

[34] Atkins, A. G. (2003) 'Modelling metal cutting using modern ductile fracture mechanics: quantitative explanations for some longstanding problems'. *International Journal of Mechanical Sciences*, 45:373–396.

[35] Atkins, A. G. (2006) 'Toughness and oblique cutting'. *Journal of Manufacturing Science and Engineering*, Transactions of the ASME, 128(3):775–786.

[36] Cook, N.H., Finnie, I., Shaw, M.C. (1954) 'Discontinuous chip formation'. *Journal of Manufacturing Science and Engineering*, Transactions of the ASME, 76:153–162.

[37] Connolly, R. and Rubenstein, C. (1968) 'Mechanics of continuous chip formation in orthogonal cutting'. *International Journal of Machine Tool Design and Research*, 8(3):159–187.

[38] Merchant, M.E. (1945) 'Mechanics of the metal cutting process. I. orthogonal cutting and a Type 2 chip'. *Journal of Applied Physics*, 16(5):267–275.

[39] Thomsen, E.G., Lapsley, J.T., and Grassi, R.C. (1953) 'Deformation work absorbed by workpiece during metal cutting'. *Journal of Manufacturing Science and Engineering*, ASME Transactions, 75:591–598.

[40] Subbiah, S., and Melkote, S. N. (2006) 'The constant force component due to material separation and its contribution to the size-effect in specific cutting energy'. *Journal of Manufacturing Science and Engineering*, Transactions of the ASME, 128(3):811–815.

[41] Childs, T.H.C. (2010) 'Surface energy, cutting edge radius and material flow stress size effects in continuous chip formation of metals'. *CIRP Annals*, 3(1):27–39.

[42] Lawn, B.R. and Evans, A.G. (1977) 'A model for crack initiation in elastic/plastic indentation fields'. *Journal of Materials Science*, 12:2195.

[43] Blake, P.N., and Scattergood, R.O. (1990) 'Ductile-regime machining of germanium and silicon'. *Journal of the American Ceramic Society*, 73(4):949–957.

[44] Venkatachalam, S., Xiaoping, Li, and Liang, S.Y. (2009) 'Predictive modelling of transition undeformed chip thickness in ductile-regime micro-machining of single crystal brittle materials'. *Journal of Materials Processing Technology*, 209(7):3306–3319.

[45] Brinksmeier, E., Preuss, W. and O. Riemer (1995) 'From friction to chip removal. an experimental investigation of the microcutting process'. Proc. 8th Intern. Prec. Eng. Seminar, 335.

[46] Paul, E., Evans, C.J., Mangamelli, A., McGaluffin, M.L. and Robert S. Polvani (1996) 'Chemical aspects of tool wear in single point diamond turning', *Precision Engineering*, 18:4-19.

[47] Sugano, T., Takeuchi, K., Goto, T., and Yoshida, Y. (1987) 'Diamond turning of an aluminium alloy for mirror'. *CIRP Annals*, 36(1):17–20.

[48] Revel, P., Khanfir, H., and Fillit, R-Y. (2006) 'Surface characterization of aluminium alloys after diamond turning'. *Journal of Materials Processing Technology*, 178(1–3):154–161.

[49] Lucca, D.A., and Seo, Y.W. (1993) 'Effect of tool edge geometry on energy dissipation in ultraprecision machining'. *CIRP Annals*, 42(1):83–86.

[50] Wang, Z.Y., Sahay, C., and Rajurkar, K.P. (1994) 'Micro-turning of copper with monocrystal diamond tool'. Proceedings of the First S.M. Wu Symposium on Manufacturing Science, 1:97–100.

[51] Moronuki, N., Liang, Y., and Furukawa, Y. (1994) 'Calculations of the effect of material anisotropy on microcutting processes'. *Precision Engineering*, 16:132–138.

[52] Komanduri, R. and Brown, R.H. (1972) 'the formation of microcracks in machining a low carbon steel'. Metals and Materials, 6: 531–533.

[53] Fang, F.Z., and Venkatesh, V.F. (1998) 'Diamond cutting of silicon with nanometric finish'. *CIRP Annals*, 47(1):45–49.

[54] Liu, K. and Li, X.P. (2001) 'Ductile cutting of tungsten carbide'. *Journal of Materials Processing Technology*, 113(1–3):348–354.

[55] Beltrao, P., Geeb, A.E., Corbett, J. and Whatmore, R.W. (1999) 'Ductile mode machining of commercial PZT ceramics' . *CIRP Annals*, 48(1):437–440.

[56] Arif, M., Rahman, M., Wong, Y. S., and Neha, D. (2010) 'An experimental approach to study the capability of end-milling for microcutting of glass'. *International Journal of Advanced Manufacturing Technology*, 53(9–12):1063–1073.

[57] Ueda, K. and Manabe, K. (1992) 'Chip formation mechanism in microcutting of an amorphous metal'. *CIRP Annals*, 41(1):129.

[58] Bakkal, M., Shih, A.J., Scattergood, R.O. and Liu, C.T. (2004) 'Machining of a Zr-Ti-Al-Cu-Ni metallic glass'. *Scripta Mater*, 50:583.

[59] Jayasena, B. and Subbiah, S. (2011) 'A novel mechanical cleavage method for synthesizing few-layer graphenes'. *Nanoscale Research Letters*, 6:95.

[60] Lipomi, D.J., Ramses, V., Martinez, R.M., Rioux, L.C., Reus, W.F. and Whitesides, G.M. (2010) 'Survey of materials for nanoskiving and influence of the cutting process on the nanostructures produced'. *ACS Applied Materials Interfaces*, 2(9): 2503–2514.

[61] Xu, Q., Rioux, R.M., and Whitesides, G.M. (2007) 'Fabrication of Complex Metallic Nanostructures by Nanoskiving'. *ACS Nano*, 1(3):215–227.

6

Modelling and Simulation of Micro Cutting

Ying-Chun Liang, Qing-Shun Bai and Jia-Xuan Chen
Harbin Institute of Technology, Harbin, China

In recent years, the demands for miniature and micro products, such as micro radiator, micro sensors, micro pump, micro mirrors, micro fibre optics connectors and micro inertial MEMS devices have greatly increased. Micro cutting technology plays an important role in the production of all these micro products. With its extensive contents, micro cutting has the advantage of fabricating complex real three-dimensional (3D) features which can be as small as several microns with a high efficiency. This technology can also be used for processing various kinds of materials. Modelling and simulation of micro cutting can show the ability to simulate industrial processing, which has enabled great advances in these processes to take place. Typically, finite element (FE), molecular dynamics (MD) and multiscale simulation (MS) are the typical methods for the process of modelling and simulation of micro cutting. Table 6.1 shows the contrasts between the three simulation techniques used in the modelling and simulation of micro cutting.

Table 6.1 Developments and applications for various simulation techniques

Techniques	FE	MD	MS
Simulation Scale	Big	Small	Medium
Basic unit	Node	Atom	Atom
Generated time (Approx.)	1940s	1950s	1990s
Basic principle	Constitutive Equation	Potential function and Newton's second law	_
General tools	ANSYS, Abaqus, Deform 3D, …	Lammps, GROMOS, MDynaMix, …	QC, MAAD, CADD, CGMD, …

Micro-Cutting: Fundamentals and Applications, First Edition. Edited by Kai Cheng and Dehong Huo.

6.1 FE Modelling and Analysis

The finite element method (FEM) is today without doubt one of the major numerical tools in computational science. This includes the results on the theory of finite elements that have been produced by mathematicians since the 1960s, particularly regarding error estimation; and last but by no means least, the inexorable increase in computer power over last half century and the corresponding increase in the expectations of users [1]. The result of these simultaneous factors has been that theory for finite elements is now available for a very large range of applications, and that a large quantity of commercial software is now available for a very large range of applications, and for implementing finite element techniques. The mechanism of micro cutting is a relatively complex process, in which the micro tool plays an important role. The FE has provided a useful tool for simulating micro cutting in the discrete node of FE, which gains the benefits from the development of computer hardware and commercial software. Besides, most of the current simulation models are two dimensional (2D), which are based on the hypothesis of orthogonal cutting. Three dimensional analysis of micro cutting offers more realistic simulation of the process by avoiding the assumptions accompanied with 2D models. So, finite element simulation can be thought of as a key technology for understanding the behaviour of micro cutting in the view of continuous mechanics.

6.1.1 Finite Element Model

Micro milling is a typical complex process for material removal. The micro milling modelling using the FE method is built to investigate the burr formation, cutting force, breakage and thermal distribution. As is shown in Figure 6.1, the 3D micro-burr formation finite element model of micro-end-milling operations has been developed by using commercial FE code (Deform -3D™), which is typically used in the simulation of traditional cutting [2]. The micro tool used in the present milling process is a micro-diameter cutter with an edge diameter of 0.5 mm. The SEM morphology of a tungsten carbide micro cutter is shown in Figure 6.1(a). It can be seen that the cutter has two extended helix flutes to the cone section of the tool. Based on parameterized design, the geometry model of the micro-diameter cutter was first designed in Pro/Engineer® software as shown in Figure 6.1(b). The geometry parameters of the micro-diameter cutter are listed in Table 6.2.

In order to improve the simulation's efficiency, the geometry model of the micro-diameter cutter was simplified and just the effective edge section was kept, that is, 0.1 mm from tool top end (Figure 6.1(c)). This geometry model was imported into the finite element (FE) environment Deform-3D™ and embedded in a peripheral milling environment with the workpiece. The meshed model of micro tool and workpiece is shown in Figure 6.1(d).

A brass alloy HPb63-3 is employed as workpiece material. The flow stress used in the simulation is assumed to be a function of strain, strain rate and temperate. The Johnson-Cook's constitutive equation, which is shown in Equation (6.1), is adopted to describe the material behaviour [3].

$$\bar{\sigma} = \left[A + B\varepsilon^n\right]\left[1 + C\ln\left(\frac{\dot{\varepsilon}}{\dot{\varepsilon}_0}\right)\right]\left[1 - \left(\frac{T - T_r}{T_m - T_r}\right)^m\right] \tag{6.1}$$

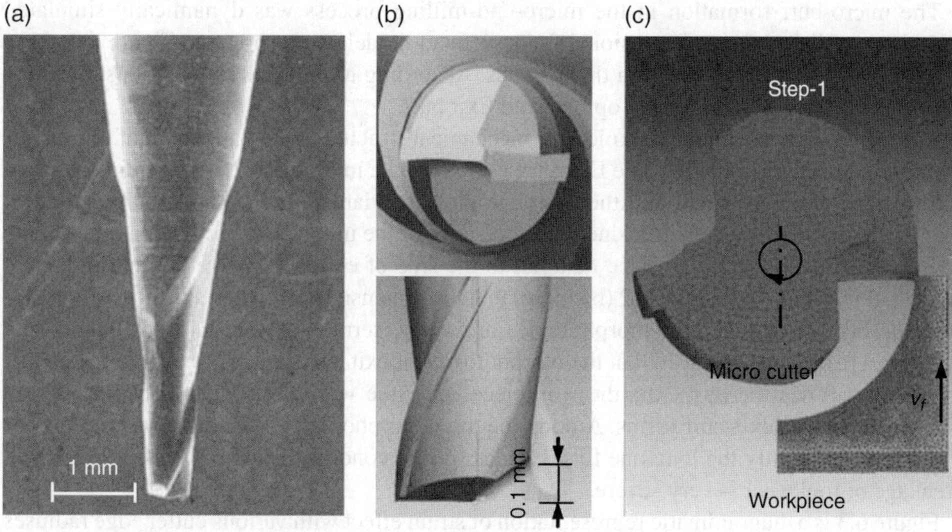

Figure 6.1 Micro cutter and the geometry model for cutter and micro milling

Table 6.2 Geometry parameters of micro cutter

Overall length	Diameter	Flute length	Rake angle	Flank angle	Helix angle
38 mm	0.5 mm	1.0 mm	5°	12°	30°

Brass alloy HPb63-3 has an excellent machining performance; therefore, it is widely used to manufacture micro parts of watches and automobiles. The Arbitrary Lagrange Euler (ALE) finite element method allows mesh movement without material dependency, which can improve the quality of large deformations. In the simulation of micro milling, the ALE method was adopted to simulate cutting chips separating from workpiece automatically. The Coulomb friction model was adopted to simulate the friction between tool and workpiece [4].

6.1.2 Simulation on Micro-burr Formation

The burr formation factors in micro-end-milling are different from those in traditional macro cutting and the burrs formed in micro-end-milling show salient size effect. Hence, investigating the burr formation mechanism on the micro-scale will improve the quality of micro features. The micro-burr formation mechanism is mainly studied using the following three methods: the analytical method, the finite element method and the experimental method, which have been used by many scholars.

The micro-burr formation in the micro-end-milling process was dynamically simulated with the established three-dimensional finite element model. As is shown in Figure 6.2, there are three types of burrs formed in this process. According to the forming positions, they can be classified into Entrance burr, Top burr and Exit burr.

Compared with machined workpiece or mechanical structure, the relative size of micro burr in micro-end-milling is larger. The Entrance burr is shown in Figure 6.2 (a). The dimension of entrance burrs is very small and the morphologies are triangular, ellipsical and pin-shaped. The formation of entrance burr is due to the presence of the initial negative shear angle and the compressed and bulged workpiece material to the side of entrance before permanent plastic deformation occurring. Figure 6.2 (b) shows that the dimension of top burrs is relatively larger than the other types and the morphologies are wavy, serration, bend and roll-over. From Figure 6.2 (c) it can be noticed that the morphologies of exit burrs are roll-over, pin-shape and strip-shape. When the tool exits the workpiece edge, the workpiece material is pushed out plastically and the exit-burr forms. Among the three morphologies of the exit burrs, the strip-shape burr, especially the long one forms under special condition in which the micro-tool-tip breakage or tool wear is very severe.

Figure 6.3 is a diagrammatic representation of strain effect with various cutter edge radiuses on burr formation. In this simulating section, cutter edge radiuses are respectively 4 μm, 6 μm and 8 μm in different simulating processes. It is found that various cutter edge radiuses have a significant influence on burr formation. As shown in Figure 6.3, when the cutting edge radius increases from 4 μm to 8 μm, the dimension of micro burr increases greatly. The average value of the effective rake angle can be derived by Equation (6.2). [5, 6]

$$\gamma_{ave} = -\frac{\pi}{2} + \cos^{-1}\left(1 - \frac{t}{r_n}\right) \tag{6.2}$$

Where t is the undeformed chip thickness and r_n is the tool edge radius. From Equation (6.2), it can be seen that with increase of r_n, the effective rake angle would be more negative so that the sliding and ploughing processes dominate. Providing f_t is less than r_n, chip formation may not occur but the burrs occur under the effect of plastic deformation.

6.1.3 *Influence of the Tool Edge Radius on Cutting Forces*

In the micro-scale milling process, the size effect is typically characterized by a non-linear increase in the specific cutting force as the minimum chip thickness is decreased [7]. Figure 6.4 shows variations of the simulated cutting force with various tool edge radii, 3.2, 5.0 and 7.0 μm while machining aluminum alloy Al2024-T6, respectively [8]. It can be seen that the cutting forces increase rapidly as the tool advances into the workpiece. When the micro-cutter cutting edge radii are between 3 and 5 μm, the cutting force did not change significantly. However, as the micro-cutter cutting edge radius is equal to 7 μm, the cutting force changed significantly. This simulation also showed that, when the tool edge radius is set as 7 μm, feed per tooth f_z of 0.1 μm, the ratio of feed to tool edge radius (f_t/r_n) is less than the critical chip thickness. Thus, under this cutting condition, the cutting force increased significantly.

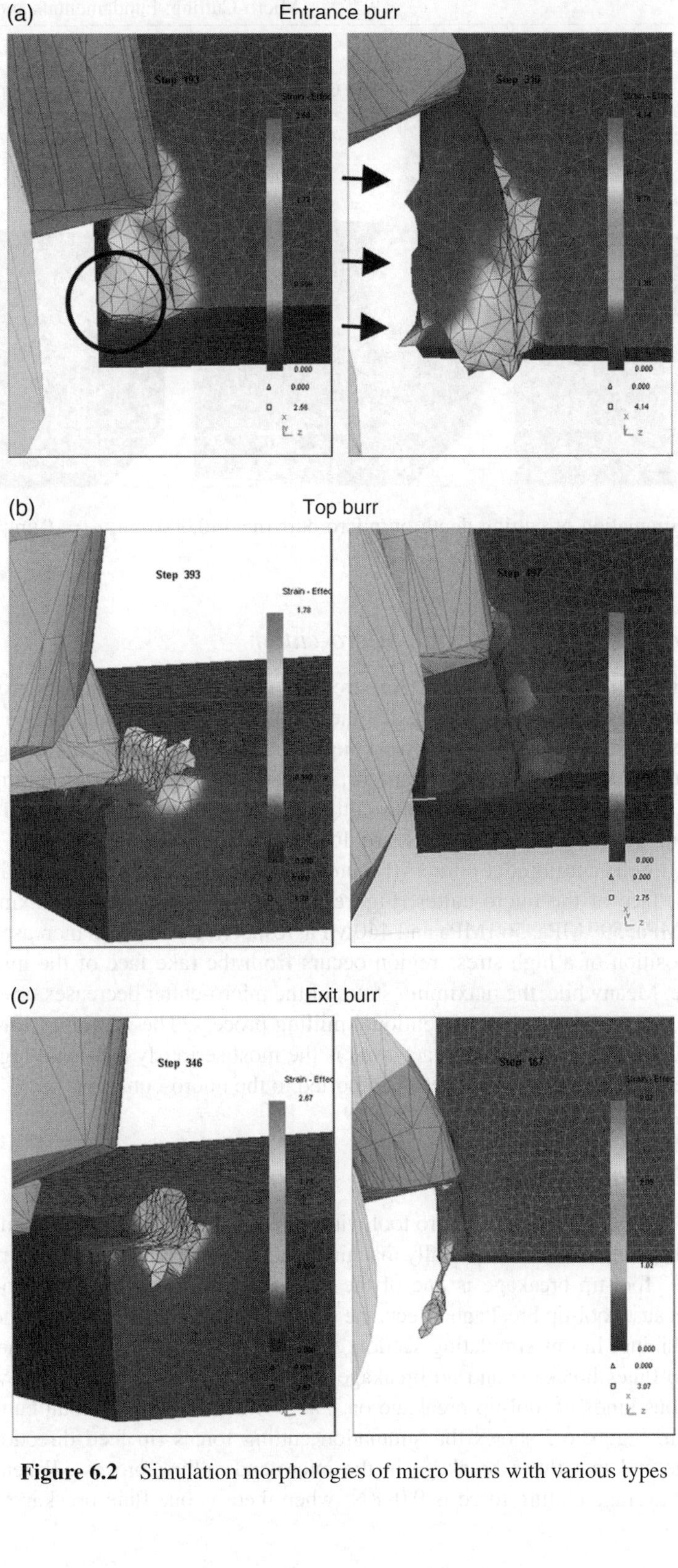

Figure 6.2 Simulation morphologies of micro burrs with various types

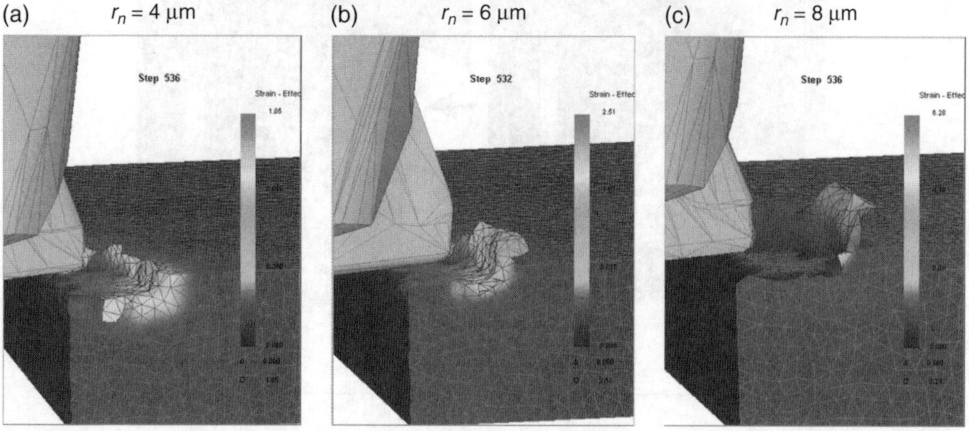

Figure 6.3 Simulation of cutting depth on micro burr (n = 140,000 rpm, $a_p = 10\ \mu m$, $a_e = 0.25\ \mu m$, $f_z = 1.0\ \mu m$)

6.1.4 *Stress Distribution on the Micro-cutter*

Figure 6.5 shows images of the effective stress of chip, workpiece and micro-cutter with various cutting edge radii. It can be seen that the maximum effective stress only occurs in very small regions of the micro-cutter and workpiece. Figure 6.5(a) shows a state generated by continuing cutting chips; Figure 6.5(b) and (c) are states that resulted from interrupted chips, and Figure 6.5(d) shows a state without any cutting chips. With an increase of tool edge radius, the maximum effective stress regions are located at the rake face of the micro-cutter (Figure 6.5(a)), helix cutting edge corners (Figure 6.5(b)), the end cutting edge (in Figure 6.5(c)), and the flank face of the micro-cutter (Figure 6.5(d)), respectively. The maximal effective stress is 647 MPa, 599 MPa, 560 MPa and 460 MPa, respectively. With an increase of tool edge radius, the position of a high stress region occurs from the rake face of the micro-cutter to the flank face. Meanwhile, the maximum stress of the micro-cutter decreases obviously, but it is still much higher than that of conventional milling process. The maximum stress distribution micro-cutter and workpiece contact zone is the most seriously frictional region, and the heat generated by friction will be directly imported to the micro-cutter.

6.1.5 *Micro-tool-tip Breakage*

In micro-end-milling process, the micro tool with very small flute diameter is commonly used. Tool-tip breakage often occurs, especially machining hard-to-machine materials or material with hard particles. Tool-tip breakage is one of the most influential aspects of micro-end-milling operations because tool-tip breakage affects the quality of the machined surface and the cutting force of machining. In this simulating section, three micro cutters were established (one flute breakage, two flutes breakage and no breakage micro-cutter) [9]. Figure 6.6 shows the strain effect of various kinds of tool-tip breakage on burr formation when the total cutting distance reaches 60 μm. Figure 6.7 shows the simulation cutting forces (in feed direction) with one flute breakage and two flutes breakage in the micro-end-milling process. When there is no breakage, the average cutting force is 0.098 N; when there is one flute breakage, the average

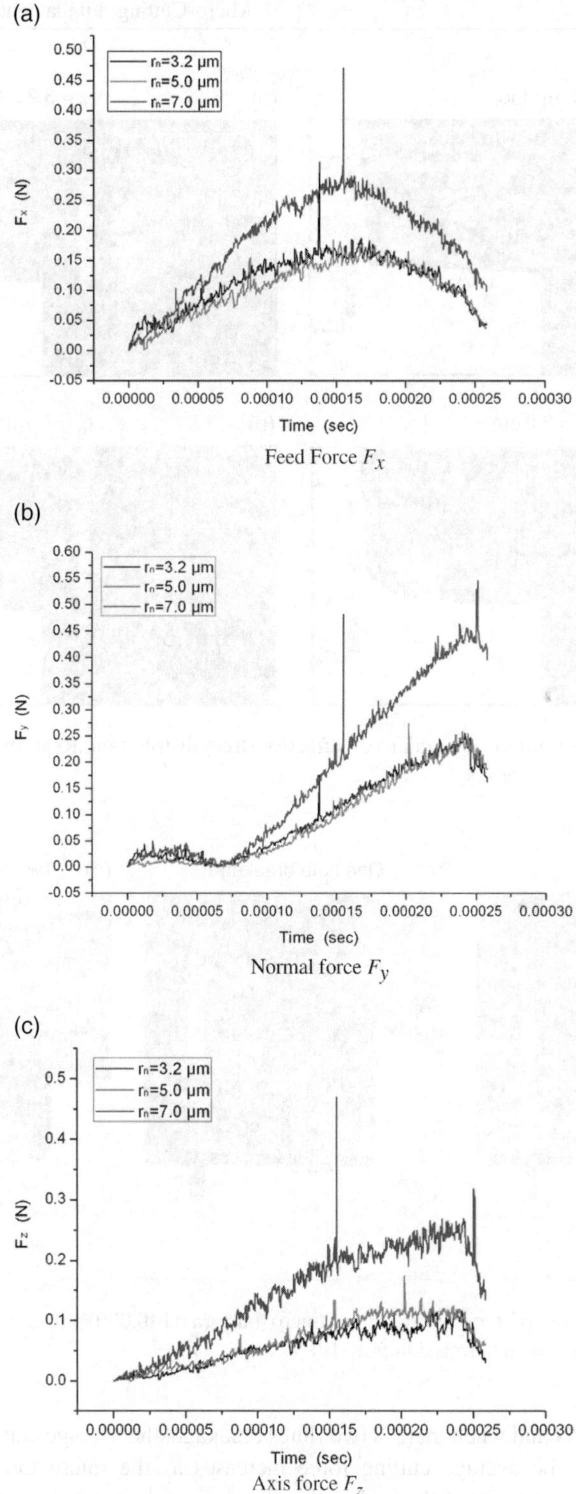

Figure 6.4 Effect of tool edge radius on three different cutting forces. Reproduced with permission from [8]. Copyright 2011 Springer

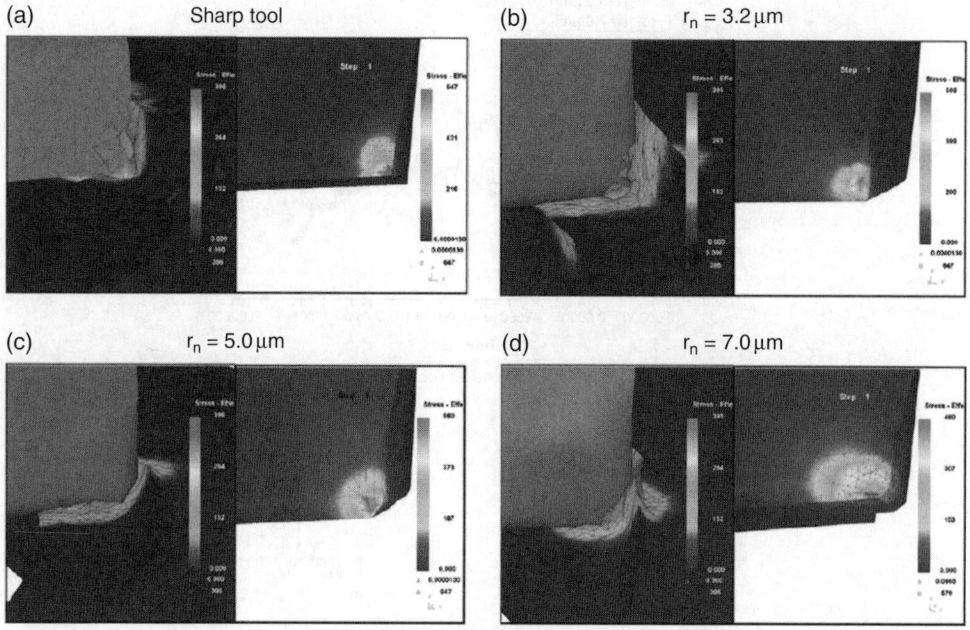

Figure 6.5 The effect of tool edge radius on effective stress distribution. Reproduced with permission from [8]. Copyright 2011 Springer

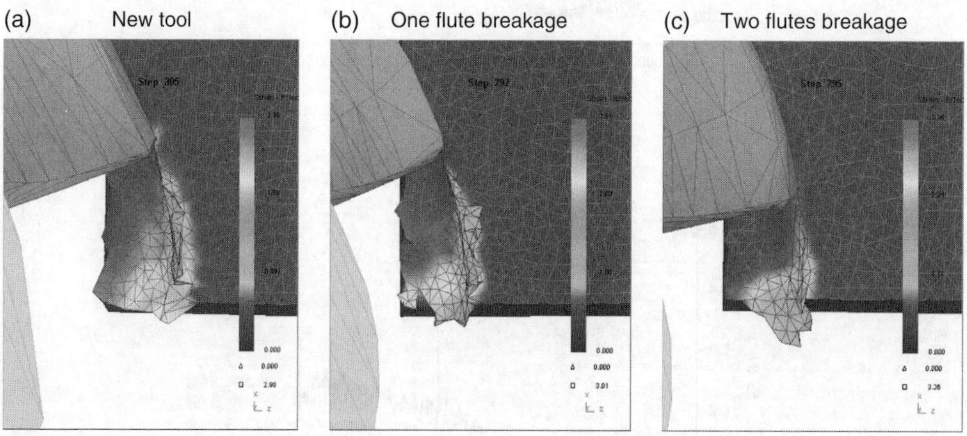

Figure 6.6 Simulation of tool breakage on micro burr (n = 140,000 rpm, a_p = 10 μm, a_e = 0.25 μm, f_z = 1.0 μm). Reproduced with permission from [9]

cutting force is 0.12 N and when there is two flutes breakage, the average cutting force is 0.18 N. It can be seen that the average cutting force increases as the micro tool breakage occurs. Meanwhile, with the increase of the tool-tip breakage, the rake angle becomes more negative, which also results in the increased dimension of the exit-burrs.

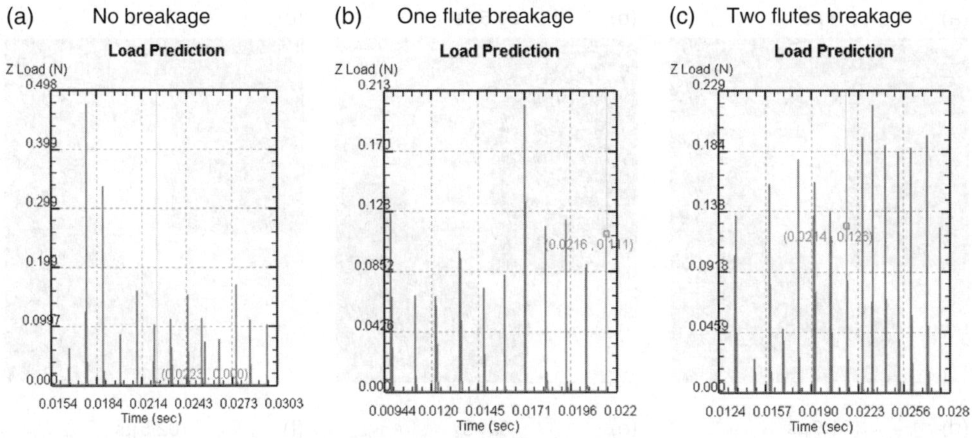

Figure 6.7 Simulation of feed direction cutting force on micro burr (n = 140,000 rpm, a_p = 10 μm, a_e = 0.25 μm, f_z = 1.0 μm). Reproduced with permission from [9]

6.1.6 *Thermal Analysis on Micro Cutting*

In a peripheral milling process, the main thermal-mechanical coupling occurs in the period from the tool workpiece contact until the separation of tool from workpiece. In this process, the stresses and strains of tool and workpiece are very complex and the typical characteristics of micro milling can be observed. A 3D coupled thermal–mechanical finite element mesh model was used for the analysis of thermal analysis of micro cutting [10]. Figure 6.8 shows the simulated temperature distributions of the micro milling process in the various cutting distances. In the initial stage of milling, the small deformation of the workpiece can be produced by micro-diameter cutter, which may be converted into thermal energy. The heat produced by the workpiece is not conducted into the micro tool in time, which will lead to a low cutter temperature increase. With the proceeding of micro milling, the temperature on the micro cutter rises greatly and minor temperature distinction between tool and workpiece can be observed. From the isotherms of Figure 6.8(e) and Figure 6.8(f), it can be seen that the temperature of micro tool and workpiece are approximately equal.

The temperature distribution in certain simulation steps (Step No.1530) is shown in Figure 6.9. It can be seen that the thermal effect focuses only on a small region between tool and workpiece. The effective region is so small that there is no medium for tool and workpiece to cool. Figure 6.9(b) and (c) show the temperature distribution of micro tool and workpiece in the simulation step separately. It is also found that the highest temperature in the workpiece locates in the root of the cutting chip. Figure 6.9(c) illustrates that the distribution of temperature on the rake face of micro cutter is variable. The maximal stress doesn't occur at the end cutting edge and helix cutting edge but rather in the rake face leaving a certain distance from the end cutting edge. The temperature distribution curve in the rake face of tool tip along the cutter axis is shown in Figure 6.9(d), in which the measured distance of the temperature peak is about 1.6 μm.

With the proceeding of cutting, the cutting chips flow from the rake face and the heat which accumulates gradually from friction and deformation is dissipated through chips, tool body

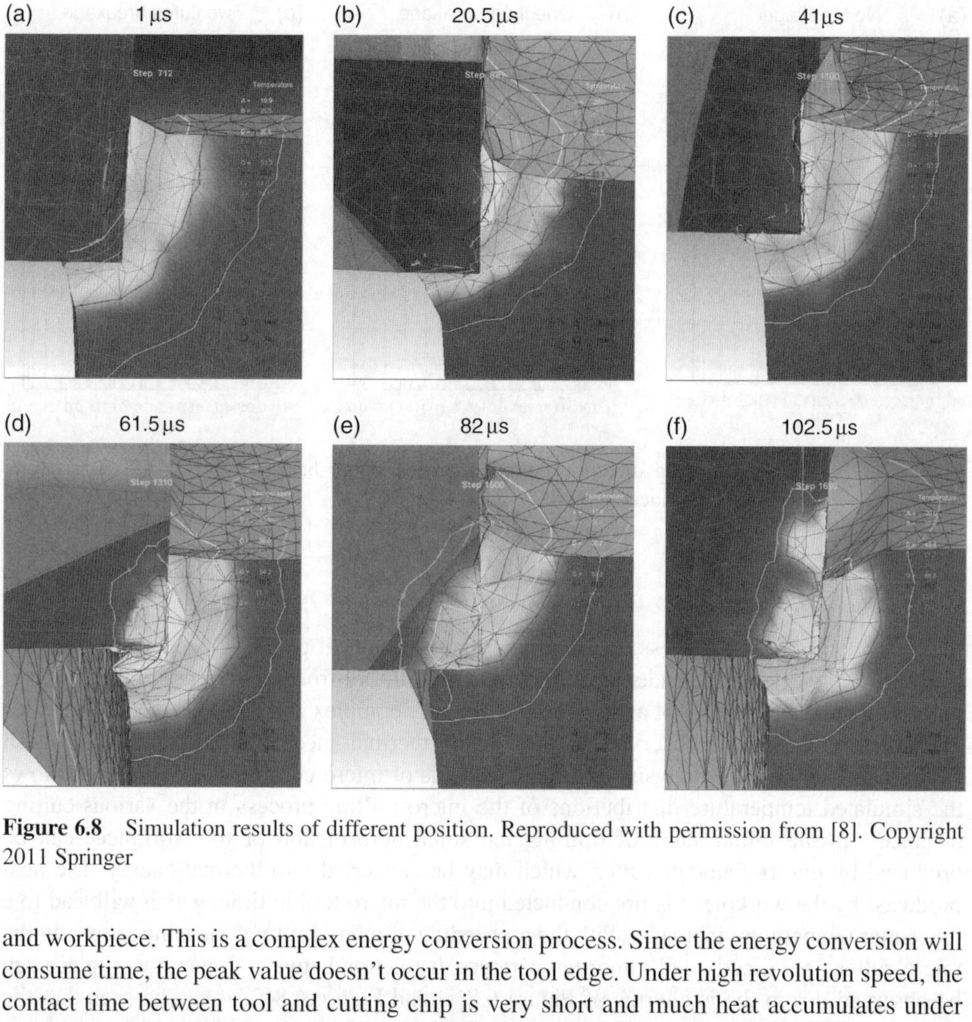

Figure 6.8 Simulation results of different position. Reproduced with permission from [8]. Copyright 2011 Springer

and workpiece. This is a complex energy conversion process. Since the energy conversion will consume time, the peak value doesn't occur in the tool edge. Under high revolution speed, the contact time between tool and cutting chip is very short and much heat accumulates under the cutting chip. So the position with the highest temperature is very close to the end cutting edge. The highest temperature in the simulation can exceed 90°, which may cause heat stress concentrations related to size-effects on the micro-diameter cutter.

6.2 Molecular Dynamics (MD) Modelling and Analysis

6.2.1 MD Modelling Process and Simulation

With the development of science and technology, the surface finish has increased greatly, even researched to the nanometer scale. The mechanisms and principle of micro cutting in nano-scale has created an increased interest for scholars. Molecular dynamics (MD), which can model the micrometer and nanometer structure, is an effective method to explore the phenomenon of the micromachining in atomic view. It was in the late 1950s that Alder and Wainwright initially investigated the equation of state (EOS) of liquid and atmosphere by

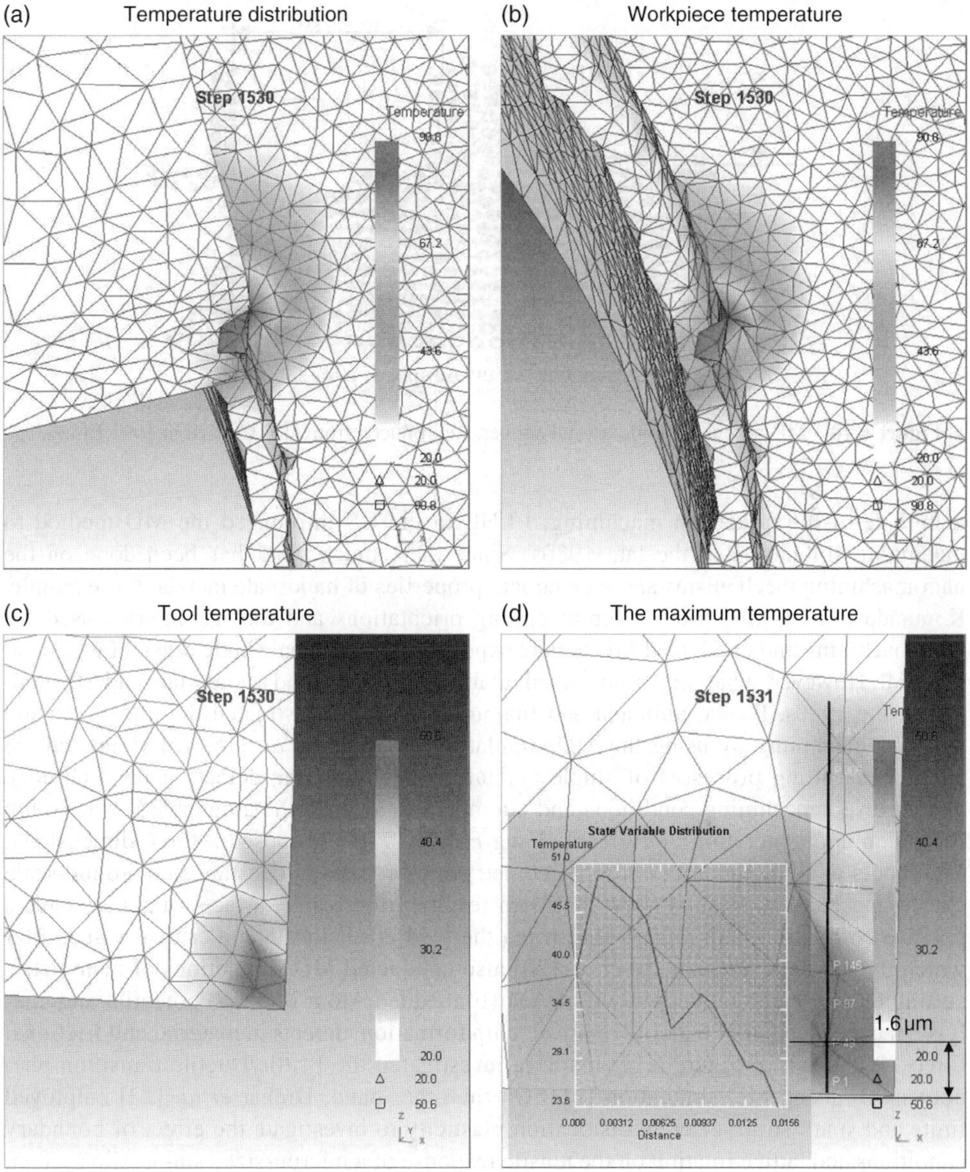

Figure 6.9 Temperature distribution of micro milling. Reproduced with permission from [10]

using molecular dynamics [11]. Hoover *et al.* explored the micro friction on the surface of single crystal copper by using MD in 1989 [12]. After that, Belak and Stowers built the MD model for the indentation and cutting of single crystal copper with traditional Morse potentials for calculating the interact force between two atoms, as shown in Figure 6.10 [13]. Also this result was shown in CIRP assembly in 1990, which promoted the simulation

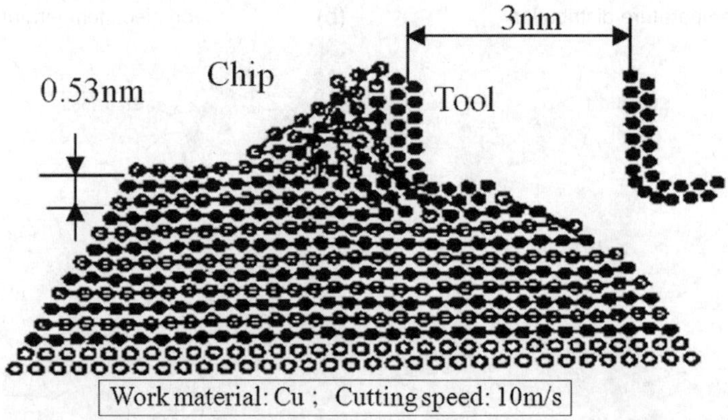

Figure 6.10 MD model of single crystal copper. Reproduced from [13] Copyright 1991 Elsevier

technique of ultraprecision machining. LLNL in the US introduced the MD method to machining simulation in the late 1980s. Since then, much work has been done on the nanomachining mechanisms and mechanical properties of nanoscale metals. For example, Komanduri *et al.* [14] tried different cutting orientations and tool geometries used for nanomachining and conducted MD tensile experiments with ideal single crystal FCC metal material. However, what they used for calculations were two-body potentials and 2D models. In the 1990s, Ikawa, Shimada and Inamura in Japan investigated the ultra-precision and nanomachining by using the MD simulation. Shimada *et al.* [15, 16] simulated the nanometer-cutting processes of single crystal copper with free defects using fine edge under perfect machining conditions and attempted to analyze the attainable surface, chip formation and generation of surfaces. Fang *et al.* [17-19] proposed a three-dimensional MD model to study the effect of tool geometry and processing resistance on atomic-scale cutting mechanism. Results show that both resultant force and surface roughness have a positive correlation with rate of feed when the feed is smaller than a critical value, after which they remain constant. Ye *et al.* [20] also conducted MD simulations of nanometric cutting of single crystal copper with EAM (Embedded Atom Method) potential and analyzed the mechanism of material removal, chip formation, defects in material and frictional forces. As shown in Figure 6.11, Inamura investigated the Brittle/Ductile transition phenomena by using MD simulation [21]. On the other hand, Dreher *et al.* [22] employed finite and small strain discrete dislocation plasticity to investigate the effect of boundary conditions and lattice rotation on the tensile response of a micron-sized single-slip system. Horstemeyer *et al.* [23, 24] simulated the shearing process of an ideal single crystal metal, and found that yield stress increases as size scale decreases, and the results of MD simulation are similar to the results of a finite element simulation or a physical experiment. It can be seen by comparing the three numerical frameworks that show that maximum dislocation nucleation occurs where there is maximum stress and strain gradients. They suggested that the yield stress in the specimen was related to the aspect ratio of the specimen. Fang *et al.* [25] found that a single crystal has the smallest Bauschinger effect, and high angle grain boundary (GB) has the strongest Bauschinger effect. Dislocation nucleation occurred earlier in low angle GB than in single crystal and high angle GB cases. This finding was

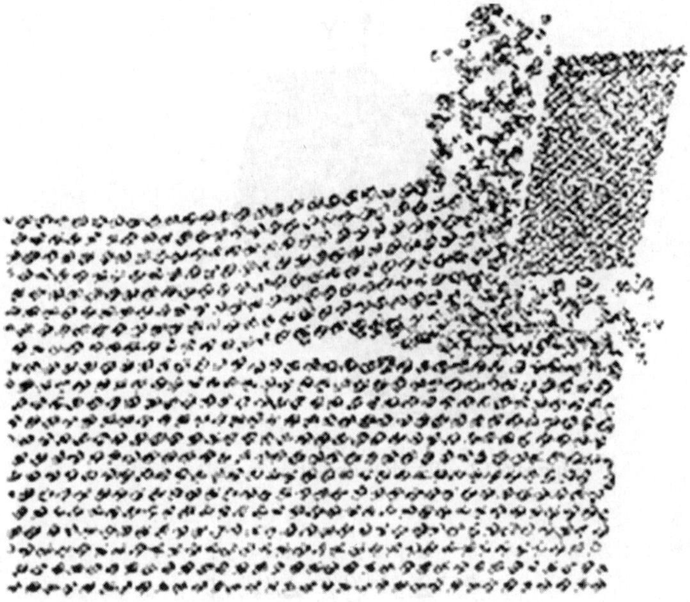

Figure 6.11 Crack during nanomachining. Reproduced from [21]. Copyright 1997 Elsevier

true for both fixed-end and flexible-end boundary conditions. Diao *et al.* [26] attempted to study the asymmetry of tensile/compressive yield stress and considered the effect of surface stress and different slip systems active in tensile and compressive yielding. Lin *et al.* [27, 28] also analyzed different multishell structures made of ideal Au nano wires for their effect on Young's modulus. Heino *et al.* studied the mechanical properties of copper with different defects [29].

6.2.2 *Modelling Analysis of Micro Cutting*

Morse potential model is used to predict the interaction between diamond atoms and Cu atoms in the calculation process. The potential model has been used to effectively depict the interaction between two molecules combined by covalent bonding in several similar studies.

The Morse potential function is adopted for the diamond atoms and the Cu atoms of the calculation process. It is

$$\mu(r_{ij}) = D[\exp(-2A(r_{ij} - r_0) - 2\exp(-A(r_{ij} - r_0)))] \tag{6.3}$$

Here $u(r_{ij})$ is a pair potential energy function. D, A, r_0 and r_{ij} correspond to the cohesion energy, the elastic modulus, the atomic distance at equilibrium and the atomic distance between atom i and atom j, respectively.

The EAM potential model is applied to the calculation process of between Cu and Cu atoms, it is

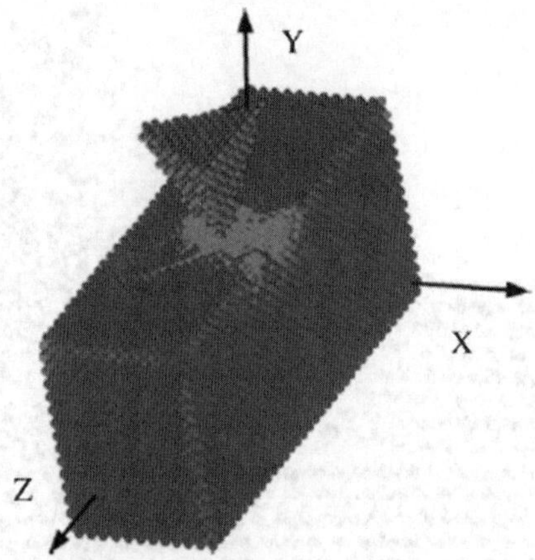

Figure 6.12 3D MD simulation model of scratch

$$E = \sum_{i}^{N} \left[F(\rho_i) + \sum_{j>i}^{N} u(r_{ij}) \right] \tag{6.4}$$

$$\rho_i = \sum_{j} \phi(r_{ij}) \tag{6.5}$$

Where E is the sum of the energy, F, $u(r_{ij})$ and $\phi(r_{ij})$ correspond to the sum of the embedding energy, the total potential energy and the embedding function the electron density around an atom.

Initially, the molecular model is limited to a 2D model down to the capability of computer hardware and software. Now 3D MD is common for most simulations of micro cutting, which can be used for more visual observation of simulation. Also the verities of tool and workpiece can be much more complex than 2D model. Figure 6.12 is a 3D molecular dynamics model for scratching. In Figure 6.13 the oblique cutting is simulated by using the 3D MD method [30].

6.2.3 Scratching Simulation by Using MD

The diamond tool was used to scratch the {010} surface of specimen along [001] direction. The initial displacement of the specimen results from the FCC copper lattice, the coordinate system is selected to be axis x, y and z in the [100], [010] and [001] directions, respectively. The size of the specimen is 12-17-20 a_0 and 12-17-35 a_0 from axis x, y and z, respectively, where a_0 is the equilibrium lattice constant (for FCC Cu, $a_0 = 0.362\,nm$). The tool is a tri-pyramid diamond tip with a rake angle of 60° and the angle of diamond tip is 60°, and the depth of scratch is 0.5, 1, 2 and 3 a_0, respectively.

At the initial stage of simulation, the atomic arrangement is composed of a FCC crystal which relaxes at 293 K to minimum energy, and the corner of specimen tends to be a curved surface to decrease the surface energy, as an indication of size effect. And then, the diamond tool starts moving along [001] from +X to –X. In the process of scratching, due to the ploughing of the tool, the atoms of the specimen near the tool are subjected to a strong shear force, and a shear zone is set up in front of the tool in the specimen. When the shear stress is up to

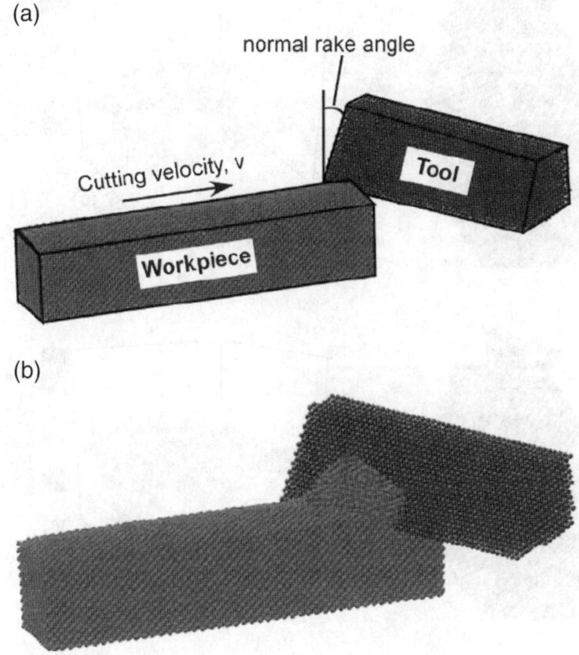

Figure 6.13 Oblique cutting of aluminium crystal. Reproduced from [30]. Copyright 2004 Elsevier

the threshold, the lattice of the specimen is buckled, and some atoms in the specimen start leaving the specimen and are adsorbed onto the surface of the tool under the shear and attraction forces of the tool, which is the initial stage of chip formation. As the tool moves forwards, the numbers of pile-up atoms ahead of the tool increase as the atoms gradually climb up the tool's front surface, and form a chip. As the scratch depths increase, the numbers of pile-up atoms in front of the tool increase and the atoms in the surface and subsurface are subject to an increasing shear force.

The distributions of shear stress in the second layers (xz plane) of atoms under tool tip at different scratch depths during machining are shown in Figure 6.14. It can be seen that the atomic shear stresses in the subsurface increase remarkably with the increase of scratch depths. In terms of both area and intensity, when the scratch depth is shallow, scratching has a little effect on the shear stress in the subsurface of the specimen, the effect of scratching on the shear stress in the subsurface of the specimen increases as the scratch depth increases. In the scratching process, the lattices of the specimen deform under the action of shear stress. When the shear stress exceeds a specific level, the atoms begin to rearrange and some dislocations are generated in the specimen. As the cutting edge moves forward, some dislocations are successively generated near the tool-specimen interface. As the dislocations are intrinsically linked to the stress level of atoms in the subsurface of specimen, the high atomic stress in the subsurface resulting from the cutting effect of tool is high enough to cause the nucleation of dislocations and drive the dislocations downwards along different slip planes in the specimen as shown in Figure 6.15(a). When the level of shear stress in the subsurface is low while the scratch depth is shallow, there is not enough drive force to move the dislocations downwards, and the dislocations do not penetrate into the specimen under the cutting edge, and propagate

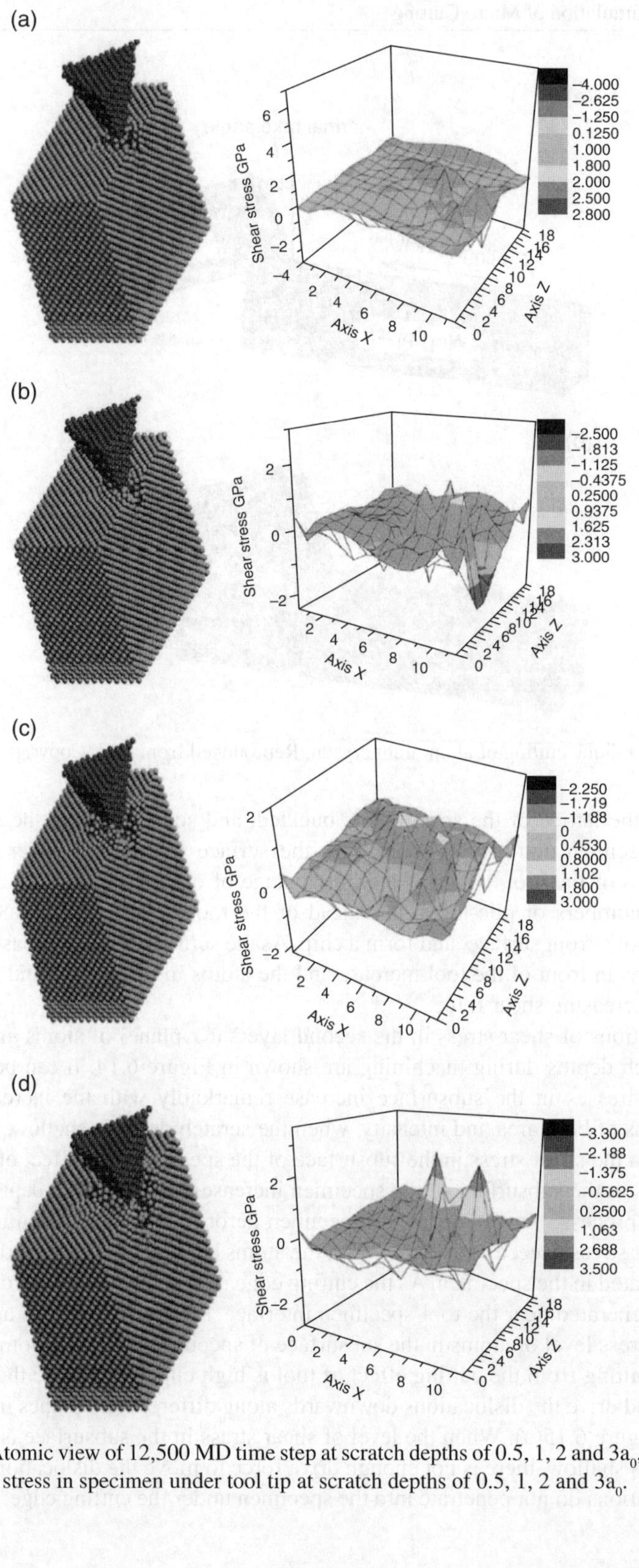

Figure 6.14 Atomic view of 12,500 MD time step at scratch depths of 0.5, 1, 2 and 3a_0, and distribution of atomic shear stress in specimen under tool tip at scratch depths of 0.5, 1, 2 and 3a_0.

(a) (b) (c)

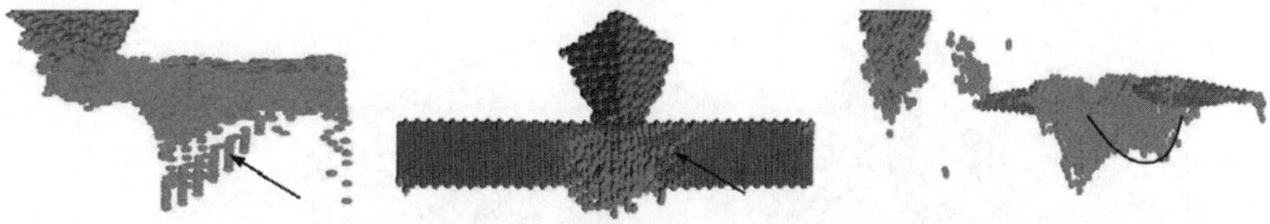

Figure 6.15 Effect of tool scratching specimen (010) plane at (a) scratch depth of $3a_0$ with arrow denoting downward propagation of dislocations; (b) scratch depth of 0.5 a_0, arrow denotes dislocations propagating in subsurface and (c) with changes in the subsurface work-affected layer below tool during scratching; atoms shown with slip vector index >0.3.

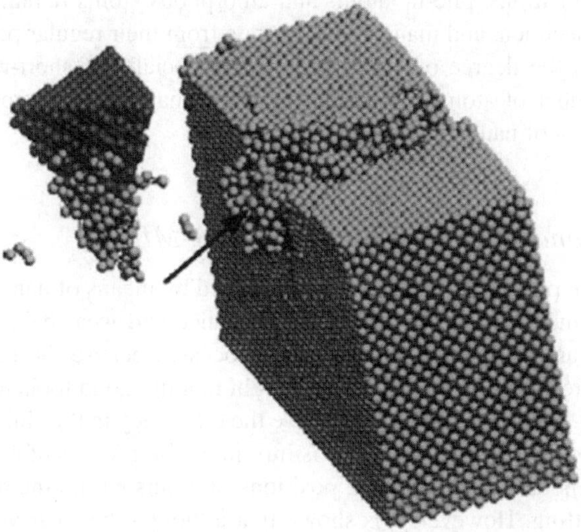

Figure 6.16 Effect of tool withdrawn from the specimen, the arrow denotes burr and surface recovery

in the surface and subsurface only as shown in Figure 6.15(b). Dislocations disappear when they reach the free boundary. With the passage of the tool, the stress of affected atoms in the subsurface are partially released, and some atoms do not rearrange but move upwards to cause the partial recovery of deformed areas as shown in Figure 6.15(c). Some defects remain in the subsurface because the stress is not fully released, which is a hidden trouble, and has its inevitable effect on the mechanical properties of a nanostructure. After the tool scratches the surface of the specimen, some atoms are taken away by the tool and rearranged under the action of interactive force, and the groove left in the surface restores to a certain extent.

When the tool moves out from the specimen, some pile-up atoms in the specimen are attracted backwards, and form a protrusion on the specimen, which is a burr as shown in Figure 6.16. Other pile-up atoms are adsorbed onto the tool surface and taken away by the tool. The number of defects in the surface and subsurface of the specimen increases as the scratch depth increases.

It can be seen from radial distribution function (RDF) that the scratch depth has an important effect on the crystal formation in the surface and subsurface of the specimen. The atomic

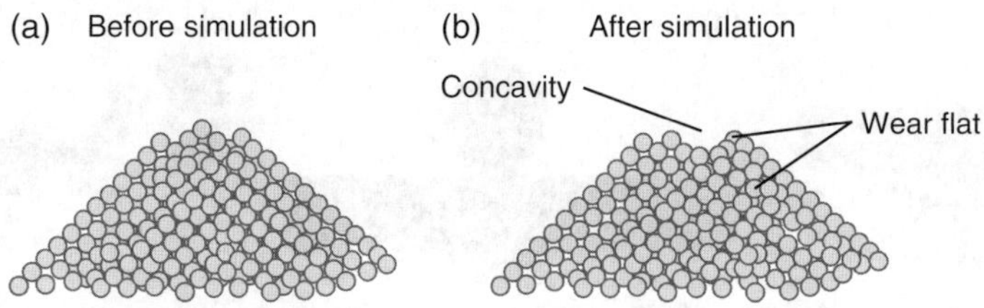

Figure 6.17 Simulation of tool wear in nano cutting. Reproduced from [33]. Copyright 2003 Elsevier

distribution is regular at its initial position and shows several discrete peaks. After scratching, such defects as dislocations, pile-up atoms and amorphous atoms remain in the surface and subsurface of the specimen, and many atoms deviate from their regular positions. There is an obvious decrease in the degree of lattice ordering, especially of short-range ordering. The decrease in the number of atoms in a regular position may have an important effect on the mechanical properties of nanostructures.

6.2.4 Friction and Wear Simulation by Using MD

Currently, nanoscale precision surface can be achieved by means of nanometric cutting with a single crystal diamond tool. The cutting performance and wear behaviour of a diamond tool play a significant role in the nano-cutting process. Because the diamond is has high hardness and wear resistance, it tends to be thought that diamond tools used in nano cutting will hardly be worn. Indeed, in order to improve the efficiency in the simulation, most scholars assume that the diamond tool's rigidity is infinite in the process of establishing the nano cutting simulation model. So the relative positions of atoms within the tool have no change in simulation of cutting. However, it is shown that although the wear volume is so tiny, its impact on the surface accuracy is significant in nano cutting. Maekawa and Komanduri took advantage of the MD method to study the friction and wear behaviour of the diamond tool in machining the single crystal copper and single crystal aluminum [31, 32]. Also, as shown in Figure 6.17, Cheng and Luo analyzed the wear behaviour of the diamond tip when it scratched the surface of a single crystal silicon, and concluded that thermo chemistry behaviour, especially friction between the workpiece atoms and the diamond tip atoms, plays the leading role in the wear process of tip [33]. Figure 6.18 shows the simulated images of the AFM diamond tips before and after the cutting, respectively. A few diamond tip atoms have been removed from the cutting edge near the rake face and, consequently, an obvious wear flat appears there.

Figure 6.19 shows the atomic position of the tool and the workpiece in different molecular dynamics time steps in the process of single crystal silicon nano cutting [34]. From Figure 6.19 it can be seen that the integral atomic structure had no obvious change in the initial cutting stage. But some atoms of tool fell off. Subsequently, the interior atoms turned into exterior ones. The lattice atoms on the external surfaces of the tool had the characteristic of unsaturated bonds. As shown in Figure 6.19 (b) and (c), individual atoms entered into the diamond lattices because of the extrusion effect, and occupied the position of carbon atoms whose bonds are weak.

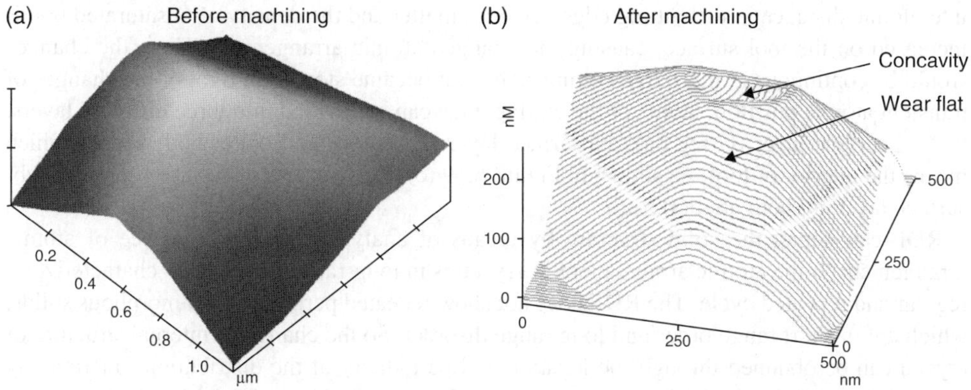

Figure 6.18 Wear of AFM tip in nano cutting. Reproduced from [33]. Copyright 2003 Elsevier

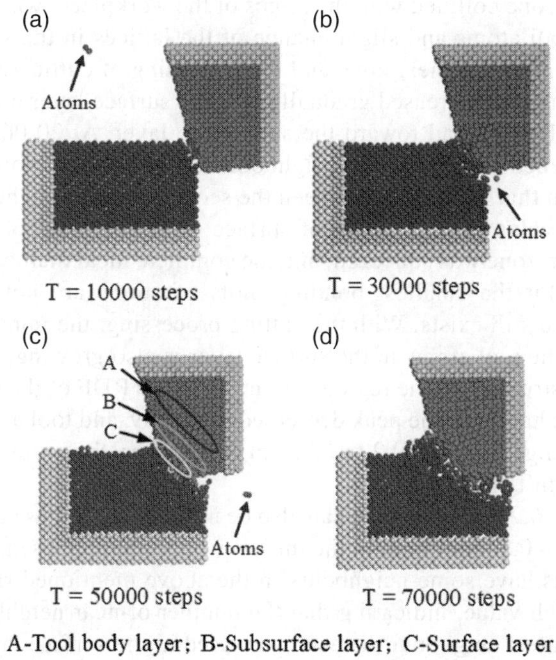

Figure 6.19 Atomic position in the simulation of diamond. Reproduced with permission from [34]

It can also be noticed from Figure 6.19(b) and (c) that in the process of nano cutting, the falling of the tool surface atom becomes weak, and the cutting edge radius tended to increase. After that, diamond atoms couldn't be found in the cutting chips, and no significant changes happened to the interior lattices of the tool. The cutting edge of the tool became smoother. It was because the falling off of atoms made the nose of the tool blunter. Meanwhile, the interatomic distance in the cutting edge became smaller and the density of unsaturated bonds increased on the tool surface, causing the changed atomic arrangement. Then, the changes would be continuous in the stable cutting stage, but became slower. In light of the changes of transient atomic position, atoms of the cutting edge can be divided into three different layers, that is, surface layer (C region), subsurface layer (B region) and cutter body region which means the interior of tool. As shown in Figure 6.19 (c) the thickness of surface layer and subsurface layer are both about 0.5 nm.

RDF can depict the solid structure by means of analyzing the order degree of atomic arrangement. If the atomic arrangement in crystal is in long-range order, RDF characterizes a regular and repeated cycle. The RDF does not show repeated periodicity in amorphous solids, which are in short range order and long range disorder. So the change of internal structure of crystal can be obtained through the location and periodicity of the distribution function. By calculating the RDF, the lattice atoms' position can be analyzed, and then the changes of the tool's crystal structure can be determined.

Figure 6.20 shows the changes of the RDF curves of three different areas in various molecular dynamics time steps. It can be seen that the atomic lattices of the cutting edge in the deformation zone collided with the atoms of the workpiece, which resulted in position deviation of tool atoms and slight change of the lattices in the surface layer. Then, with the enlargement of the shear zone and the increasing of cutting chips, the extrusion force suffered by the tool increased gradually, and the surface layer has a relatively obvious deformation which spread toward the subsurface layer. At 20 000 steps, dispersion phenomenon happened to the nearest neighbours of the surface atoms, but not significantly. In the region that is distance between the second nearest neighbour and the lattice constant, the peak value of RDF curve of surface layer decreases obviously, indicating that the deformation zone was squeezed, and the complete ideal diamond crystal structure has disappeared. But the smallest bonding units of atoms are not damaged, that is, α-diamond structure still exists. With the cutting processing, the impact from workpiece atoms suffered by the tool atoms in the surface layer is also growing, causing significant changes of crystal structure in the region. As shown in the RDF of the surface layer in the following steps, the height of the peak decreased gradually, and tool atoms are in the scattered state in the range of about 0.2 to 0.35 nm. Thereafter, the minimum bonding structure of diamond didn't exist at all.

From both Figure 6.20 (a) and (b), it can also be noticed that atoms have very few neighbours in about 0.17 ~ 0.24 nm range in the initial cutting stage. Subsequently, in the stable cutting stage, atoms have some neighbours in the above mentioned range, and the RDF shows relatively high value, indicating that the number of near neighbour atoms outside the range of the nearest neighbour distance increased in the surface layer. The atom density increased, resulting in induration of the diamond tool. So, in the whole cutting process, some atoms of surface layer fall off, the crystal structure changes significantly and is distributed in the periphery of the subsurface layer. Meanwhile, the crystal structure of the subsurface layer changes partly, causing the density to increase, which protects the diamond tool from further wear.

In the whole nano-cutting process of single crystal silicon, the RDF in the area of the tool body layer almost did not have any change, indicating that the internal structure of the crystal did not change, and the key effects of wear were on the atomic lattice layer of tool surface and subsurface layers.

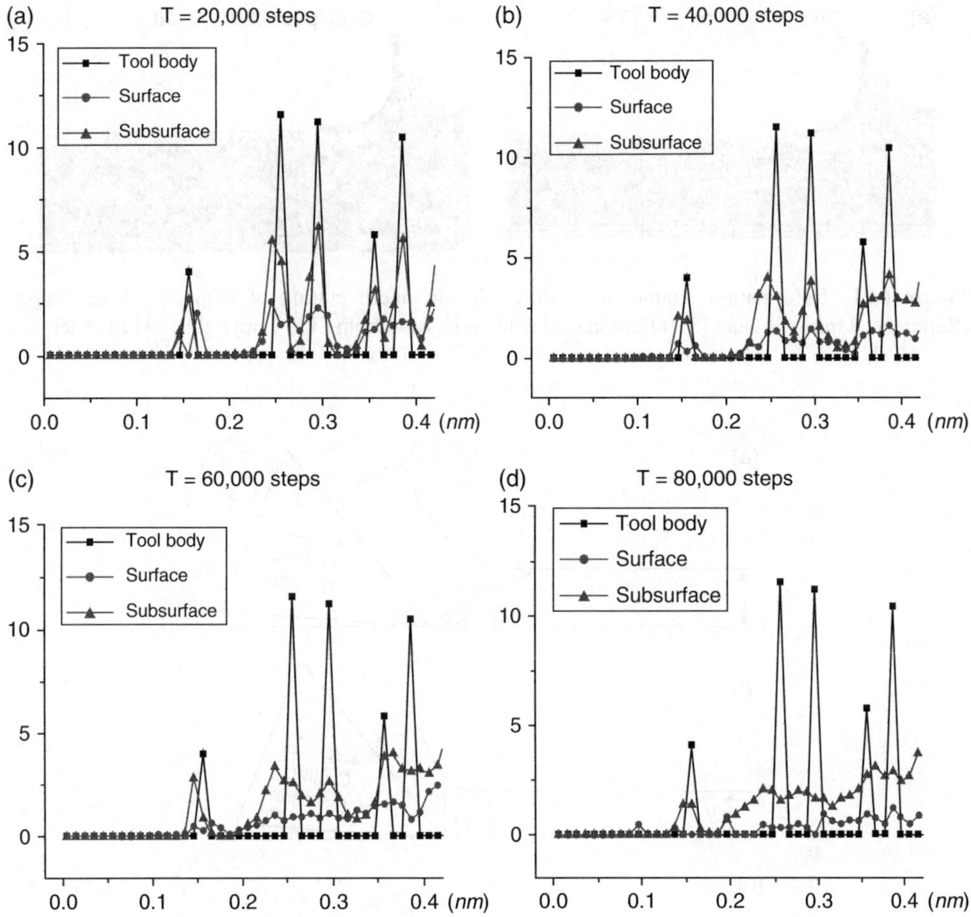

Figure 6.20 The RDFs in various regions of diamond tool. Reproduced with permission from [34]

6.2.5 *Effect of the Crystal Plane of Single Crystal and Multicrystalline*

The mechanics and crystal character are quite important for the performance of micro machining. As for single crystal material machining, the combination of crystal plane and cutting direction, will ensure material defects both contribute to the machining quality and efficiency. As for the multicrystalline, the factors may be more complex than single crystal material for micro cutting.

6.2.5.1 The Effects of Combination of Cutting Direction and Crystal Plane

Cutting direction and crystal plane of workpiece will inevitably influence the cutting behaviour for the reason of anisotropy. Shimada and Inamura considered the effect of the crystal plane and cutting direction on cutting at first [35, 36]. As shown in Figure 6.21, with a 2D

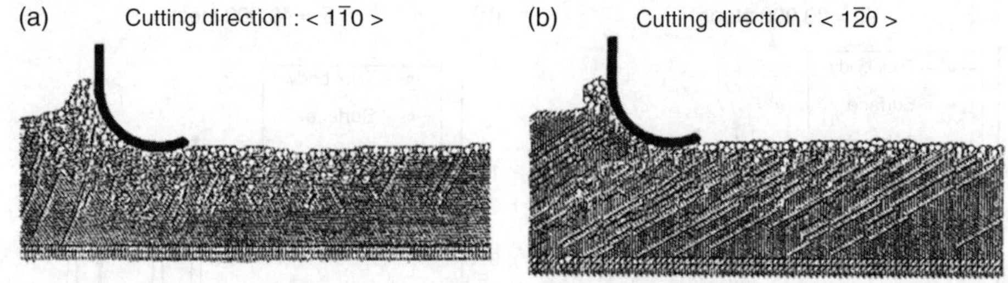

Figure 6.21 Deformation around the cutting edge in micro cutting of monocrystalline copper. (Reproduced from [34] and [35]) Reproduced with permission from [35]. Copyright 1994 Elsevier

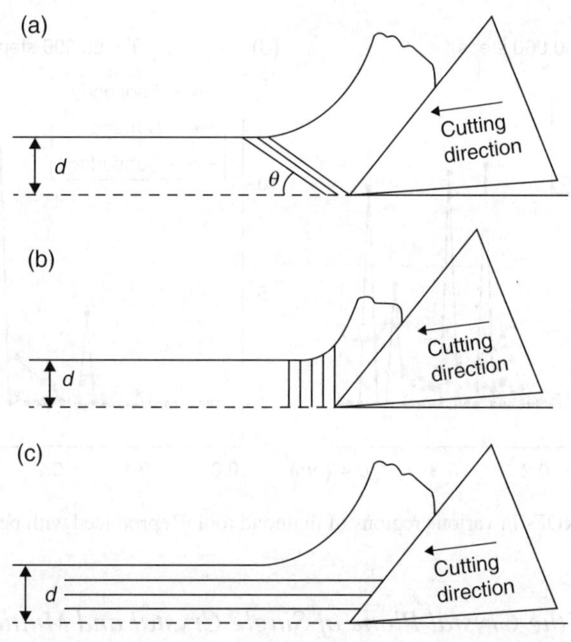

Figure 6.22 Three modes of plastic deformation in the shear zone. Reproduced from [37]. Copyright 1999 Elsevier

MD model, they investigate the machining process of the single crystal copper. Moreover, the surface roughness, defect layer and dislocation were examined in their researches. Komanduri and Chandrasekaran analyzed the deform mechanism in various combinations of cutting direction and crystal plane (Figure 6.22) [37]. It was found that the combination would affect the cutting force, the ratio of thrust force and tangential force, and deformation mechanism. Three plastic modes in nanometric scale cutting were summarized as 'parallel to cutting direction', 'perpendicular to cutting direction' and 'certain degree to cutting direction' [37, 30]. From the MD simulation, the combination of cutting direction and

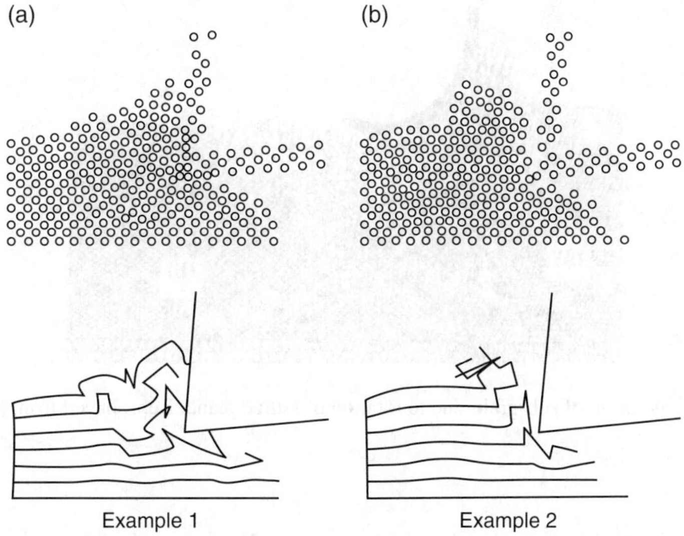

Figure 6.23 Chip formation observed in the experiment (upper: particle representation; lower schematic representation). Reproduced from [36]. Copyright 1992 Elsevier

crystal plane can lead to a certain surface finish and quality of workpiece and can affect the energy consumption. So it is an interesting project to investigate the effects of crystal plane and cutting direction.

6.2.5.2 Cutting Mechanism of Multicrystalline Material

The MD simulation of micro cutting mainly focused on the ideal single crystal material. However, various multicrystaline materials were used in the engineering area. Many crystal grains, growing in various directions, connected each other with complex grain boundaries. It may be useful for understanding the micro-cutting process in atomic views to investigate the cutting mechanism of multicrystalline material. In 1992, Inamura built a simple model with two single crystals in different crystal directions, by which the plastic deformation is investigated in the MD simulation (Figure 6.23) [36]. As shown in Figure 6.24, Shimada also contrasted the cutting of single crystal copper and multicrystalline copper [38]. Residual deformation inevitably occurred in the boundary of crystal grains in multicrystalline material, whereas it cannot be found in single crystal material [38]. Though some research results have been grained, the model is relatively simple and it is difficult to verified it by experiments. So to investigate the cutting simulation of multicrystalline material further is a promising topic for the future.

6.2.6 Improvement of the MD Simulation Capability

In order to improve the model scale and computational efficiency of MD simulation, the key technique is through the improvement of hardware or software. In the aspect of hardware, supercomputers can be assumed to improve the computational capability; however, it is not

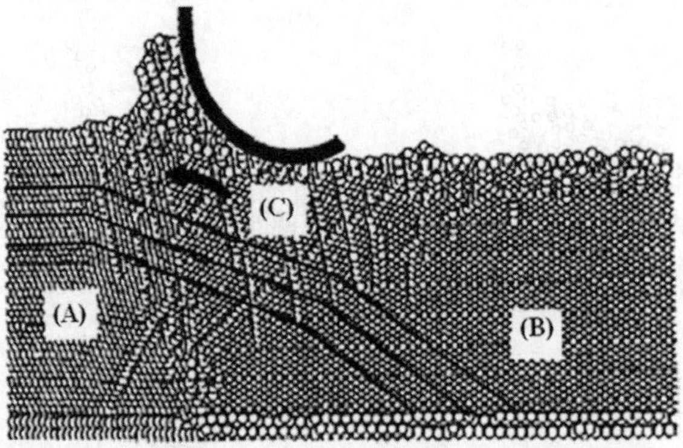

Figure 6.24 Generation of sub-grain due to rotation of lattice plane. Reproduced from [38]. Copyright 1994 Elsevier

the most cost effective way to provide the simulation. With regard to software, new adapted MD methods are also investigated to improve the efficiency, such as book-keeping method, linked-list method, ARMD (Area restricted molecular dynamics) and LRMD (Length restricted molecular dynamics) [39, 40]. High efficiency and extendable Parallel MD, which characterize the properties of hardware and software, can be adopted for large-scale MD simulation in the environment of cluster parallel computers and parallel algorithm.

Figure 6.25 is a diagram for the micro cutting of single crystal silicon with a parallel algorithm. Spatial Decomposition Algorithm is used for the simulation process, and the whole process is built on the Message Process Interface (MPI) simulation environment of a Linux operating system. About 120 000 atoms of workpiece were built in the parallel simulation and the depth of cutting is about 2.7 nm. Compared with the serial MD process, the parallel MD can be an effective way for large-scale and high-efficiency simulation of micro cutting.

6.3 Multiscale Modelling and Analysis

From the analysis of the above section, it can be seen that the MD method has almost become a dominant tool for investigating the mechanism of micro cutting at nanoscale. It is convenient to obtain some important information in the machining process, such as cutting forces, chip morphology, residual stresses, subsurface damaged layer, temperature distribution and energy variation in the atomic scale. However, MD simulations have their own limitations. For example, the typical MD simulations are restricted to much smaller systems with several millions of atoms or even less and time scales in the order of picoseconds. Thus the MD method can't be used to calculate much larger structures, and it is also difficult to compare the results attained from simulations with experimental data. In the past few decades, the FE method was a key numerical analysis tool in the machining process. It can predict the macroscopic changes of forces, stresses, strains and temperatures in micro cutting, but can't accurately predict those parameters at atomic scale. Although the fine elements can improve the accuracy, constitutive equations have significant errors when the

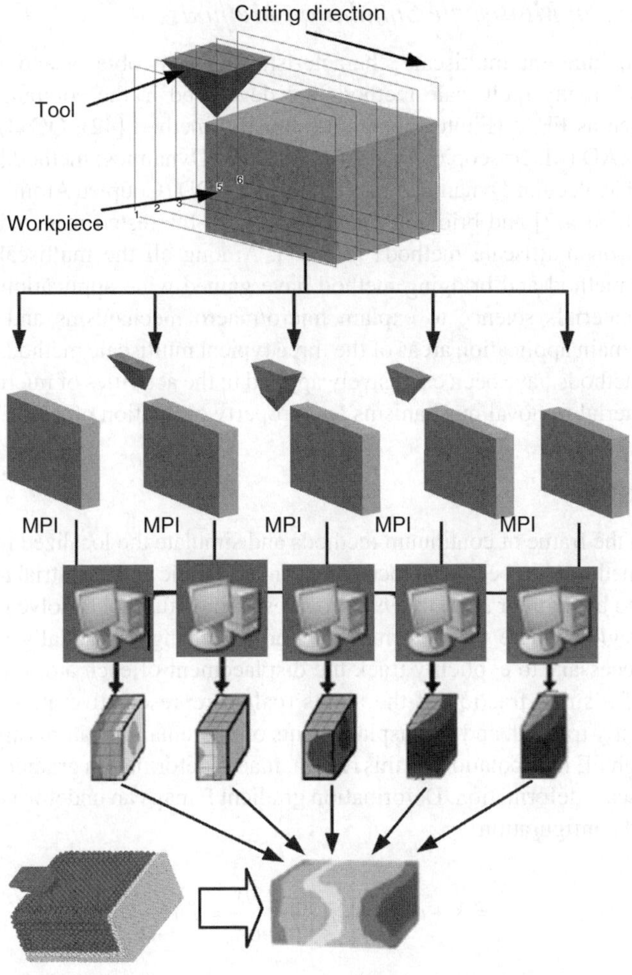

Figure 6.25 Parallel MD simulation of micro cutting

elements are meshed to atomistic size [41]. Because of the limitation of governing equations, further mesh refinement cannot help to solve the problem.

So it is necessary to develop multiscale models which link atomistic and continuum scales to simulate the micro cutting processes. The multiscale methods which combine atomistic and continuum scales are discussed in this section. An atomistic simulation method is used in the localized region concerned, such as crack tip, in which it is important to capture the individual atomistic dynamics accurately. A continuum method is used in most of the other parts in which the deformation is considered to be homogeneous and smooth. Thus the physical parameters gained from microscale are conveyed to macroscale through some equivalent methods, and macroscale calculations supply some boundary conditions for microscale. This not only saves computational time but captures micro deformation mechanisms of objects accurately.

6.3.1 *Advance in Multiscale Simulation Methods*

Recently, due to inherent multiscale characteristics of real objects and improvement in computing power, many multiscale methods are developed in the computational materials science field, such as FEAt (Finite Element-Atomistic) method [42], QC (Quasicontinuum) method [43], MAAD (Macroscopic, Atomistic, Ab initio Dynamics) method [44, 45], CGMD (Coarse-Grained Molecular Dynamics) method [46], CADD (Coupled Atomistic and Discrete Dislocation) method [47] and bridging method [48, 49]. Interested readers can refer to two overview papers on multiscale methods [50, 51]. Among all the multiscale methods, QC method, CADD method and bridging method have gained wide applications in the field of computational materials science to explain micro/macro mechanisms and solve scientific issues. From the main application areas of the three typical multiscale methods, it can be noted that multiscale methods have been extensively applied in the activities of micromanufacturing, for example, material removal mechanisms and property evaluation of micro structures.

6.3.1.1 QC Method

How to maintain the frame of continuum methods and simulate the localized interesting region using atomistic methods has become a focus of recent academic and industrial research. The QC method developed by Tadmor *et al.* in 1996 provides a powerful tool to solve this problem.

In the local region where the deformation gradient changes gradually on the atomistic scale, it is not necessary to explicitly track the displacement of each atom. Instead, only the displacements of a small fraction of the atoms (called representative atoms or 'repatoms') need to be explicitly tracked, and the displacements of the remaining atoms approximately are calculated through FE interpolation. In this region, macro deformation gradient F is considered to be fit for fine scale deformation. Deformation gradient F maps an undeformed configuration X onto deformed configuration

$$x = X + u(X), \quad F(X) = \frac{\partial x}{\partial X} = I + \frac{\partial u}{\partial X} \tag{6.6}$$

where I is the unit tensor and $u(X)$ is the displacement tensor. The first Piola-Kirchhoff stress tensor P and Lagrangian stiffness tensor C are defined respectively in Equation (6.7)

$$P = \frac{\partial W}{\partial F^T}, \quad C = \frac{\partial^2 W}{\partial F^T \partial F^T} \tag{6.7}$$

where W is the strain energy density function that is calculated directly from the ineratomic potentials by using the Cauchy-Born rule. Thus the non-linear finite element technique can be used to solve this problem.

In the non-local region such as dislocation cores, surfaces and interfaces, it is necessary to describe the potential energy accurately. In the energy-based formulation, the energy of model is calculated accurately by explicitly computing only the energy of repatoms ($N_{\text{rep}} << \text{N}$)

$$E^{tot,h} \approx E^{tot,h'} = \sum_{\alpha=1}^{N_{rep}} n_\alpha E_\alpha(u_h) \tag{6.8}$$

where n_α is the weight function for repatom α, which will be high for repatoms in regions of low repatom density and vice versa.

In the transition region where atomistic and continuum models are combined, the total potential energy is approximated through summing the energy of local and non-local repatoms as

$$E^{tot,h} \approx \sum_{\mu=1}^{N_{loc}} n_\mu E_\mu(u_h) + \sum_{\alpha=1}^{N_{nl}} n_\alpha E_\alpha(u_h) \tag{6.9}$$

where nμ, nα are the weight functions for local and non-local regions respectively, Eμ is the energy of local elements and Eα is the energy of non-local repatoms.

The QC method essentially is an adaptive finite element method. So the simulation size of atomistic area can be controlled easily following the change of deformation gradient. The QC method has already been applied in micro defect [43, 52], nano indention [53] and crack evolvement [54], and has been extended from 2D to 3D simulations [55, 56]. At present, the main development of the QC method is a complete dynamics theory at finite temperature [57] and new research results about non-zero temperature QC models have come forth [58–60]. These expand the application range of QC method extensively.

6.3.1.2 CADD Method

Recently, the CADD method [61, 47] was presented by Curtin, Miller and Shilkrot. Actually, the method combines the QC method with continuum defect models such as the discrete dislocation (DD) method. This method combines the continuum region with defects and the atomistic region. The defects (such as dislocations) that are nucleated in the atomistic region are allowed to pass through the atomistic/continuum border into the continuum where they are presented by discrete dislocation mechanics.

This method is viewed as a general boundary value problem. A body is divided into one or more atomistic regions Ω_A and one or more linear elastic continuum regions Ω_C, and combined by interfaces $\partial\Omega_T$. Tractions $T=T_0$ are prescribed on boundary $\partial\Omega_T$ and displacements $u=u_0$ on $\partial\Omega_u$. Region Ω_C contains N continuum discrete dislocations with Burgers vectors b^i and positions d^i.

In order to make sure that real atoms at and near the interface are properly coordinated, a pad of atoms which overlay with continuum region are added. Due to the presence of dislocations in the continuum, the total stresses, strains and displacements in the whole continuum regions are written as functions of the contribution from the discrete dislocations and a correction term

$$u = \tilde{u} + \hat{u}, \sigma = \tilde{\sigma} + \hat{\sigma}, \varepsilon = \tilde{\varepsilon} + \hat{\varepsilon} \tag{6.10}$$

According to the decomposition above, the continuum energy is written as the energy of element plus boundary work

$$E^c = \frac{1}{2}\int_{\Omega_C} (\hat{\sigma} + \tilde{\sigma}) : (\hat{\varepsilon} + \tilde{\varepsilon}) dV - \int_{\partial\Omega_T} T_0(\tilde{\mu} + \hat{\mu}) dA \tag{6.11}$$

where T_0 is the prescribed traction on the continuum boundary $\partial\Omega_T$.

The energy functional for the entire system is expressed as

$$\Psi = \frac{1}{2}\int_{\Omega_C} (\hat{\sigma}+\tilde{\sigma}):(\hat{\varepsilon}+\tilde{\varepsilon})dV - \int_{\partial\Omega_T} T_0(\hat{\mu}+\tilde{\mu})dA + E_{at}(r_A, r_I, r_p) - f_A \cdot \mu_A \tag{6.12}$$

where E_{at} is the atomistic energy, r_A are the positions of atoms, μ_A are the displacements of atoms and r_p are the positions of pad atoms.

The approach has been validated by 2D problems, including fracture, nano indentation and atomic scale void growth. It can also be applied to simulate mechanical behaviour of crystalline solids under mechanical loads, such as materials removal processes that will be described later.

6.3.1.3 Bridging Method

Different from the QC and CADD models, bridging methods don't need to mesh elements in the continuum region down to the atomistic size, thus atomistic and continuum regions can calculate using their natural time scales.

The bridging domain method is one of the bridging methods. It was proposed by Xiao and Belytschko in 2003, in which the overlapping region is called the bridging domain. The total Hamiltonian in the overlapping region is written to be a linear combination of the molecular and continuum Hamiltonians.

$$H = (1-\alpha)H^M + \alpha H^C \tag{6.13}$$

where scaling parameter α represents the contribution of each domain to the total Hamiltonian. Both of the regions are constrained on the overlapping subdomain by a Lagrange multiplier

$$g_I = \left\{ \sum_J N_J(X_J)u_{iJ} - d_{iI} \right\} = 0 \tag{6.14}$$

where g_I are the Lagrange multipliers, u_{iJ} are the nodal displacements and d_{iI} are the MD displacements.

The coupling equations of motion are defined to be

$$\bar{M}_I \ddot{u}_I = f_I^{extC} - f_I^{intC} - f_I^{LC} \tag{6.15}$$

$$\bar{m}_I \ddot{d}_I = f_I^{ext} - f_I^{int} - f_I^{L} \tag{6.16}$$

where f_I^{LC} and f_I^{L} are due to the constraints enforced by the Lagrange multipliers.

This approach was validated by 2D wave and crack propagation problems. Recently it has been extended to investigate dynamics, heat conduct and the fracture of defected single-layer atomic sheets and nanotubes [62].

The bridging scale method is the other bridging method, which was presented by Wagner and Liu for couple atomistic and continuum simulations in 2003. The fundamental idea is to decompose the total displacement field into fine and coarse scales

$$u(x) = \bar{u}(x) + u'(x) \tag{6.17}$$

where $\bar{u}(x)$ and $u'(x)$ are coarse and fine scale displacements respectively.The coarse is defined to be

$$\bar{u}(X_\alpha) = \sum_I N_I^\alpha d_I \tag{6.18}$$

where $N_I^\alpha = N_I(X_\alpha)$ is the shape function of node I at the initial atomic position X_α and d_I are the finite element node displacements associated with node I.

The fine scale is that part of total displacement that the coarse scale cannot represent. Thus, the fine scale is defined to be the projection of the total displacement onto the FE basis functions subtracted from the total solution

$$\sum_\alpha m_\alpha | q_\alpha - \sum_I N_I^\alpha w_I |^2 \tag{6.19}$$

So the fine scale displacement is defined as

$$u' = q - Nw \tag{6.20}$$

or

$$u' = q - Pq \tag{6.21}$$

The equation of total displacement is written as

$$u = Nd + q - Pq \tag{6.22}$$

The bridging scale method inherently provides for solving finite temperature and dynamic problems [63] and can be easily applied to static problems. It has been applied in the research of wave propagation, dynamic crack propagation [64, 65] and bending of CNTs [66].

6.3.2 *Applications of Multiscale Simulation in Micro Cutting Processes*

Han *et al.* [67] tried to link the micro cutting and nanometric cutting using the hierarchical coupling methods. Figure 6.26 shows the coupling model of the micro-cutting process. The whole model of micro cutting was set up by FE method, in which the dimension of workpiece is the micrometer order of magnitude. The area for extracting node forces is re-meshed to fine size, which ensures the volumes of re-meshed element have the same order of magnitude as the MD area for applying the node forces. Through applying the forces extracting from the FE calculation to the MD area, the movement of atoms like cutting was observed.

Aly *et al.* [68] established a hierarchical coupling to simulate the micro cutting of single crystal Si, in which the parameters extracting from the MD simulation acted as the input of the

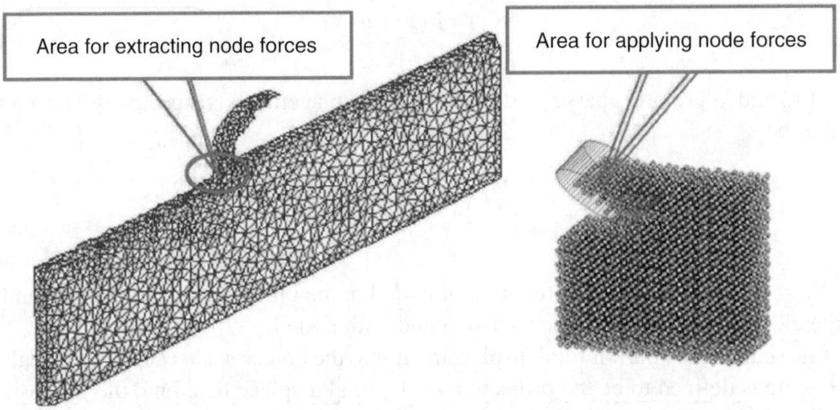

Figure 6.26 The hierarchical coupling model for micro cutting. Reproduced with permission from [67]

FE model. They first simulated the uniaxial tension process of single crystal Si, then attained the strength, yield stress and elastic modulus of Si from the stress-strain curve extracted from the MD simulation. Those parameters were inputted into the FE model as the material constitutive equations. The good agreement of cutting forces between simulation and micro cutting tests verified the validation of the model. The hierarchical coupling micro cutting model doesn't need to cope with the complicated linking problems between atomic and continuum regions. But the simulation results still depend on the scales of input ends.

Lin and Huang [69] proposed a method based on combining the concepts of shape functions of the FEM and the MD technique to evaluate the chip formation and strain and stress distribution in the cutting of single crystal Cu comparatively early. In the model, the atom is regarded as a node and the lattice is regarded as an element. This method can directly attain the stress and strain of atoms in the MD zone during nanoscale cutting.

Shiari *et al.* [70] researched the features of dislocation distribution in nanoscale material removal processes of single crystal Al using the CADD method. The material is modelled using the EAM potential. Figure 6.27 shows the dislocation distribution for two different time periods which can capture the compression and relaxed states in the machined surface, respectively. As shown in Figure 6.27(a), at least five of the dislocations penetrate very deeply into the workpiece in a compression state. Figure 6.27(b) is a snapshot of the dislocation distribution after the tool passes further along the top of the region. As can be seen, with the tool advancement, four out of five numbered dislocations in Figure 6.27(a) glide back from continuum to the atomistic region and disappear into the free surface. The fifth dislocation is forced to move deeper into the workpiece because of the tool advancement. The use of a multiscale model avoids the requirement of 300 000 or more atoms in full atomistic simulations and the simulation is performed with only 24 000 atoms and 2500 nodes.

It can obviously be a meaningful attempt for the introduction of the QC method into the micro cutting of single crystal Cu [71]. As shown in Figure 6.28, the feasibility was validated by comparing the cutting forces with the MD results. As a typical multiscale simulation tool, the QC method was adopted to investigate the influence of cutting speeds on the deformation mechanism in micro cutting of single crystal Cu [72]. In the machining model, a diamond tool

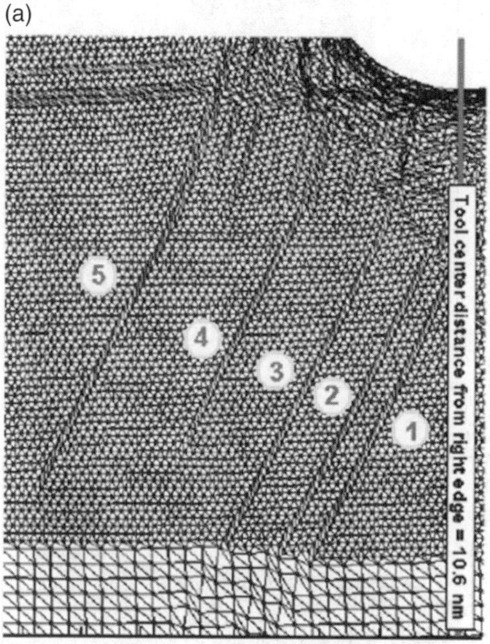

Figure 6.27 Distribution of dislocations in two different states: (a) Compression state, (b) Relaxed state. Reproduced from [70]. Copyright 2007 Elsevier

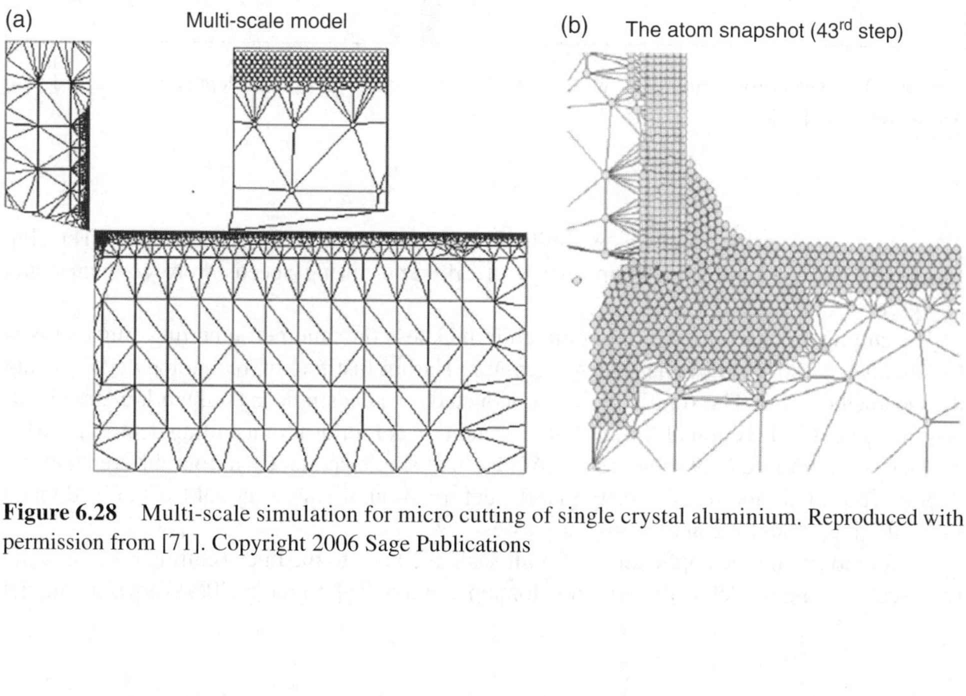

Figure 6.28 Multi-scale simulation for micro cutting of single crystal aluminium. Reproduced with permission from [71]. Copyright 2006 Sage Publications

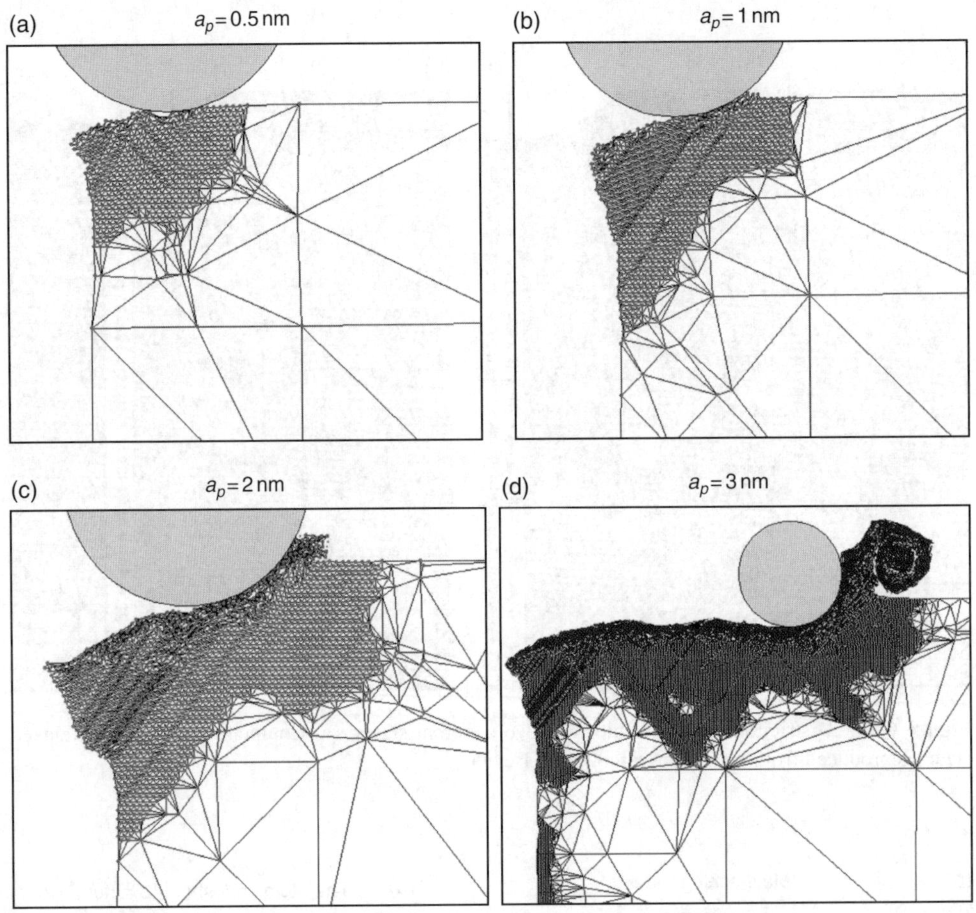

Figure 6.29 The deformation of single crystal Cu at four different cutting depths. Reproduced with permission from [73]

was simplified as a rigid circle, as is similar to the tool tip with certain edge radius. The chip formation, cutting force and strain energy at different cutting speeds were compared and analyzed.

The cutting mechanism and minimum chip thickness on nanometric cutting single crystal Cu are also investigated by the research group in Harbin Institute of Technology (HIT) using the QC method [73]. The simulation was conducted on a Cu single crystal work material cutting along [−110] direction of (001) crystal orientation at four different cutting depths (a_p = 0.5, 1, 2 and 3 nm). Figure 6.29 shows the deformation of single crystal Cu at four different cutting depths. The simulation results verify that the deformation of single crystal Cu can be divided into adhering, ploughing and cutting processes.

In order to extend the applications of multiscale methods to the micro cutting process, a 3D multiscale cutting model by the bridging domain method [74]. Figure 6.30(a) shows the model

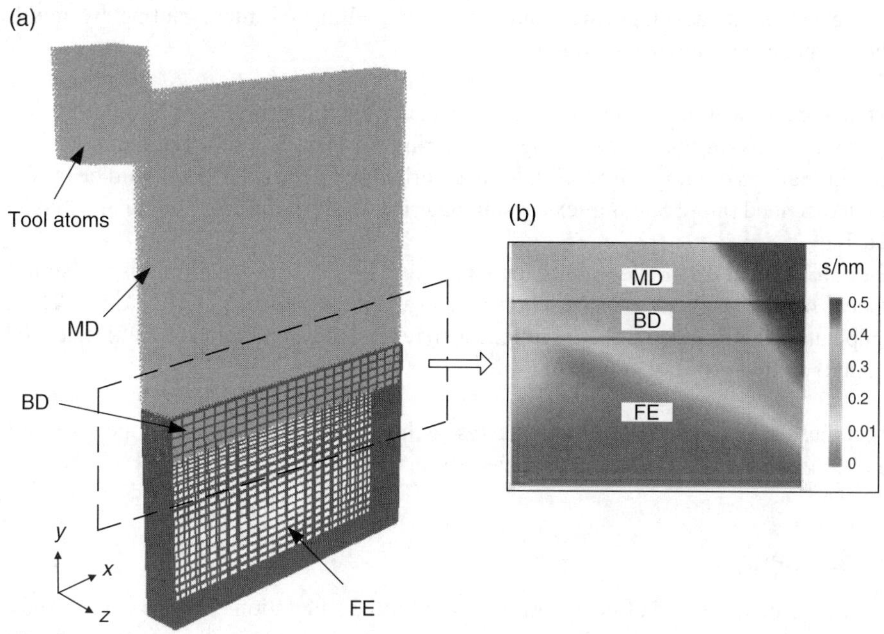

Figure 6.30 The bridging domain simulation model of nanometric cutting: (a) Multiscale model and (b) Close-up snapshot for the displacement distribution of model from the lateral view of workpiece materials. Reproduced with permission from [75]. Copyright 2011 IOP Publishing

for nanometric cutting of single crystal Cu [75]. The top surface of the workpiece was the (100) crystal surface. The directions X, Y and Z corresponded to the crystal directions [100], [010] and [001] respectively. There were 257 600 atoms, 9216 tetrahedron elements and 10 625 nodes in the workpiece. The tool consisted of 35 602 atoms. The EAM potential was adopted for the interaction between Cu atoms in the workpiece. The Morse potential was used for interactions between C atoms in the tool, atoms of the diamond tool and atoms of the Cu workpiece. In order to test the efficiency of the coupling state between MD and FE in the present model, the displacement distribution of workpiece materials under certain machining parameters is shown in Figure 6.30(b). Under the highly plastic deformation in the shear area, the displacement from the MD region to the FE region has a continuous variation. It shows that the high frequency waves in the atomistic region pass by the bridging domain into the continuum as a low frequency wave.

6.3.3 Research Challenges and Future Trends

From these examples above we can see multiscale methods have been applied to research the cutting mechanisms of micro cutting and a lot of interesting deformation mechanisms have been revealed. It is a new research field to apply multiscale methods to material removal processes. Owing to the complexity of machining processes, many aspects of existing multiscale methods, such as the complicated modelling and computational efficiency need to be improved.

At present, main development trends in the modelling of micro cutting by multiscale methods can be summarized as follows:

- Extending and scaling up 3D multiscale models of micro cutting.
- Dynamics analyzing of micro cutting considering the influence of temperature.
- The simulations of simple Bravais lattice materials (e.g., fcc and bcc) should be performed for an extended range of complex crystal materials (e.g., diamond cubic Si and hcp metals like Ti).
- The development of parallel multiscale simulation codes. The size of multiscale simulation models has been close to or reached real available experimental conditions. However, computational efficiency is still a bottleneck to restrict the application of multiscale methods to micro cutting processes.

Most of these are challenging tasks for the researchers, but the authors still hope some of them might be attempted and tackled in the near future.

6.4 Summary

The finite element method offers a powerful tool for the simulation of micro cutting and more valuable results have been gained in recent years. However, dislocation, size effect phenomena and multiscale behaviour in micro cutting may not be easily explained by using the FE method. Also the discrete atomic and chemical phenomenon between tool and workpiece are hard to describe using the continuous mesh method. So the potential of FEM used in micro cutting should be explored further by researchers.

Molecular dynamics simulation can present the scientific understanding from an atomic view. The technique now experiences a developing stage, which will inspire a novel idea and method to explain the phenomenon of micromachining. Though the MD simulation for micro cutting is limited by experience potentials, small-scale model, time consumption and simple model, it will contribute to the scientific understanding for the process of micromachining.

As a viable and effective simulation method, multiscale simulation can build a bridge between the macroscale and the microscale, or between the microscale and the nanoscale. Though there might be some problems that should be solved, such as large deformation, non-linear behaviour and computational efficiency, it can be confirmed that multiscale simulation is the potential research and development direction for understanding the complicate phenomenon of micro machining.

References

[1] Babuska I., Whiteman J., and Strouboulis T. (2011) *Finite Element – an introduction to the method and error estimation*. Oxford University Press.

[2] Aurich, J.C., and Bil, H. (2006) 3D finite element modelling of segmented chip formation, *Annals of the CIRP*, 55, 47–50.

[3] Özel, T., Liu, X., and Dhanorker, A. (2008) Modelling and simulation of micro-milling process, *Int. J. Mechatron. Manuf. Syst.*, 1, 23–42.

[4] Marusich, T., and Ortiz, M. (1995) Modeling and simulation of high-speed machining. *Int. J. Numer. Meth. Eng.*, 38, 3675–3694.

[5] Fang, F.Z., and Liu, Y.C. (2004) On minimum exit-burr in micro cutting. *J. Micromech. Microeng.*, 14, 984–988.

[6] Bai, Q.S., Yang, K., Liang, Y.C., Yang, C.L., and Wang, B. (2009) Numerical simulation and experiment on micro-diameter cutter wear considering tool run-out in micro-end-milling. *J. Vac. Sci. Technol.* B, 27: 1566–1572.

[7] Kai Yang, Qingshun Bai, and Yingchun Liang (2010) Numerical simulation and experimental investigation of tool edge radius effect on micro-cutter wear in micro-end-milling process. *Advanced Materials Research.* 97–101: 2542–2545.

[8] Yang, K., Liang, Y.C., Zheng, K.N., Bai, Q.S., and Chen W.Q. (2011) Tool edge radius effect on cutting temperature in micro-end-milling process. *Int. J Adv Manuf Technol.*, 52, 905–912.

[9] Liang, Y.C., Yang, K., Bai, Q.S., Chen, J.X., and Wang, B. (2009) Modeling and experimental analysis of micro-burr formation considering tool edge radius and tool-tip breakage in microend milling. *J. Vac. Sci. Technol.* B, 27, 1531–1535.

[10] Bai, Q.S., Liang, Y.C., Yang, K., and Luo, Z. (2009) Simulation and experimental research on thermal effects in micro-milling of micro parts. 3rd International Conference of Asian Society for Precision Engineering and Nanotechnology (ASPEN2009), Nov. 11–13, Kitakyushu, Japan.

[11] Alder, B.J., and Wainwright, T.E. (1957) Phase transition for a hard system. *J. Chem. Phys*, 27, 1208–1209.

[12] Hoover, W.G., Hoover, C.G., Stower, I.F., and Siekhaus, W.J. (1989) Interface tribology via nonequilibrium molecular dynamics. *Mater Res. Soc. Symp.*, 140: 119–124.

[13] Ikawa, N. *et al.* (1991) Ultraprecision metal cutting – the past, the present and the future. *Annals of the CIRP*, 40, 587–594.

[14] Komanduri, R., Chandrasekaran, N., and Raff, L.M. (2000) MD simulation of nanometric cutting of single crystal aluminum-effect of crystal orientation and direction of cutting. *Wear*, 242, 60–88.

[15] Shimada, S. (1995) Molecular dynamics analysis of nanometric cutting process. *Int. J. Japan Soc. Prec. Eng.*, 29, 283–286.

[16] Shimada, S., Inoue, R., Uchikoshi, J., and Ikawa, N. (1995) Molecular dynamics analysis on microstructure of diamond-turned surfaces. *Proc. SPIE*, 2576, 396–405.

[17] Fang, T.H., and Weng, C.L. (2000) Three-dimensional molecular dynamics analysis of processing using a pin tool on the atomic scale. *Nanotechnology*, 11, 148–153.

[18] Fang, T.H., Weng, C.L., and Chang, J.G. (2002) Molecular dynamics simulation of nano-lithography process using atomic force microscopy, *Surf. Sci.*, 501, 138–147.

[19] Fang, T.H., Chang, W.J., and Weng, C.I. (2006) Nanoindentation and nanomachining characteristics of gold and platinum thin films. *Mater. Sci. Eng. A*, 430, 332–340.

[20] Ye, Y.Y., Biswas, R., Morris, J. R., Bastawros, A., and Chandra, A. (2003) Molecular dynamics simulation of nanoscale machining of copper. *Nanotechnology*, 14, 390–396.

[21] Inamura, T., Shimada, S., Takezawa, N. *et al.* (1997) Brittle/Ductile transition phenomena observed in computer simulations of machining defect – free monocrystalline silicon. *Annals of the CIRP*, 46: 31–34.

[22] Dreher, M., Pauly, F., Heurich, J., Cuevas, J.C., Scheer, E., and Nielaba, P. (2005) Structure and conductance histogram of atomic-sized Au contacts. *Phys. Rev.* B, 72, 075435.

[23] Horstemeyer, M.F., Baskes, M.I., and Plimpton, S.J. (2001) Computational nanoscale plasticity simulations using embedded atom potentials. *Theor. Appl. Fract. Mech.* 37, 49–98.

[24] Horstemeyer, M.F., Baskes, M.I., Prantil, V.C., Philliber, J., and Vonderheide, S. (2003) A multiscale analysis of fixed-end simple shear using molecular dynamics, crystal plasticity, and a macroscopic internal state variable theory. *Modelling Simul. Mater. Sci. Eng.* 11, 265–286.

[25] Fang, H., Horstemeyer, M.F., Baskes, M.I., and Solanki, K. (2004) Atomistic simulations of. Bauschinger effects of metals with high angle and low angle grain boundaries, *Comput. Meth. Appl. Mech. Eng.*, 193, 1789–1802.

[26] Diao, J.K., Gall, K., Dunn, M.L., and Zimmerman, J.A. (2006) Atomistic simulations of the yielding of gold nanowires. *Acta Mater.*, 54, 643–653.

[27] Ju, S.P., Lin, J.S., and Lee, W. (2004) A molecular dynamics study of the tensile behaviour of ultrathin gold nanowires. *Nanotechnology*, 15, 1221–1225.

[28] Lin, J.S., Ju, S.P., and Lee, W.J. (2005) Mechanical behavior of gold nanowires with a multishell helical structure, *Phys. Rev.* B., 72, 085448.

[29] Heino, P., Kinen, H., and Kaski, K. (1998) Molecular-dynamics study of copper with defects under strain, *Physical Review B: Condensed Matters and Material Physics*, 58, 641–652.

[30] Komanduri, R., Lee, M., and Raff, L.M. (2004) The signification of normal rake in oblique machining. *Int. J. Mach. Tool Manu.*, 44: 1115–1124.

[31] Maekawa, K., and Itoh, A. (1995) Fiction and tool wear in nano-scale machining–a molecular dynamics approach. *Wear*, 188, 115–122.

[32] Komanduri, R., Chandrasekaran, N., and Raff, L.M. (2000) Molecular dynamics simulation of atomic-scale friction. *Phys. Rev. B*, 61, 14007–14019.

[33] Cheng, K., Luo, X., Ward, R. *et al.* (2003) Modeling and simulation of the tool wear in nanometric cutting. *Wear*, 255, 1427–1432.

[34] Bai, Q. S., Liang, Y.C., and Li, D.G. (2010) Atomic Level Simulation on Wear Mechanism of Single Crystal Diamond Tool. Proceedings of 4th CIRP International Conference on High Performance Cutting, 24–26 October, GiFu, Japan.

[35] Rentsch, R., and Inasaki, I. (1994) Molecular dynamics simulation for abrasive processes. *Annals of the CIRP*, 43: 327–330.

[36] Inamura, T., and Takezaw, N. (1992) Atomic-scale cutting in a computer using crystal model of copper and diamond. *Annals of the CIRP*, 41: 121–124.

[37] Komanduri, R., Chandrasekaran, N., and Raff, L.M. (1999). Orientation effects in nanometric cutting of single crystal materials: an MD simulation approach [J]. *Annals of the CIRP*, 48: 296–302.

[38] Shimada, S., Ikawa, N., Tanaka, H., and Uchikoshi J. (1994) Structure of micromachined surface simulated by molecular dynamics analysis. *Annals of CIRP*, 43: 51–54.

[39] Maekawa, K., and Itoh, A. (1995) Friction and tool wear in nano-scale machining–a molecular dynamics. *Wear*, 188: 115–122.

[40] Chandrasekaran, N., Khajavi, A.N., Raff, L.M., *et al.* (1998) A new method for molecular dynamics simulation of nanometric cutting. *Philosophical Magazine B*, 77: 7–26.

[41] Rudd, R.E., and Broughton, J.Q. (2005) Coarse-grained molecular dynamics: Nonlinear finite elements and finite temperatures, *Phys. Rev.* B, 72, 1441041.

[42] Kohlhoff, S., Gumbsch, P., and Fischmeister, H.F. (1991) Crack propagation in b.c.c crystals studied with a combined finite-element and atomistic model. *Philos. Mag. A*, 64, 851–878.

[43] Tadmor, E.B., Ortiz, M., and Phillips, R. (1996) Quasicontinuum analysis of defects in solids. *Philos. Mag.* A, 73, 1529–1563.

[44] Abraham, F.F., Broughton, J.Q., Bernstein, N., and Kaxiras, E. (1998) Spanning the continuum to quantum length scales in a dynamic simulation of brittle fracture. *Europhys. Lett.*, 44, 783–787.

[45] Broughton, J.Q., Bernstein, N., Kaxiras, E., and Abraham, F.F. (1999) Concurrent coupling of length scales: methodology and application, *Phys. Rev.* B, 60, 2391–2403.

[46] Rudd, R.E., and Broughton, J.Q. (1998) Coarse-grained molecular dynamics and the atomic limit of finite elements. *Phys. Rev.* B, 58, 5893–5896.

[47] Shilkrota, L.E., Curtina, W.A., and Miller, R.E. (2002) A coupled atomistic/continuum model of defects in solids. *J. Mech. Phys. Solids*, 50, 2085–2106.

[48] Belytschko, T., and Xiao, S.P. (2003) Coupling methods for continuum model with molecular model. *Int. J. Mult. Comp. Eng.*, 1, 115–126.

[49] Wagner, G.J., and Liu, W.K. (2003) Coupling of atomistic and continuum simulations using a bridging scale decomposition. *J. Comput. Phys.*, 190, 249–274.

[50] Curtin, W.A., and Miller, R.E. (2003) Atomistic/continuum coupling in computational materials science, *Model. Simul. Mater. Sci. Eng.*, 11, 33–68.

[51] Liu, W.K., Karpov, E.G., Zhang, S., and Park, H.S. (2004) An introduction to computational nanomechanics and materials, *Comput. Methods Appl. Mech. Engrg.*, 193, 1529–1578.

[52] Tadmor, E.B., and Phillips, R. (1996) Mixed atomistic and continuum models of deformation in solids. *Langmuir*, 12, 4529–4534.

[53] Tadmor, E.B., Miller, R., Phillips, R., and Ortiz, M. (1999) Nanoindentation and incipient plasticity. *J. Mater. Res.* 14 2233–2250.

[54] Miller, R., Tadmor, E.B., Phillips, R., and Ortiz, M. (1998) Quasicontinuum simulation of fracture at the atomic scale. *Model. Simul. Mater. Sci. Eng.*, 6, 607–638.

[55] Rodney, D., and Phillips, R. (1999) Structure and strength of dislocation junctions: An atomiclevel analysis. *Phys. Rev. Lett.*, 82, 1704–1707.

[56] Knap, J., and Ortiz, M. (2001) An analysis of the quasicontinuum method. *J. Mech. Phys. Solids*, 49, 1899–1923.

[57] Miller, R.E., and Tadmor, E.B. (2002) The quasicontinuum method: Overview, applications and current directions. *J. Comput. Aided Mater. Des.*, 9, 203–239.

[58] Diestler, D.J., Wu, Z.B., and Zeng, X.C. (2004) An extension of the quasicontinuum treatment of multiscale solid systems to nonzero temperature, *J. Chem. Phys.*, 121, 9279–9282.

[59] Dupuy, L.M., Tadmor, E.B., Miller, R.E., and Phillips, R. (2005) Finite-temperature quasicontinuum: MD without all the atoms. *Phys. Rev. Lett.*, 95, 060202- (1–4).

[60] Tang, Z., Zhao, H., Li, G., and Aluru, N.R. (2006) Finite-temperature quasicontinuum method for multiscale analysis of silicon nano-structures. *Phys. Rev. B*, 74, 064110.

[61] Shilkrot, L.E., Miller, R.E., and Curtin, W.A. (2004) Multiscale plasticity modeling: Coupled atomistic and discrete dislocation mechanics. *J. Mech. Phys. Solids*. 52, 755–787.

[62] Zhang, S., Khare, R., Lu, Q., and Belytschko, T. (2007) A bridging domain and strain computation method for coupled atomistic/continuum modelling of solids. *Int. J. Numer. Meth. Engng*. 70, 913–933.

[63] Park, H.S., Karpov, E.G., and Liu, W.K. (2004) A temperature equation for coupled atomistic/continuum simulations, *Comput. Methods Appl. Mech. Engrg*. 193, 1713–1732.

[64] Park, H.S., Karpov, E.G., Klein, P.A., and Liu, W.K. (2005) Three-dimensional bridging scale analysis of dynamic fracture. *J. Comput. Phys*. 207, 588–609.

[65] Park, H.S., Karpov, E.G., Liu, W.K., and Klein, P.A. (2005) The bridging scale for two-dimensional atomistic/continuum coupling. *Philos. Mag. A*, 85, 79–113.

[66] Park, J.Y., Cho, Y.S., Kim, S.Y., Jun, S., and Im, S. (2006) A quasicontinuum method for deformations of carbon nanotubes. *Comput. Model Eng. Sci.*, 11: 61–72.

[67] Han, X. S. *(2004) Investigation of multiscale numerical simulation of ultraprecision cutting technology*. Tianjin: Tianjin University, 2004.

[68] Aly, M.F., Ng, E., Veldhuis, S.C., and Elbestawi, M.A. (2006) Prediction of cutting forces in the micro-machining of silicon using a "hybrid molecular dynamic-finite element analysis" force model. *Int. J. Mach. Tool Manu.*, 46, 1727–1739.

[69] Lin, Z.C., and Huang, J.C. (2004) A nano-orthogonal cutting model based on a modified molecular dynamics technique. *Nanotechnology*, 15, 510–519.

[70] Shiari, B., Miller, R.E., and Klug, D.D. (2007) Multiscale simulation of material removal processes at the nanoscale. *J. Mechan. Phys. Solids*, 55, 2384–2405.

[71] Sun, X., Chen, S., Cheng, K., Huo, D., and Chu, W. (2006) Multiscale simulation on nanometric cutting of single crystal copper. *Proc. IMechE Part B: J.* Engineering Manufacture, 220, 1217–1222.

[72] Pen, H.M., Bai, Q.S., Liang, Y.C., and Chen, M.J. (2009) Multiscale simulation of nanometric cutting of single crystal copper effect of different cutting speeds. *Acta Metall. Sin. (Engl. Lett.)*, 22, 440–446.

[73] Liang, Y.C., Pen, H.M., and Bai, Q.S. (2010) Multiscale simulation of minimum chip thickness on nanometric cutting process of single crystal diamond tool. Proceedings of 4th CIRP International Conference on High Performance Cutting, 24–26 October, GiFu, Japan.

[74] Anciaux, G. (2008) Available from http://libmultiscale.gforge.inria.fr/index.html

[75] Liang, Y.C., Pen, H.M., Bai, Q.S., and Lu, L.H. (2011) Multiscale simulation of nanometric cutting of single crystal Cu based on bridging domain method. *Acta. Phys. Sin.*, 60, 100205-1-6 (In Chinese).

Part Two
Applications

7

Diamond Turning and Micro Turning

Dehong Huo[1] and Kai Cheng[2]
[1] *School of Mechanical and Systems Engineering, Newcastle University, UK*
[2] *School of Engineering and Design, Brunel University, UK*

7.1 Introduction

Mechanical micro machining is an important micro manufacturing process, which offers many unique advantages such as broad spectrum of workpiece material choices, high machining accuracy and surface finishes, high aspect ratio and complex shaped components, and so on. Micro cutting fabricates precision micro components or micro features through direct machining or through micro injection moulding or micro embossing with micro machined precision moulds or replication masters.

This chapter will focus on micro turning operations for micro cutting, and classify micro turning operations into two categories – diamond turning for surface structuring through commercially available ultra-precision machine tools as the platform and micro turning for producing 3D high aspect ratio components typically through micro turning machines. Although there is abundant literature relating to conventional diamond turning, only the work concerning machining micro structures or micro features will be reviewed in this chapter. Both micro tooling and machine tools are discussed with the support of perspective applications.

7.2 Ultra-precision Diamond Turning

Precision applications like optical mirrored lenses for optical systems require surface finish in the range of tens of nanometers and form accuracies in the sub-micron range. Ultra-precision machining is one of the machining processes meeting this requirement. For some applications, for example, soft metals such as copper and aluminium which are difficult to machine by

Micro-Cutting: Fundamentals and Applications, First Edition. Edited by Kai Cheng and Dehong Huo.

grinding, lapping and polishing, diamond turning becomes an economic and sometimes unique process for these specific components to meet the requirement since it produces optical surface finish without the need for additional grinding and/or polishing operations.

The cutting tool for most of the ultra-precision machining applications is exclusively single-crystal diamond (natural or synthetic). Therefore, the process is also called diamond turning or single point diamond turning (SPDT).

7.2.1 A Historical Perspective of Diamond Turning

The use of natural diamond to cut materials and generate precision surfaces and surface structures can date back centuries. Fasdolt claimed to have a ruling engine that could generate one million lines per inch equating to a line spacing of 25 nm in the 1880s and as early as the 1920s mirror-like surfaces had been achieved by diamond turning [1].

Research and development of ultra-precision machining was mainly driven by defence applications before the 1970s. A number of national laboratories were working on manufacture and inspection of precision components with optical surface finishes. Modern diamond turning machine tools are central to the ultra-precision machining technology. Thanks to the success of precision coordinate measurement machines developed since the 1950s and the invention of laser and laser interferometry in the 1960s which has become the metrology standard of the machine tool of this kind, several large diamond turning machines were developed in the 1960s and 1970s. Examples include Diamond Turning Machine (DTM-1) and Large Optics Diamond Turning Machine (LODTM) designed and built by Lawrence Livermore National Laboratory (LLNL). LODTM was developed for diamond machining large optical components with a surface finish of a few nanometers (Ra) and a relative accuracy of the order of 10^{-6}. Figure 7.1 shows the schematic of LODTM.

In the past decades, single point diamond turning performed in ultra-precision machines has reached an extremely high level of performance through scientific and engineering efforts.

With the advances in several enabling technologies such as aerostatic and hydrostatic bearings, computer numerical control (CNC), optical displacement measuring, many commercial diamond turning machines have been developed since the 1980s, and these machines have been employed in both the defence industry and consumer product sectors. Figure 7.2 shows a modern commercial diamond turning machine. These commercial diamond turning machines normally feature hydrostatic bearing slideways, one or more hydrostatic or aerostatic bearing spindle, granite structure, high accurate displacement sensors (optical encoders or laser interferometer) with nanometric resolution. Using diamond tools with controlled cutting tool edge waviness, sub-micron accuracy and nanometer surface roughness over various engineering materials have been achieved by those commercially available machine tools in the past decades.

Since the 1990s there has been an increasing demand on micro components and micro-structured surfaces. Research efforts, therefore, have been put on fabrication of micro-structured surfaces using diamond turning machines. Research has shown that single point diamond turning extends the design freedom of the optical system with complex free form surface and micro structure features. Section 7.2.3 discusses the development of machining micro structures through single point diamond turning technology.

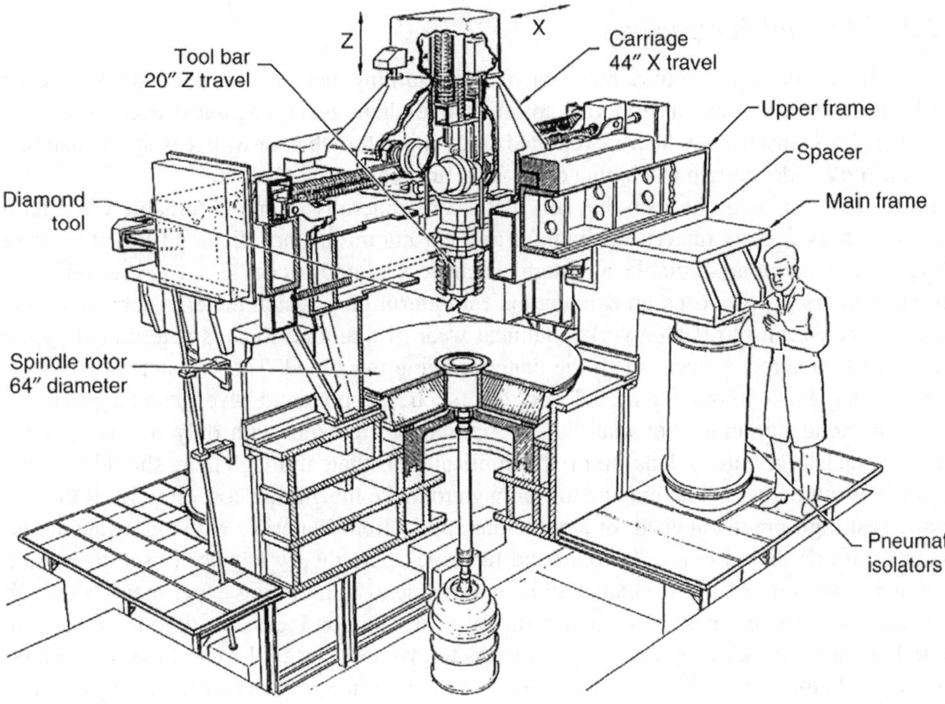

Figure 7.1 Large optics diamond turning machine (LODTM)

Figure 7.2 Nanotech 500FG (Courtesy of Moore Nanotechnology Systems) Copyright Moore Nanotech

7.2.2 *Material Perspectives*

The workpiece materials for ultra-precision diamond turning include both soft and hard, ductile and brittle materials. Some materials are not suitable for diamond turning due to excess chemical tool wear. Those materials, which are reported as 'diamond turnable' or with reasonable diamond machinability under certain machining conditions, are summarized below.

From Table 7.1 it can be seen that soft metals are generally diamond turnable while hard metals such as ferrous materials, titanium and chromium are not diamond turnable. Pure nickel is not diamond turnable although electroless plated nickel is an exception. Paul and co-workers [2] proposed an explanation for determining which materials are 'diamond turnable'. According to Paul's work, chemical wear of diamond tools is determined by the presence of unpaired d electrons in the material being machined. Those materials are composed of elements without unpaired d electrons (such as copper and silver) and are predicted to be 'diamond turnable'. For materials composed of elements with only one unpaired d electron (such as cerium and platinum) the implication is that, although there should be some chemical wear, the wear rate may be low enough to make machining cost effective if there is a particular high-precision need for a particular application. Materials with many unpaired d electrons are predicted not to be 'diamond turnable'. The theory has been used to explain diamond turnability of most metals and is in agreement with cutting experiments although there are some exceptions. Silicon and germanium are diamond turnable although great care should be taken on selecting cutting parameters to avoid excess tool wear. Less research on polymers' diamond turnability is available, but the acrylic, polycarbonate and nylon have been found to be diamond turnable without difficulty.

Diamond turned components from soft materials are widely used for electro-optics products, however precision moulds still require hard metals such as carbon steel. Research has been carried out to improve diamond machinability of hard metals. One of the most effective methods is ultrasonic vibration assisted machining which has been outlined in Chapter 2.

Table 7.1 Common materials with good diamond machinability

Material family	Materials
Metals	copper alloys
	aluminum alloys
	silver
	gold
	electroless plated nickel
	Zinc
	magnesium
Polymers	Acrylic
	Polycarbonate
	Nylon
Glass and ceramics	silicon
	germanium
	glass

7.2.3 *Micro Structuring by Diamond Turning*

High precision micro features and micro-structured surfaces are widely used in various fields ([3–5]). The increasing demand for high precision products stimulates the development of the micro fabrication technology. Micro-structured surfaces are considered to be sensible deterministic surfaces with high aspect ratios and determine the function of many products. The function of such micro-structured surfaces is very different from traditional surface finish parameters, such as surface roughness and waviness. These traditional surface finish parameters are statistically determined and cannot provide information for some functions such as diffraction grating.

Micro-structured surfaces are capable of performing many different functions and have found applications in optics, MEMS, biomedical devices, heat transfer and cooling devices, and so on. Table 7.2 lists some application areas and examples of micro-structured surfaces.

Micro-structured surfaces can be produced using a number of manufacturing processes including materials-removing methods such as machining, knurling and laser cutting; replication methods such as injection moulding and casting; and material deposition, and so on. Among these manufacturing processes, the cutting process, especially diamond cutting, is unique and cost effective and hence central to the production of a large proportion of micro-structured surfaces either through direction production of the part or through making a mould [6]. Figure 7.3 shows the precision micro-structured surface fabrication process generated by diamond micro cutting. Precision micro-structured surfaces with sub-micron accuracy and nanometre level surface finishes are commonly directly machined by diamond cutting, or through micro injection moulding or micro embossing with diamond cut micro moulds.

Table 7.2 Application of micro-structured surfaces

Application areas	Examples
Optics	Diffraction gratings Fresnel lenses Diffractive optics Retro-reflective road sign Filters Wavelength specific mirrors Backlight display
Automotive	LED car front light Head-up display
Mechatronics	Vacuum chuck Hard disk surfaces Air bearing recessed surface Velcro moulds
MEMS	Micro sensors and actuators
Biomedical	Cell culture system Intraocular lenses Capillary electrophoresis
Thermal Engineering	Heat transfer fins

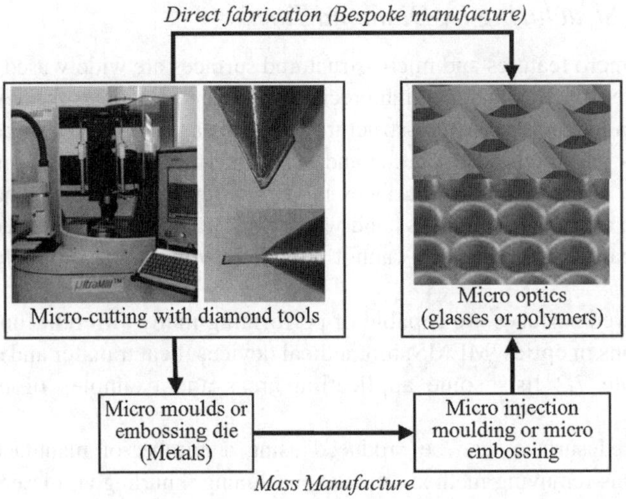

Figure 7.3 Fabrication of precision micro-structured surfaces by diamond cutting

Diamond turning is capable of producing micro features or complex micro structures on the surface. There are a number of diamond cutting approaches to producing precision micro structure, namely, direct single point diamond turning, diamond fly cutting, diamond turning with fast tool servo, and diamond micro milling. Diamond milling and its applications will be discussed in Chapter 8, and the other three approaches and their applications will be introduced in this section.

7.2.3.1 Direct Single Point Diamond Turning

The most obvious application of diamond turning to manufacture micro structures is rotationally symmetric surfaces. Spindles are not position controlled for this application. Because most precision machines have limited dynamic response (at the order of 100Hz is common), direct single point diamond turning is not suitable for producing micro-structured surface with near wavelength or even sub-wavelength optical micro structures.

Various rotationally symmetric micro structures have been produced using direct diamond turning. Brown *et al.* [7] produced Twyman sphere with groove depths of 6.33 microns using diamond turning as shown in Figure 7.4. Brinksmeier and his co-workers produced micro Fresnel lens and calibration standards for surface topography measuring systems with features of sine wave and triangle wave respectively as shown in Figure 7.5, and surface roughness was reported as small as 3.8 nm. Yan *et al.* [8] produced Infrared Fresnel lens moulds using germanium as shown in Figure 7.6. Weck presented several micro structures machined by diamond turning performed in the Fraunhoffer Institute for Production Technology IPT as shown in Figure 7.7 [9].

7.2.3.2 Diamond Turning with Fast Tool Servos

Fast Tool Servos (FTS) refer to auxiliary servos that are specially adopted to actuate the diamond tool with fine resolution, high stiffness and fast dynamic response [10]. FTS is usually used to create non-rotationally symmetric surfaces and micro surface features on a cutting

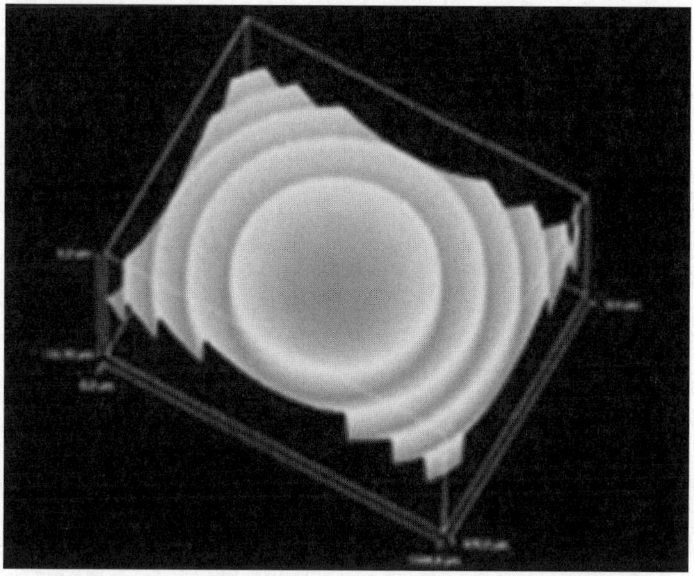

Figure 7.4 Twyman sphere with groove depths of 6.33 microns. Reproduced with permission from [7]

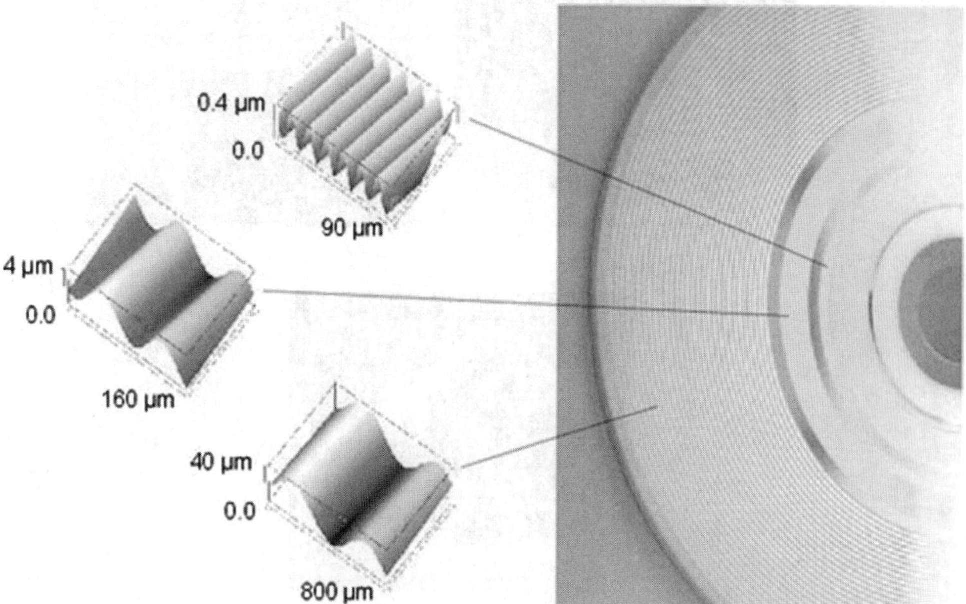

Figure 7.5 Calibration standards for surface topography measuring systems with features of sine wave. Reproduced with permission from [3]. Copyright 2001 EUSPEN

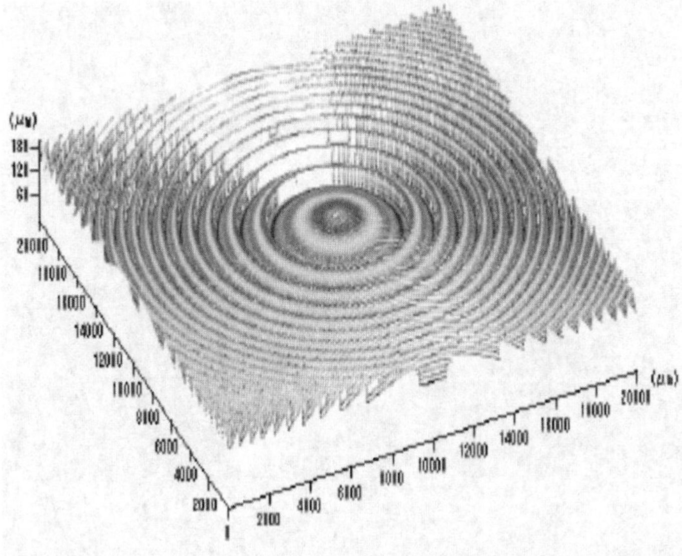

Figure 7.6 Infrared Fresnel lens produced in Tohoku University. Reproduced with permission from [8]. Copyright 2005 IOP Publishing

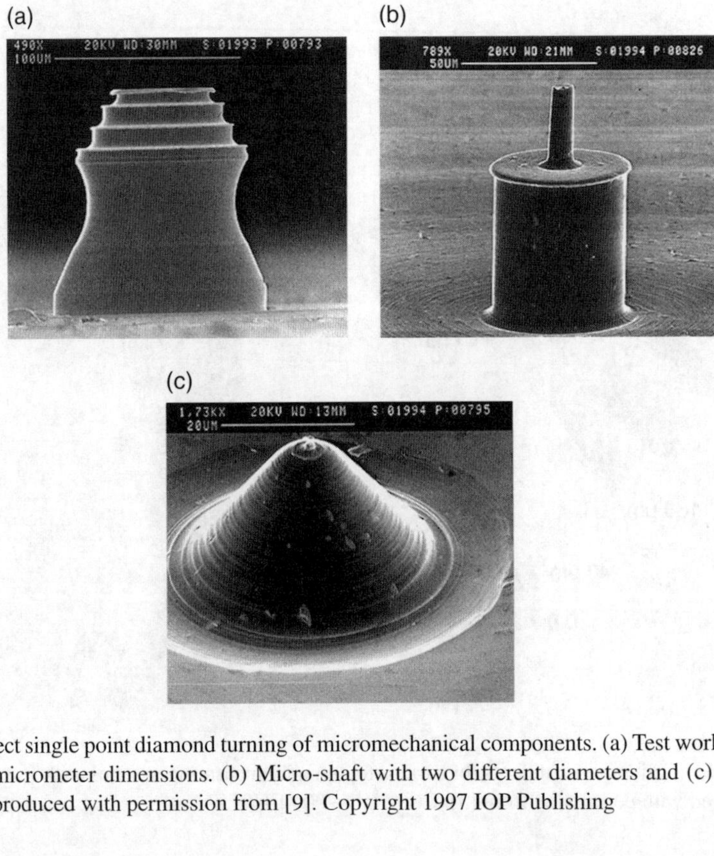

Figure 7.7 Direct single point diamond turning of micromechanical components. (a) Test workpiece for NC-lathes with micrometer dimensions. (b) Micro-shaft with two different diameters and (c) spherical press mould. Reproduced with permission from [9]. Copyright 1997 IOP Publishing

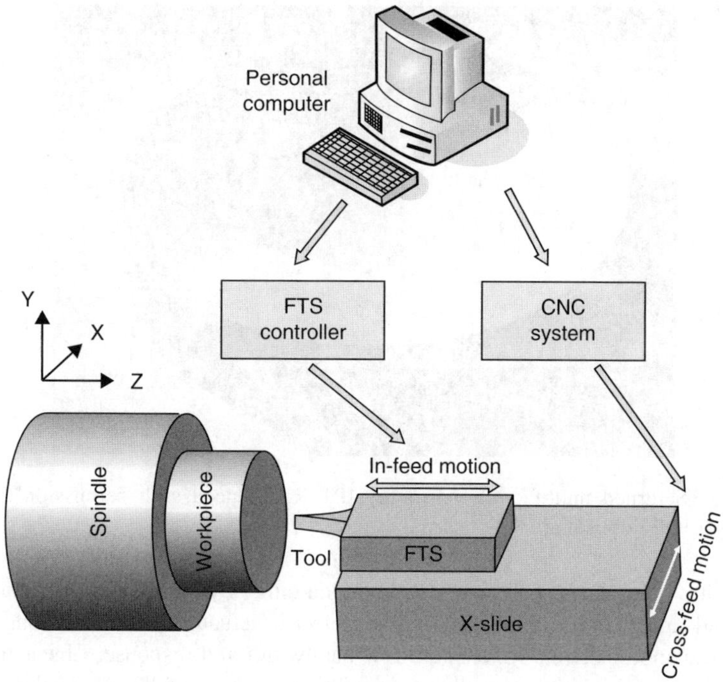

Figure 7.8 Principle of ultra-precision machining with FTS (Reproduced from [12]) Reproduced with permission from [12]. Copyright 2007 Sage Publications

surface. The application range of diamond turning machines has been extended by using the FTS. Diamond turning with a fast tool servo provides an indispensable solution for producing micro structures with sub-micron form accuracy and nanometric surface finish, without the need for any subsequent post-processing [1]and [11]. FTS diamond turning is an enabling technology to fabricate non-rotationally symmetric geometries.

Figure 7.8 presents the principle of ultra-precision machining with the aid of FTS. The workpiece is mounted on the spindle of the machine tool and rotates with precision controlled angular position. The FTS is mounted on the x-slide and is oriented in such a way that its stroke is parallel to the z-axis, perpendicular to the face of the workpiece. A single-point diamond cutting tool is mounted on the FTS. The CNC system of the machine tool controls the cross-feed motion of the x-slide, such that the tool travels from left to right relative to the workpiece, from the outer edge towards the centre of rotation. The FTS controller is independent of the CNC system. The stroke of the FTS is dependent on the angle of the spindle, θ, and the displacement of the x-slide, x. In other words, the FTS is programmed so that its stroke is a function of θ and x.

In the last two decades, various types of fast tool servos have been developed. An FTS includes an actuator with stroke ranging from tens of microns to 1 mm which holds a diamond tool. The FTS can be driven by a piezoelectric actuator, a linear motor or electromagnetic actuator with closed loop position feedback. Short stroke FTS with a travel less than 100 μm are normally driven by piezoelectric actuators. For short stroke actuation, piezoelectric actuators are superior to other methods due to their high stiffness, infinite mechanical resolution, fast response and

Figure 7.9 FTS turned multifaceted mirror by IPT Reproduced with permission from [9, 13]. Copyright 1997 IOP Publishing

high bandwidth. Long stroke FTS is normally driven either by piezoelectric actuators or linear motors. The advantage of linear motors over piezoelectric actuators lies in their long travel capability, but linear motor driven systems are low bandwidth and response. Linear motor driven FTSs are used for machining parts with larger deviations such as saddle or helical geometry [13].

Various non-rotationally symmetrical surfaces generated by FTS were reported. Figure 7.9 shows a multifaceted mirror produced by diamond turning with FTS. The surface of the mirror consists of a rotationally symmetrical part on which a non-rotationally symmetrical surface with 90 facets is superimposed. The mirror is used as an integrator optic to enhance the beam quality in high power lasers. Other micro-structured surfaces such as sinusoidal grid surface were also generated with the aid of FTS as shown in Figure 7.10 [5].

7.2.3.3 Fly Cutting with Diamond Tools

Fly cutting is another approach to generating a complex micro-structured surface. As shown in Figure 7.11, a diamond tool is mounted on a disk with a large diameter, and the disk is connected to a spindle, which is usually supported by aerostatic bearings. Workpiece is held on the slideway and fed in one direction. A groove or slot will be cut, and then the workpiece is fed in another direction, another parallel groove or slot will be cut. If this process is repeated, then an array of groove or slot will be generated. Using so called crossed fly cutting, that is, the workpiece is rotated on an angle, complex prism and pyramid structures can be generated. Figure 7.12 and Figure 7.13 show a micro-structured surface generated by diamond fly cutting.

7.2.3.4 Diamond Tools Used in Diamond Turning Micro structures

Round nosed diamond tools with a nose radius of a few mm are normally used for ultra-precision machining in order to obtain better surface finish. In addition larger nose radius tools have higher tool tip strength so that they are not easily broken or chipped. However these larger nose radius

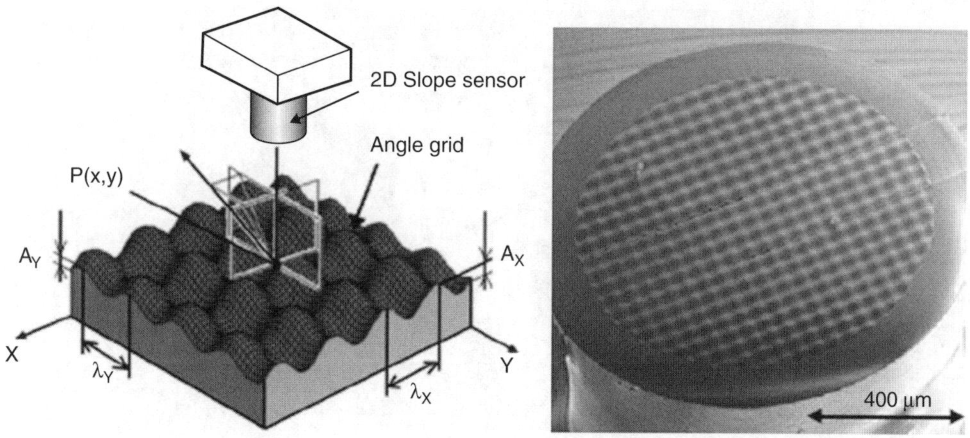

Figure 7.10 FTS diamond turned sinusoidal grid

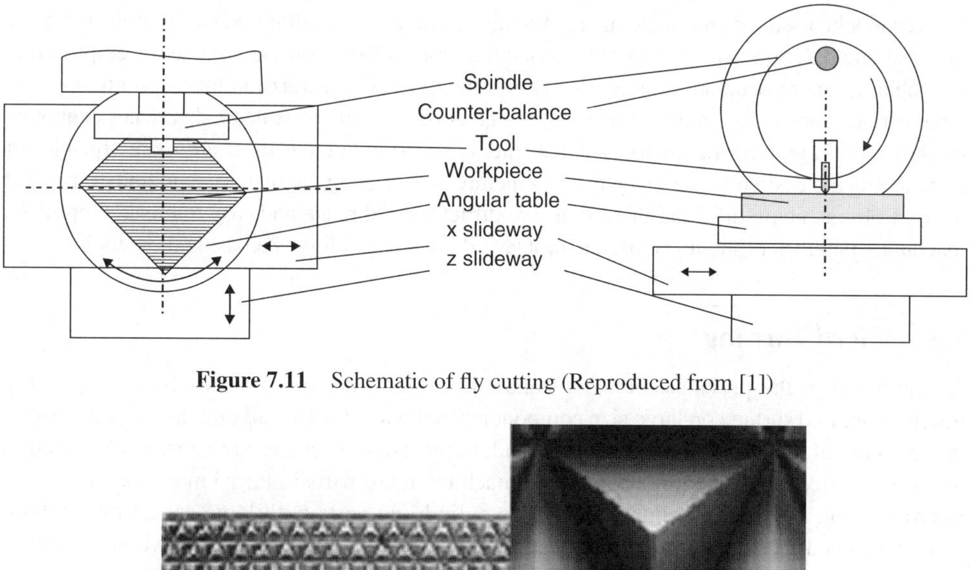

Figure 7.11 Schematic of fly cutting (Reproduced from [1])

Figure 7.12 Multi-prism structures generated by crossed fly cutting. Reproduced with permission from [3]. Copyright 2001 EUSPEN

Figure 7.13 Structured surface of a luminaire for lighting application. Reproduced with permission from [7] Copyright 2003 SPIE

tools cannot be used for producing micro structures. Therefore, small nosed radius tools with nose radius as small as 5 μm and semi-circular nosed (half radius facet) tools are frequently employed in a machining micro structure. Figure 7.14 shows the issue of transmission loss with micro-structured surfaces due to tool radius. Sharp edges and corners in micro-structured surfaces cannot be produced by large nose radius tools. Small radius and semicircular tools are more suitable for transmitting the desired shape to the micro structure and meeting the requirement, however it is at the cost of machining time and these tools are difficult to fabricate and easy to break. Optimized machining parameters should be used to elongate diamond tool life when using these tools.

7.3 Micro Turning

As discussed in the previous sections, standard precision machines are useful in machining micro-structured surface on large size components, but when the overall dimension of the part is in the region of a few millimeters or less, standard precision machines are generally believed to not be cost effective. It is more desirable to machine micro parts in micro machine tools using micro tooling. Figure 7.15 shows two micro cylindrical parts with large aspect ratio which were machined in a micro machining centre developed by Rahman and co-workers (2005). These micro pins have a diameter from tens of microns to hundreds of microns and a length up to a few millimetres. It is very challenging to machine high aspect ratio micro shafts due to their extremely small diameter hence easy to break during cutting. Success of micro turning miniaturized parts depends on many factors among which the most important factors are micro tooling and micro machine tools. In this section the development of micro turning tools and micro machines is described, and then size effect arising from micro turning is discussed.

7.3.1 Micro Turning Tool Fabrication

Although small diameter micro drills and micro end-mills down to tens of microns using various tooling materials have been developed and made commercially available, little progress has been made in the miniaturization of micro turning tools [15]. This is partly because

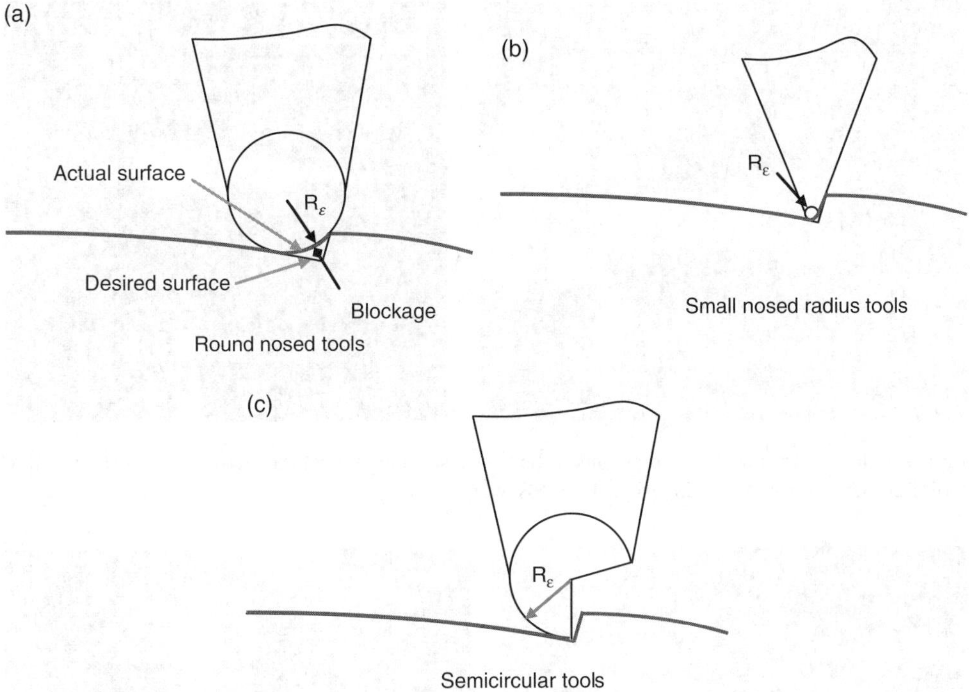

Figure 7.14 Round nosed tools and semicircular tool nose

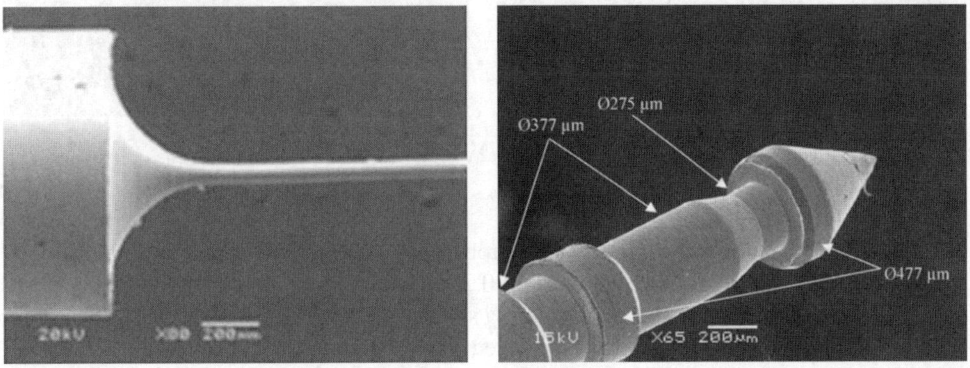

Figure 7.15 Simple and compound shaped micro pin by micro turning at NUS. Reproduced with permission from [25]. Copyright 2005 Elsevier

the micro milling operation is more universal and flexible, and dominates the mechanical micro machining so that emphasis has been put on fabrication of micro end-mills, and partly because micro peripheral turning does not necessarily require micro tools. However the size and geometry of micro turning tools determine the limit of the size and accuracy of micro features, for example, micro turning tools must be employed for micro-hole boring and face grooving of high aspect ratio micro components.

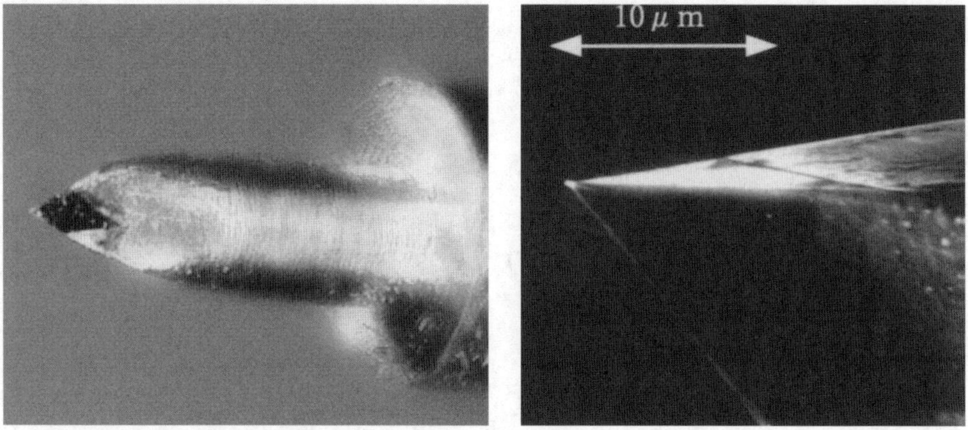

Figure 7.16 A pyramid shaped diamond probe from a scanning tunneling microscope used as a cutting tool. Reproduced from [18]. Copyright 1999 Elsevier

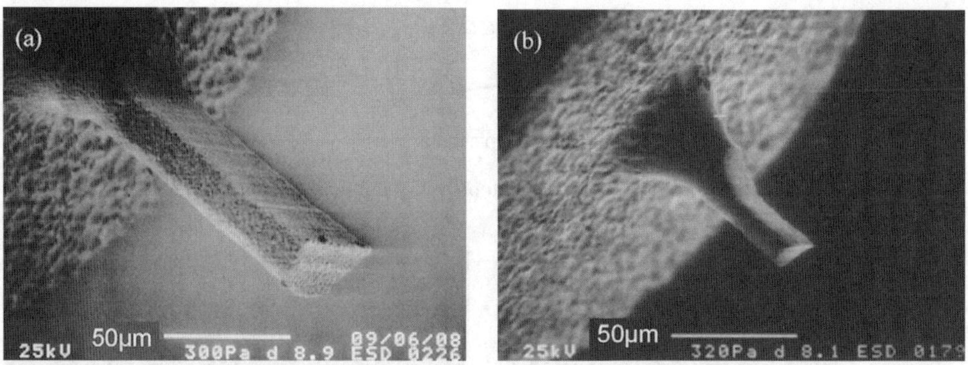

Figure 7.17 Micro turning tools fabricated by WEDG. Reproduced with permission from [15]. Copyright 2011 Elsevier

Lu and Yoneyama employed a diamond probe used in a scanning tunnelling microscope (STM) as micro point tool mounted in a rod 0.25 mm in diameter to match the micro workpiece in their in-house developed micro turning system. The single diamond tool has a trigonal pyramid shape with the rake and the included angles 0°and 60° respectively as shown in Figure 7.16 [16]. These tools are not developed and optimized for micro cutting so they are limited to certain geometries. In addition, natural diamond tools are ideal for ultra-precision machining, but they are limited to machine 'diamond turnable' materials. There is a gap to develop micro turning tools from various materials including tungsten carbide, CVD, natural diamond, and so on. The tool fabrication process must be reliable and capable of shaping micro cutting tools with the appropriate shape and dimensions, typically tens to hundreds of microns.

Egashira and co-workers fabricated a micro turning tool with a 25 and 50 μm outside diameter using wire electro-discharge grinding (WEDG) carried out on a micro EDM machine [15]. Cemented tungsten carbide was chosen as the tool material. The tool fabrication process

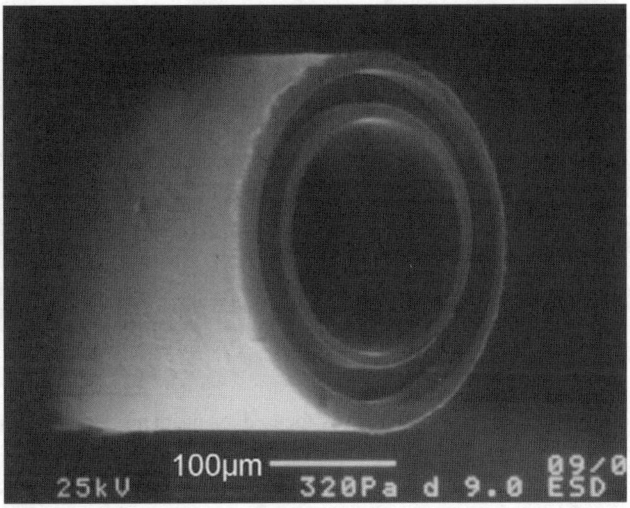

Figure 7.18 Micro face grooving on brass using micro turning tools. Reproduced with permission from [15]. Copyright 2011 Elsevier

controlled the tool geometry with a rake angle of 0° and clearance angle of 20° (Figure 7.17) Face grooving and boring trials on a brass workpiece were performed using the micro tools. Good geometric accuracy and smooth surface finish were obtained from the micro cutting experiments (Figure 7.18).

Adams and co-workers in Sandia National Laboratories have been working on micro turning tool fabrication using focused ion beam (FIB) machining (Adams 2000). A variety of micro tools used for micro turning grooves on cylinders or other rotational symmetric shapes were fabricated from tungsten carbide, high-speed steel and single crystal diamond micro tools. The micro tools have a dimension range of 15–100 μm and cutting edge radius of 40 nm. A number of tool geometry designs including rectangular, triangular and other complex shapes were reported. Figure 7.19 shows the single crystal diamond tool shaped by focused ion beam machining. Although FIB machining is a relatively slow process compared with other micro machining techniques, it is capable of shaping micro tools from both metals and ceramics with good accuracy, tool edge sharpness and surface finishes. For example, micro tools shown in Figure 7.19 have a cutting edge roughness (Ra) of 50 nm or less and a cutting edge radius of 40 nm.

These complex-shaped sharp tools were used to produce micro parts with rotational symmetry. Figure 7.20 shows micro components produced by FIB-shaped micro tools. Applications include micro parts of a micro-screw, pump threads, micro coils and micro solenoids. These 3D complex micro parts are good examples of micro turning as they are hard to produce by other micro machining methods. Due to the small tool edge radius and good surface finishes on tool edge surface, micro parts were machined with good surface finishes. Surface roughness (Ra) of 0.15 μm has been achieved from micro cutting polymers.

Development of Swiss type machines has driven the design of micro tools for machining micro features and micro parts. Swiss micro grooving and micro turning tools using the inserts with widths down to 50 μm are commercially available and these tools are capable of micro thin grooving and micro turning (GenSwiss).

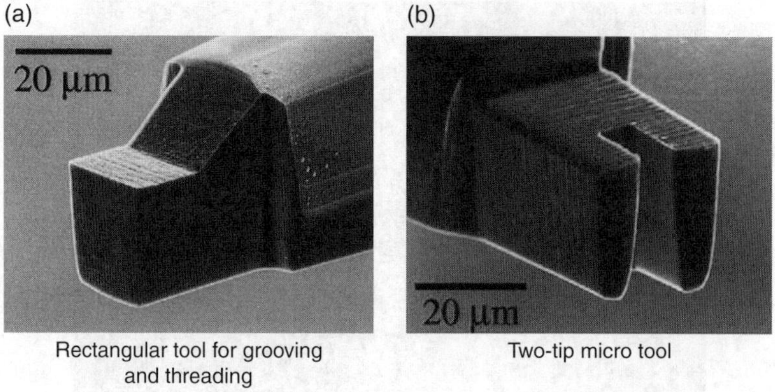

Figure 7.19 Single crystal diamond tool shaped by focused ion beam machining. Reproduced from [17]. Copyright 2003 Elsevier

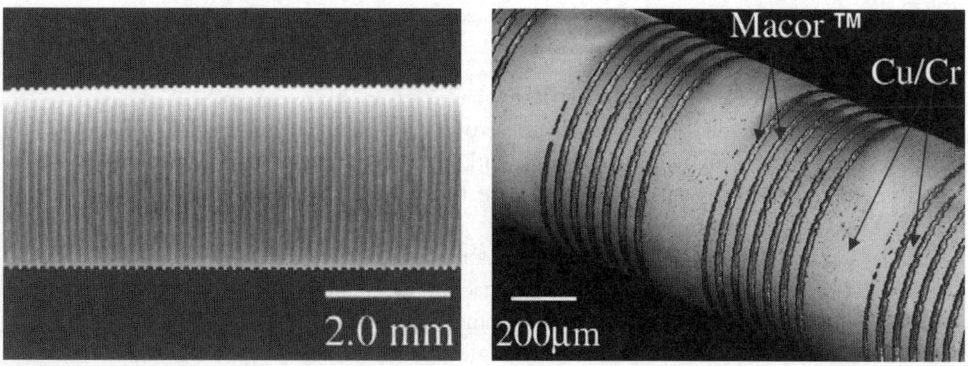

Figure 7.20 A micro screw and micro coils by micro turning tools. Reproduced from [17]. Copyright 2003 Elsevier

Fabrication of micro tools is central to the application of the micro turning process. Research on micro turning tools is still at the research stage, and there are several challenges facing the micro manufacturing community. Different to conventional tools, the micro cutting tools can be easily broken before tool wear actually takes place. Premature tool breakage can be avoided by using optimized cutting conditions, for example, optimal depth of cut and federate must be used in a precision machine with high dynamics so that the vibration is reduced to the minimum. Egashira[15] reported an experiment on the influence of depth of cut and feedrate on tool breakage. Micro tool wear and breakage is a new research area, more experimental data on various materials and cutting conditions are needed to characterize tool performance.

EDM, WEDG and FIB are being used for fabricating micro tools with good quality, but these methods are subject to machining time and not cost-effective for some applications. More micro machining methods should be utilized to fabricate micro tools, such as micro laser cutting and mechanical grinding which are widely used for shaping conventional tools.

7.3.2 Micro Machines for Micro Turning

Micro turning can be performed in conventional ultra-precision turning machines which usually have a footprint of several m2, but obviously conventional precision machines are not well suited to micro turning operations due to high investment costs and energy consumption. Small micro turning systems have many advantages over a conventional turning machine such as being less sensitive to environmental fluctuations, low production cost and high flexibility, and so on.

There has been a need to fill the gap of the micro machine tools for micro cutting. Chapter 2 of this book reviews the latest development on micro machines. The micro machines discussed in Chapter 2 are mainly micro milling machines. This section will discuss the-state-of-the-art micro turning system, both research efforts (in-house micro machines) and commercial development (industrial micro lathes).

The advantages of micro turning systems over the conventional ultra-precision turning machines can be summarized as follows:

- **Cost effective and high dynamics**. Compared with conventional machines, micro machines require a smaller amount of materials and fewer components to fabricate. It does not compromise the performance, in fact, micro machines have higher natural frequencies compared with conventional machines due to the small mass used. Higher frequencies are always desirable for machine tool design because a wide range of spindle speeds can be used without regenerative chatter instability. In addition, micro machines have much lower vibration amplitude than conventional machines.
- **Small footprint and energy-efficient**. Micro machines require a small footprint, typically less than 1 m^2 (including control system and other peripheral equipment), and only consume a small fraction of conventional machine tools' energy consumption.
- **Ease of localized environmental control and thus low operational costs**. Conventional ultra-precision machine tools require a temperature-controlled room and a sophisticated vibration isolation/control system to perform ultra-precision machining. For micro machines it is not necessary to invest such expensive facilities because a localized environmental control is enough to control temperature and vibration.
- **High modularity, flexibility and mobility**. Micro machines can be easily moved from one place to another with little setup/calibration time, and factories made up of micro machines can be easily rearranged in response to market changes.

7.3.2.1 Industrial Micro Lathes

In this section industrial micro lathes are classified into two groups: Swiss-type micro lathes and ultra-precision diamond micro turning machines.

Watchmaking industry in Switzerland has stimulated the development of precision machine tools, particularly Swiss type lathes. Figure 7.21 shows a Swiss type lathe for micro turning. A Swiss type lathe is similar to regular CNC turning machines but they are much more compact in size and are designed to produce very small, finely detailed cylindrical shaped micro parts such as screws and shafts for the Swiss watchmaking industry. Now applications of Swiss type lathes or Swiss turning exist not only in the

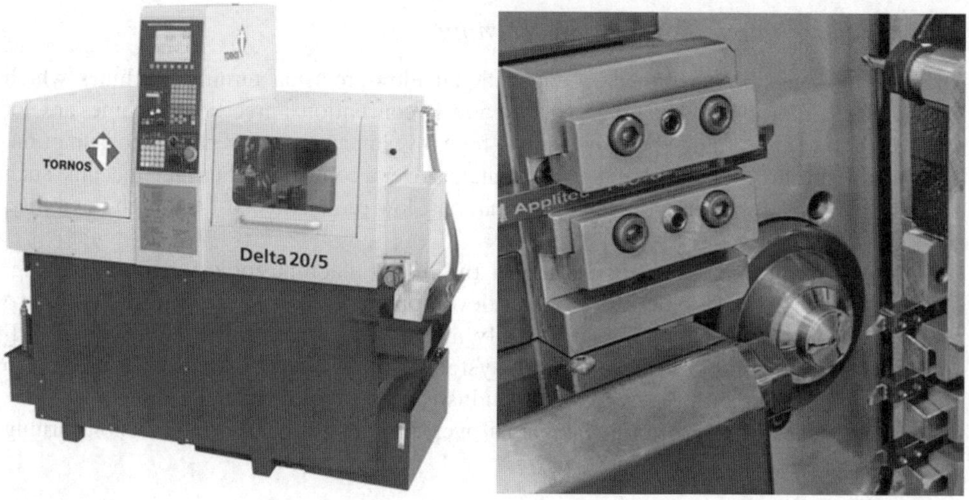

Figure 7.21 Swiss type lathe for micro turning (images courtesy of Tornos) Images courtesy of Tornos

watchmaking industry, but also in many other industries, namely automotive, electronics and medical engineering industries, and so on. Table 7.3 lists these application areas and part examples. Swiss turning is believed to be a cost effective method to mass production of small and miniature metal components or parts with the diameter as small as 0.5 mm in a tight tolerance. Figure 7.22 shows some typical micro turned parts by the Swiss type machine.

Although research and development on ultra-precision micro turning machines are still at their nascent stage, some micro turning machines have already found their way to the market. The miniature ultra-precision lathe shown in Figure 7.23 is capable of producing a nanometric surface finish required in ultra-precision components. It features a small workspace of 100 mm by 100 mm, making it extremely cost effective and suitable for the fabrication of miniature ultra-precision components, for example, inserts for micro-lens (Mikrotools). It is one of the few miniature ultra-precision lathes of this kind commercially available in the market. Important specifications of this machine include an aerostatic bearing spindle with submicron radial run-out, linear motor driven slideway and high static stiffness. Figure 7.23 shows the micro lathe and some micro components produced.

Nano Corporation has produced a series of Nanowave® micro precision CNC lathes for more than a decade (Nano Corporation) as shown in Figure 7.24. All these micro lathe products have the same design and similar specifications. The precision micro lathes have cross roller slideways and are arranged in a 'T' configuration to form high stiffness. The machines have an overall size of around 200 mm × 300 mm, and can hold a workpiece with a diameter up to 5 mm. Micro turning experiments produced a surface roughness (Ra) of 60 nm and circularity of 50 nm from stainless steel, and 20 nm Ra from brass. Asuperfine graphite pin with a diameter of 10 μm and length of 200 μm machined from MTS3 was reported.

Table 7.3 Examples of micro turned parts machined by Swiss turning

Applications	Examples
Automotive	Micro shafts and axles for motors and micro machines Micro shafts for fuel injection pump Knurled shafts Poppet valves
Electronics	Micro connectors Sensor pins Spring contacts Micro probes
Watch making	Cylindrical pins Curb pins Watch screws and studs Fastening keys Eccentrics
Biomedical Engineering	Implants e.g. for neurosurgery Micro dental tools and instruments Catheter tips Hearing aid screws, orthopedic bone screws Surgical pins

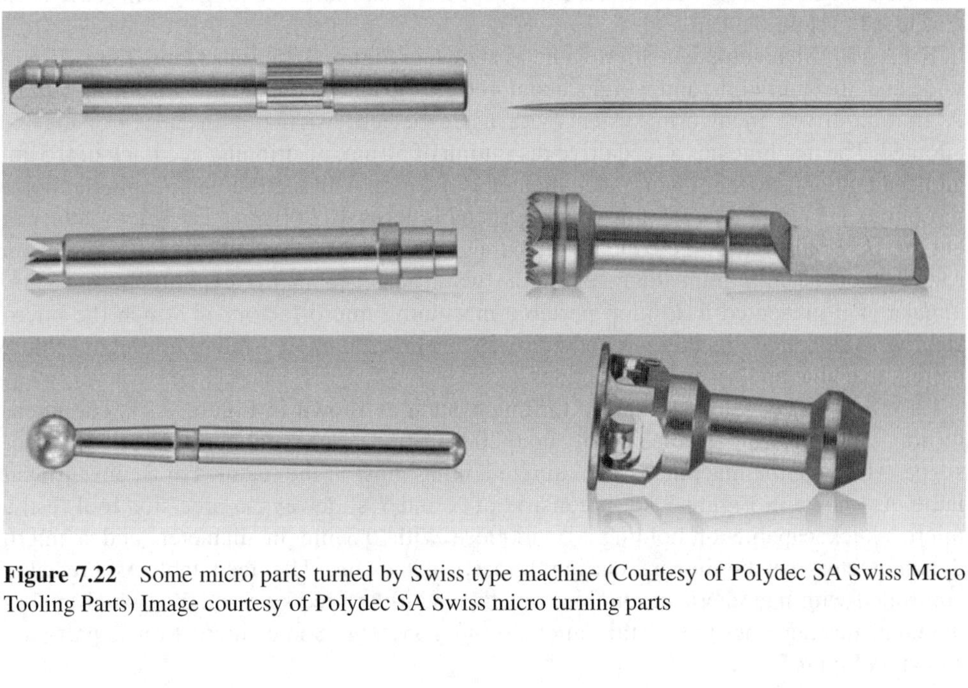

Figure 7.22 Some micro parts turned by Swiss type machine (Courtesy of Polydec SA Swiss Micro Tooling Parts) Image courtesy of Polydec SA Swiss micro turning parts

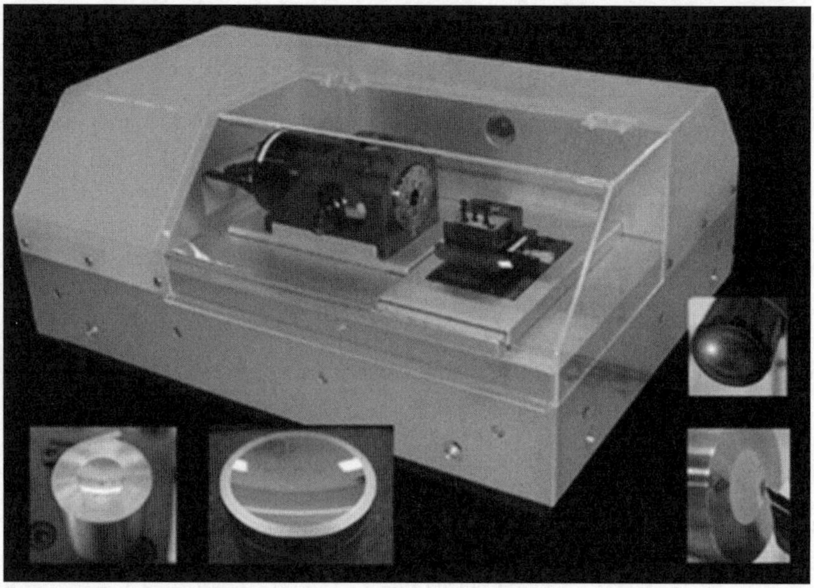

Figure 7.23 The miniature ultra-precision lathe UPL420 (Courtesy of Mikrotools) Copyright Mikrotools

7.3.2.2 In-house Micro Turning Machines

It has been believed that there is a need to produce precision small machine, that is, 'small machines for small parts'. Therefore, a number of in-house micro turning systems have been developed to close the gap.

Kitahara [18] developed a micro lathe as shown in Figure 7.25. It is 32 mm long, 25 mm wide and 30.5 mm high, and weighs about 100 g. It is one of the smallest working lathes. A friction-driven inchworm system with a piezoelectric actuator in an X–Y stage is adopted. The machine has a feed drive resolution of 50 nm, positioning accuracy of 0.5 μm and can hold a maximum workpiece diameter of 2 mm. Although the main spindle motor has only a 1.5 W rated power, the rotating speed is about 10 000 rpm. The micro lathe can cut brass with an accuracy of 1.5 μm straightness in the feed direction and 2.5 μm roundness. The minimum diameter of the work achieved in the experiment is 60 μm. Tanaka [19] presented a desktop portable machining micro factory in which the micro lathe was integrated together with other micro machines, namely a micro milling machine, a press machine and assembly machines.

Lu *et al.* [16] developed a micro turning system as shown in Figure 7.26. The micro turning system has a portable size of about 200 mm and can be set under an optical microscope. Although the machine is miniaturized, it has most of the features of a conventional lathe. As shown in Figure 7.26, it consists of X and Y slideways, a precision tool rest, a micro chuck capable of holding a workpiece with 0.3 mm in diameter, and a micro dynamometer for monitoring cutting forces, and so on. The feed table is precision controlled with nanometic resolution, together with the precision spindle, which makes diamond turning possible in this micro turning system. Some micro turned parts are shown in Figure 7.27.

Figure 7.24 Nanowave® series precision CNC micro lathes (Courtesy of Nano Corporation, Iijima 2002. Ito 2002) Image courtesy of Nano Corporation

Figure 7.25 Micro lathe developed by (Reproduced from [18])

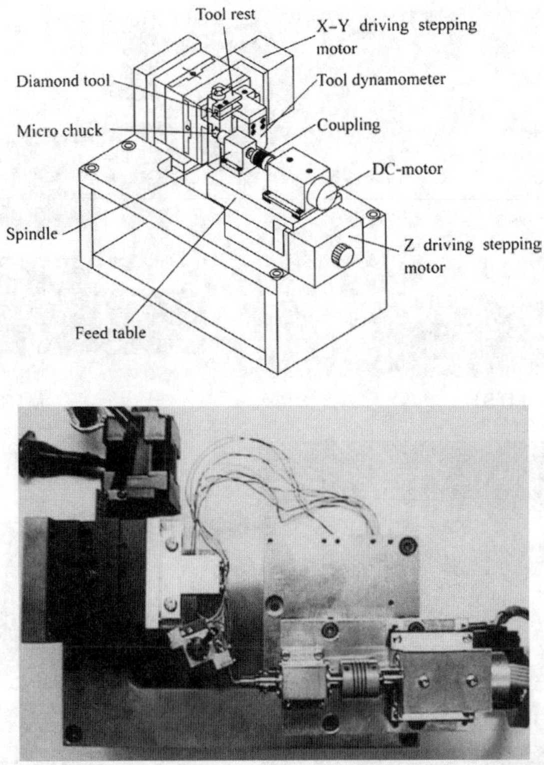

Figure 7.26 Micro turning system Reproduced from [16]. Copyright 1999 Elsevier

Huo *et al.*[20] built a small research diamond turning machine as shown in Figure 7.28. This in-house research micro machine was used for production of micro machined parts. Materials that have been machined with this system include copper, aluminium and steel. The micro turning system has a footprint of 400 mm × 600 mm and equipped with an in-house designed fast tool servo system as shown in Figure 7.28a. The machine is featured with air bearings for both the spindle and slideways and linear motor direct drive. Some micro parts with nanometric surface finishes have been machined with or without the aid of the FTS. Figure 7.28b shows an example of micro turned parts.

Compared with standard machine tools, micro turning machines have a simple design and consist of lesser components. Some micro actuation and sensors which are not suitable for standard precision machines, such as piezo-actuator and friction drive, can be found in some micro machines. Spindle, drive system and configuration commonly used in micro machines are discussed below.

Spindle is a key element of the precision machine tool because the spindle motion error will have significant effects on the surface quality and accuracy of machined components. The most frequently used spindles in precision machine tools are aerostatic spindles and hydrostatic spindles. They both have high motion accuracy and are capable of high rotational speed. Because of small cutting forces in micro turning aerostatic spindles are often used in micro turning machines. These spindles in commercial micro turning machines are reported to have a speed range of 1000–10 000 rpm which is slower than the high speed spindle used for the micro milling process for which a rotational speed higher than 100 000 rpm is not uncommon.

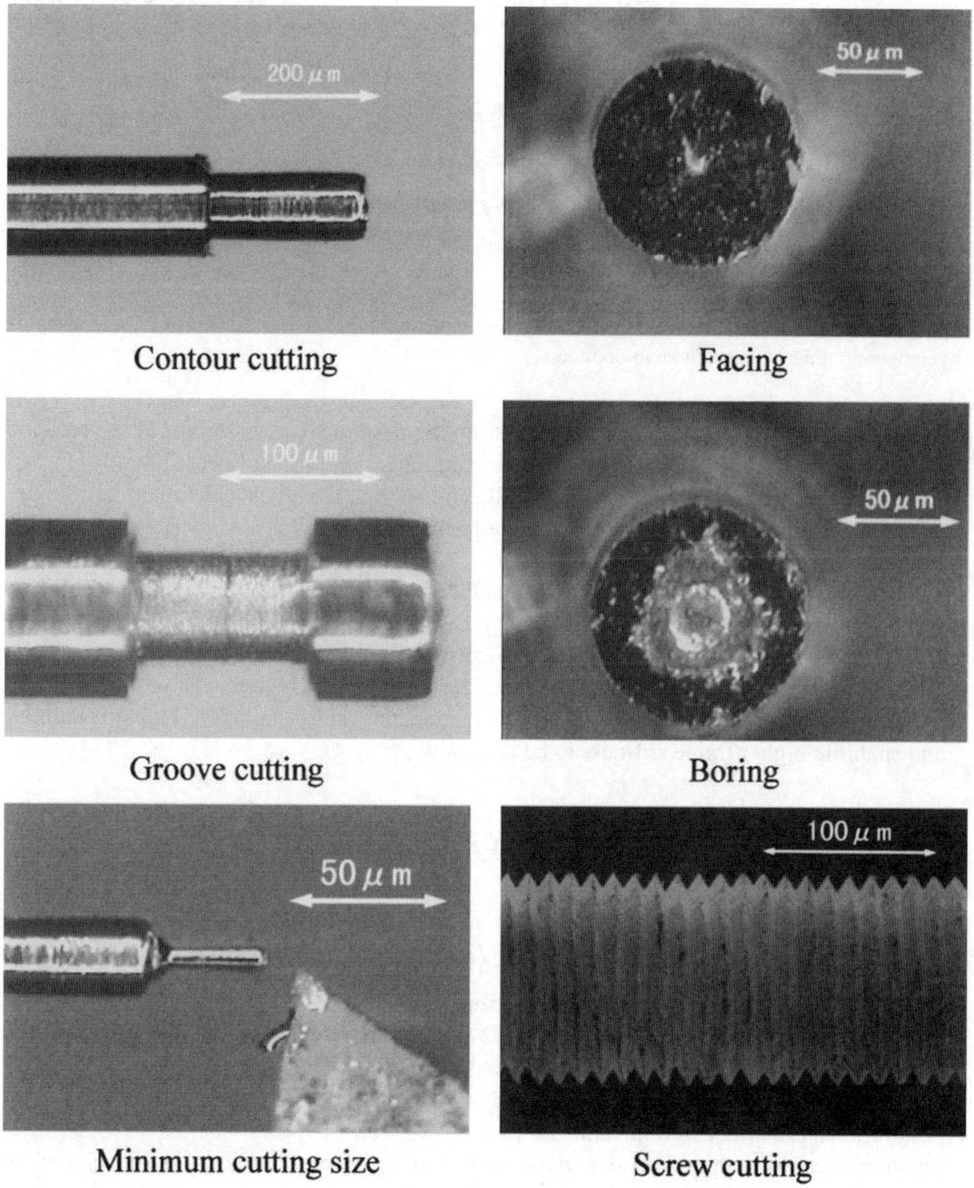

Figure 7.27 Samples micro turned by the micro turning system. Reproduced from [16]. Copyright 1999 Elsevier

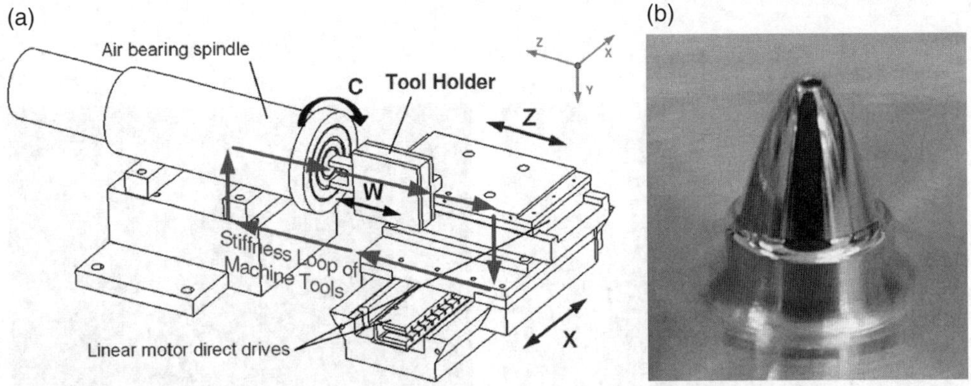

Figure 7.28 The diamond turning machine tool equipped with the FTS. (a) schematic of the machine (b) component micro turned by diamond tools

A high speed and accurate motions and positioning system are essential for ultra-precision micro machines. The direct feed drive system, using linear motors for linear slideways in particular, is becoming increasing popular in fulfilling such applications. The main advantages over the indirect feed drives can be summarized here [21–23]:

- no backlash, no lead screw error, no belt stretch and less friction, resulting in high accuracy;
- no mechanical limitations on acceleration and velocity, the velocity is only limited by the encoder bandwidth or by the power of the electronics;
- high jerks and high Kv factor which characterizes the ability of high precision at high speed;
- mechanical simplicity, resulting in ease of maintenance and installation, higher reliability and enabling higher frame stiffness to be achieved;

Although various configurations have been implemented in conventional turning machines, 'T' configuration is popularly used for most of the precision micro turning machines due to its high rigidity and simplicity [20].

7.3.3 Size Effect Arising from Micro Turning

Size effect on mechanical properties and different types of forces for micro parts or processes during scaling down was first investigated by [24]. Trimmer proposed a scaling factor *S*, which is the one dimensional linear size ratio between two components. For example if S is 10, all the dimensions of the system are decreased by a factor of 10. The scaling effect of different types of forces is summarized in Table 7.4 [25].

In order to estimate the size effect on the micro turning process, a micro shaft machined by micro turning is considered in this section. During the turning operation, high aspect ratio cylindrical parts are easily deflected by thrust force. As illustrated in Figure 7.29, during cutting, the micro shaft is loaded by two forces: the cutting force and the thrust force. For external cylindrical longitudinal turning, thrust force results in bending deflection and maximum stress at the end of the cantilever shaft, whilst cutting force is responsible for compressive deformation and buckling.

Table 7.4 Scaling effect of various forces. Reproduced from [25]. Copyright 2006 Elsevier

Kind of force	Symbol	Equation	Scaling effect	
Electromagnetic Force	F_{magc}	$\frac{B}{2\mu}S_m$	S^2	μ: permeability; B: magnetic field density; S_m: area of cross section of coil; S: scaling factor
Electrostatic Force	F_{static}	$\frac{\varepsilon S_m}{2}\frac{V^2}{d^2}$	S^0	ε:permittivity; V: applied voltage; S_m: surface area; d: gap between electrodes; S: scaling factor
Thermal expansion force	F_{ther}	$ES_m\frac{\Delta L(T)}{L}$	S^2	E: Young's modulus; L: length; ΔL: strain; T: temperature; S: scaling factor
Piezoelectric force	F_{piezo}	$ES_m\frac{\Delta L(E_e)}{L}$	S^2	E: Young's modulus; L: length; ΔL: strain; T: temperature; S: scaling factor
Inertial force	F_i	$m\frac{\partial^2 x}{\partial t^2}$	S^4	m: mass; t: time; x: displacement; S: scaling factor
Viscosity force	F_v	$c\frac{S_m}{L}\frac{\partial x}{\partial t}$	S^2	c: viscosity coefficient; x: displacement; S_m: surface area; L: length; t: time; S: scaling factor
Elastic force	F_e	$ES_m\frac{\Delta L}{L}$	S^2	E: Young's modulus;S_m: cross section area; ΔL: strain; L: length; S: scaling factor

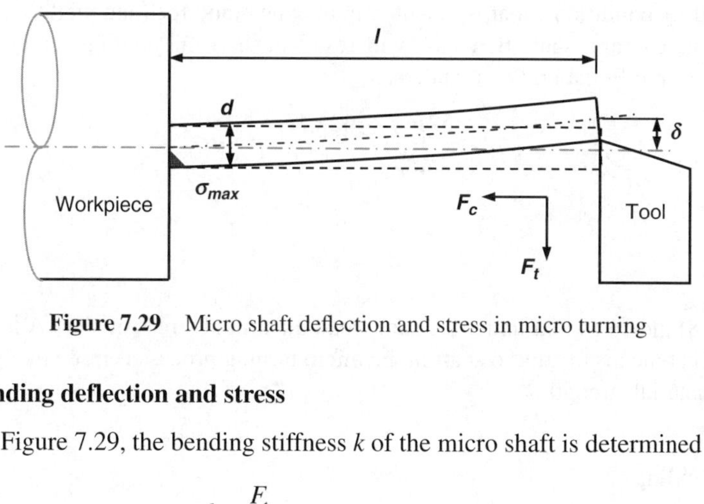

Figure 7.29 Micro shaft deflection and stress in micro turning

7.3.3.1 Bending deflection and stress

As shown in Figure 7.29, the bending stiffness k of the micro shaft is determined by:

$$k = \frac{F_t}{\delta} \tag{7.1}$$

Where F_t is the thrust force and δ is the bending deflection at tool-workpiece contact point, which can be determined by:

$$\delta = \frac{F_t l^3}{3EI} = \frac{64 F_t l^3}{3\pi E d^4} \tag{7.2}$$

Where l is the length of the shaft, d is the diameter of the shaft, E is Young's modulus, and I is the moment of inertia. Therefore the bending stiffness can be expressed as:

$$k = \frac{3EI}{l^3} = \frac{3\pi E d^4}{64 l^3} \tag{7.3}$$

Bending stress at the end of the shaft, it leads to:

$$\sigma = \frac{32 F_t l}{\pi d^3} \tag{7.4}$$

Here we consider a micro shaft A with a length of l_a and a diameter of d_a, and a conventional shaft B with a length l_b and a diameter of d_b respectively. Two shafts have same aspect ratio d/l, and scaling factor S is defined as l_b/l_a. There is:

$$E_B = E_A,\ l_B = S l_A,\ d_B = S d_A,\ I_B = S^4 I_A,\ m_B = S^3 m_A \tag{7.5}$$

Substituting Equation (7.5) into Equation (7.3) gives:

$$k_B = \frac{3E_B I_B}{l_B^3} = \frac{3E_A S^4 I_A}{S^3 l_A^3} = S k_A \tag{7.6}$$

Equation (7.6) shows that a size scaling down gives a reduction on the bending stiffness by the same factor.

Consider that the same thrust force is loaded on shaft A and B, and this is likely the case for the same cutting conditions, that is, uncut chip thickness and feedrate might be used for cutting micro shaft A and shaft B, which will result in the same cutting forces. Substituting Equation (7.5) into Equation (7.4) renders:

$$\sigma_B = \frac{32 F_t l_B}{\pi d_B^3} = \frac{32 F_t S l_A}{\pi S^3 d_A^3} = \frac{1}{S^2}\sigma_A \tag{7.7}$$

or

$$\sigma_A = S^2 \sigma_B \tag{7.8}$$

Equation (7.8) indicates that under the same load maximum stress of shaft A is S^2 times that of shaft B. That means the micro shaft in the micro turning process is more likely to reach the maximum material strength.

7.3.3.2 Buckling

When axial force (here is the cutting force F_c) acts on the shaft, simple compression occurs for low values of the force. When F_c reaches a specific value, the long shaft becomes unstable and unstable bending or buckling occurs. The critical force for the end condition as illustrated in Figure 7.29 is governed by the Euler column formula:

$$P_{cr} = \frac{\pi^2 EI}{4l^2} \tag{7.9}$$

Substituting Equation (7.5) into Equation (7.9) gives:

$$P_{crB} = \frac{\pi^2 E_B I_B}{4l_B^2} = \frac{\pi^2 E_A S^4 I_A}{4S^2 l_4^2} = S^2 P_{crA}$$

or

$$P_{crA} = \frac{1}{S^2} P_{\mathrm{crB}} \tag{7.10}$$

Equation (7.10) shows that a size reduction by a factor of S results in a decrease of critical buckling force P_{crA} by a factor of $1/S^2$, which means a size reduction causes micro shafts to be more sensitive to buckling.

Through the above theoretical analysis, we can conclude that size effect changes structural deformation in many ways and great care should be taken when choosing cutting parameters, that is, depth of cut, feedrate and spindle speed.

7.3.3.3 Eccentric force

In order to increase linear cutting speed, spindles for micro cutting normally have higher rotational speed than that used in conventional cutting processes. However the spindle errors such as radial runout are not simply scaled down. In fact miniaturized high speed spindles might have a higher level of errors than conventional spindles. Therefore it is necessary to analyze the size effect on deflection of micro shafts under eccentric force.

Consider two shafts (the same as the above analysis) are mounted in two spindles which are subjected to radial runout ε_A and ε_B respectively. We assume two spindles have the same radial runout. The rotational speed of the miniaturized spindle A is S times that of conventional spindle B, this is normally the case when the same linear speed and acceptable productivity need to be maintained:

$$\varepsilon_A = \varepsilon_B \quad \text{and} \quad \omega_A = S\omega_B \tag{7.11}$$

The eccentric force is given by:

$$F_e = m\omega^2 \varepsilon \tag{7.12}$$

and the deflection and stress under eccentric force can be estimated by the Equation (7.2) and Equation (7.4).

Substituting Equation (7.5) and (7.11) into Equation (7.12) gives:

$$F_{eB} = m_B \omega_B^2 \varepsilon_B = S^3 m_A S^{-2} \omega_A^2 \varepsilon_A = SF_{eA} \tag{7.13}$$

Substituting Equation (7.5) and Equation (7.13) into Equation (7.2) represents:

$$\delta_B = \frac{F_{eB} l_B^3}{3E_B I_B} = \frac{SF_{eA} S^3 l_A^3}{3E_A S^4 I_A} = \delta_A \tag{7.14}$$

Substituting Equation (7.5) and Equation (7.13) into Equation (7.4) leads to:

$$\sigma_B = \frac{32F_{eB}l_B}{\pi d_B^3} = \frac{32SF_{eA}Sl_A}{\pi S^3 d_B^3} = \frac{1}{S}\sigma_A \tag{7.15}$$

Equation (14) implies that if a size reduction by a factor of S and a speed increase by the same factor of S, which is close to some micro cutting applications. The deflection of shaft caused by eccentric force remains the same as that of conventional cutting before scaling down. In terms of maximum stress shaft A is S times that of shaft B. That means micro shaft would be more sensitive to eccentric force induced deflection, and the micro shaft in the micro turning process is more likely to reach material strength. We should take these into consideration when we choose spindle speed for micro cutting process.

7.4 Challenges Arising from Micro Turning

There are many benefits of applying micro cutting technology, which have been discussed and explored throughout the book. Some issues and challenges in micro turning, in addition to the size effect issues as discussed in this chapter, are continuously emerging and pushing forward along with more and more applications in the industry. Nevertheless, the essentials of the research and development in micro turning can be foreseen and highlighted as follows:

1. In the micro turning process, the rigidity or strength of the shaft decreases as the diameter of the shaft reduces. There will be a restriction of achievable minimum diameter of the shaft which can withstand the magnitude of the cutting forces for the acceptable deflection in the shaft. The micro tool and workpiece breakages can take place during the fabrication.
2. Currently, most of the research is focused on micro turning of free cutting materials, especially non-ferrous metals. As the requirements for micro turning continue to grow, other materials such as technical ceramics, hard alloys and biological materials have been receiving the attention. Micro machinability of the micro turning process will impose a challenge. Some measures such as laser assisted micro turning need to be taken into full account so as to render micro machinability in a truly predictable and highly productive manner.
3. Although micro turning is a unique micro machining process for some applications, it has limited productivity compared with other moulding based processes. Research and development efforts should be put into how to combine diamond turning and micro turning with other micro manufacturing processes such as chemical etching and embossing to extend application areas of micro turning.

References

[1] Davies, M.A., Evans, C.J., Patterson, S.R., Vohra, R., and Bergner, B.C. (2003) Application of precision diamond machining to the manufacture of micro-photonics components. *Proc. SPIE*, 5183, 94–108.

[2] Paul, E., Evans, C.J., Mangiamelli, A., McGlauflin, M.L., and Polvani, R.S. (1996) Chemical aspects of tool wear in single point diamond turning, *Precision Engineering*, 18, 4–19.

[3] Brinksmeier, E., Preuss, W., and Gessenharter, A. (2001) Manufacturing of calibration standard by diamond machining. In Proceedings of the 2nd Euspen International Conference, Turin, Italy, 2001, 680–683.

[4] Wallrabe, U., Ruther, P., Schaller, T., and Schomburg, W.K. (1998) Microsystems in medicine. *International Journal of Artificial Organs*, 21, 137–146.
[5] Gao, W., Araki, T., Kiyono, S., Okazaki, Y., and Yamanaka, M. (2003) Precision nano-fabrication and evaluation of a large area sinusoidal grid surface for a surface encoder. *Precision Engineering*, 27, 289–298.
[6] Evans, C.J. and Bryan, J.B. (1999) 'Structured', 'Textured' or 'Engineered' Surface, *Annals of the CIRP*, 48(2), 541–556.
[7] Brown, N.J., Donaldson, R.R., and Thompson, D.C. (1983) Fabrication of Machined Optics for Precision Applications, Proc. of the SPIE, Presented at the SPIE Symp. on Optical Surface Technology, GarmischPartenkirchen, April 1983.
[8] Yan, J., Maekawa, K., and Tamaki, J. (2005) Micro grooving on single-crystal germanium for infrared Fresnellenses. *Journal of Micromechanics and Microengineering*, 15, 1925–1931.
[9] Weck, M., Fischer, S., and Vos, M. (1997) Fabrication of microcomponents using ultraprecision machine tools, *Nanotechnology* 8, 145–148.
[10] Kim, H.S. and Kim, E.J. (2003) Feed-forward control of fast tool servo for real-time correction of spindle error in diamond turning of flat surfaces, *International Journal of Machine Tools and Manufacture*, 43, 1177–1183.
[11] Yang, Y., and Chen, S. (2006) Application of ultra-precision machining technique in micro machinery manufacturing. *Journal of Aviation Precision Manufacturing Technology*, 42, 1–4.
[12] Yang, Y., Chen, S., Huo, D., and Cheng, K. (2007) Performance analysis and optimal design of fast tool servo used for machining microstructured surfaces, *Proc. IMechE, Part C: J. Mechanical Engineering Science*, 222, 1541–1546.
[13] Weck, M., Klocke, F., Oezmeral, H., Hennig, J., Ruebenach, O., Ehl, M., Grosser, N., Leier, R., Henning, T., Unnebrink, L.,and Bernges, J. (1999) Manufacturing and applications of non-rotationally symmetric optics, *SPIE*, 3739, 94–107.
[14] Rahman, M.A., Rahman, M., Kumar, A.S., and Lim, H.S. (2005) CNC microturning: an application to miniaturization, *International Journal of Machine Tools & Manufacture*, 45, 631–639.
[15] Egashira, K., Iwata, M. and Nomura, Y. (2011) Boring and face grooving using micro turning tools, *Annals of the CIRP*, 60, 81–84.
[16] Lu, Z.N., and Yoneyama, T. (1999) Micro cutting in the micro lathes turning system, *International Journal of Machine Tools and Manufacture*, 39, 1171–1183.
[17] Picard, Y.N., Adams, D.P., Vasile, M.J. and Ritchey, M.B. (2003) Focused ion beam-shaped microtools for ultra-precision machining of cylindrical components, *Precision Engineering*, 27, 59–69.
[18] Kitahara, T., Ishikawa, Y., Terada, T., Nakajima, N., and Furuta, K. (1996) Development of micro-lathe. Mechanical Engineering Lab. Report, *AIST/ MITI*, 50, 117–123.
[19] Tanaka, M. (2001) Development of desktop machining microfactory *RIKEN Review*. 34, 46–49.
[20] Huo, D., and Cheng, K. (2008) A dynamics-driven approach to precision machines design for micro-manufacturing and its implementation perspectives. *Proceedings of the IMechE, Part B, Journal of Engineering Manufacture*, 222, 1–8.
[21] Brecher, C., Klar, R., and Wenzel, C. (2007) Development of a high precision miniature milling machine. *Proceedings of the 3rd International Conference on Multi-Material Micro Manufacture, 4M2007*, 327–330.
[22] Otten, G., De Vries, T.J.A., Van Amerongen, J., Rankers, A.M., and Gaal, E.W. (1997) Linear motor motion control using a learning feedforward controller, *IEEE/ASME Transactions on Mechatronics*, 2(3): 179–187.
[23] Denkena, B., Tönshoff, H.K., Li, X., Imiela, J., and Lapp, C. (2000) Analysis and control/monitoring of the direct linear drive in end milling, *International Journal of Production Research*, 42(24), 5149–5166.
[24] Trimmer, W.S.N. (1997) Micromechanics and MEMS, Classic and Seminal papers to 1990, *IEEE*, New York.
[25] Kussul, E., Baidyk, T., Ruiz-Huerta, L., Caballero-Ruiz, A., and Velasco, G. (2006) Scaling down of microequipment parameters, *Precision Engineering*, 30, 211–222.

Other Sources

Mikrotools Pte Ltd. http://www.mikrotools.com
Moore Nanotechnology System. website: http://www.nanotechsys.com/
Nano Corporation, website http://www.nanowave.co.jp/english
Multidec 1600 series micro tools http://www.genswiss.com/multidec1600.pdf
Polydec.Website:http://www.polydec.ch/

8

Micro Milling: The State-of-the-art Approach Towards Applications

Tao Wu and Kai Cheng
School of Engineering and Design, Brunel University, UK

8.1 Introduction

During the last couple of decades, the demand for high-accuracy miniature and micro features or components has experienced a phenomenal growth in various markets including aerospace, medical, automotive, electronics, telecommunication, optics, and so on. Some specific applications, such as micro channels for lab-on-chips, micro fuel cells, fluidic micro-chemical reactors, micro-nozzles for high-temperature jets, micro-holes for fibre optics, micro-moulds and deep X-ray lithographic masks, have already been reported [1–3]. For Micro Electro Mechanical Systems (MEMS) only, Yole Development's analysis shows that the overall MEMS market is expected to grow from $8 billion in 2010 to $18 billion in 2015, while the emerging MEMS markets have the potential to add a further $2.2 billion to the overall MEMS market by 2015 [4]. This emerging trend and the reported developments indicate micro parts and systems are becoming more and more important, both for enhancing product performance and diversity as well as for enhancing industrial economic growth.

To cope with the different requirements of a wide range of applications, there is a growing need for a fast, direct, reliable and repeatable approach to fabricate freeform and complex three dimensional features out of a diverse range of engineering materials. However, some traditional MEMS manufacturing techniques, such as chemical etching and LIGA, are mainly applicable to silicon-like materials and 2D features or components [2]. Micro milling on the other hand has the capability of producing freeform micro-structures and micro components with complex 3D shapes out of materials that include metals, polymers, composites, glass and ceramics. Figure 8.1 shows some high quality or high aspect ratio micro features produced by micro milling various materials [5–10].

Potential applications as well as advantageous process capabilities drive researchers to enhance the understanding of micro milling technology. Micro milling is a newly emerging

---

Micro-Cutting: Fundamentals and Applications, First Edition. Edited by Kai Cheng and Dehong Huo.

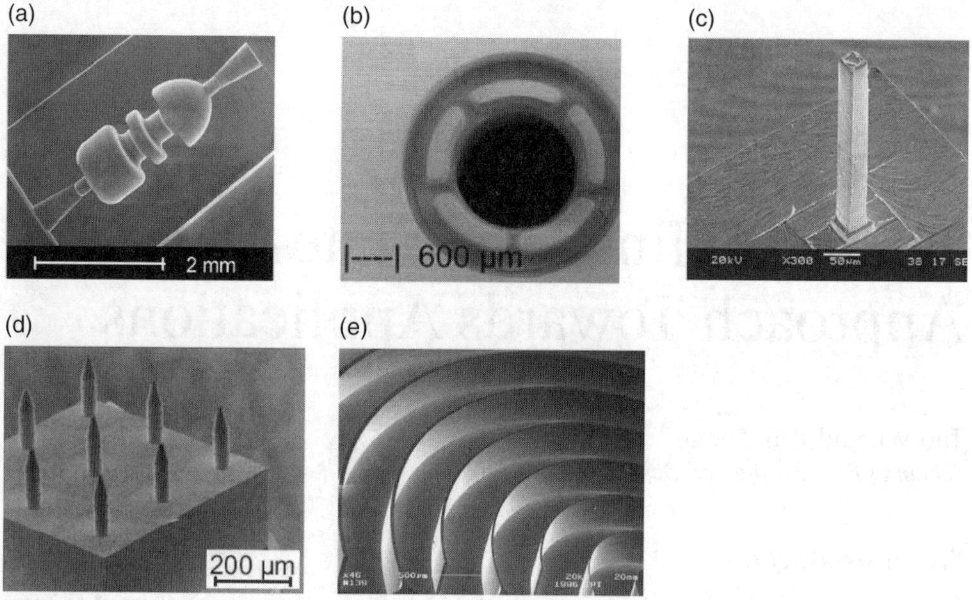

Figure 8.1 High quality or high aspect ratio micro features: (a) and (b) Micro moulds (steel). Reproduced from [5, 6], (c) and (d) Micro columns (brass). Reproduced from [7, 8], (e) Micro walls (nickel silver). Reproduced with permission from [9]. Copyright 2001 SPIE

technology but is directly scaled down from conventional milling. Therefore, the two cutting processes have similar kinematics, with material removal achieved by mechanical interaction between a rotating cutter and workpiece, causing shearing of the material along defined tool paths and removal of unwanted material in the form of chips [11]. Nevertheless, a number of critical issues arise on transition of mature macro-domain knowledge into the micro-domain. Issues such as size effect and tool failures make such technology considerably challenging compared to macro-scale milling. Consequently, a comprehensive insight into micro milling is essential for extending its practical applications. The chapter aims to provide a critical review of the state-of-the-art approach, covering machines, tooling, processes and their holistic integration for applications.

8.2 Fundamental Elements in Micro Milling

The main objective of micro milling is to fabricate material oriented miniature components and products accurately and economically according to specific customer requirements. Relationships between fundamental elements for the approach are illustrated in Figure 8.2. Ultra-precision micro milling machines and cutting tools, together with appropriate process conditions play an important role for implementing the technology for industrial applications.

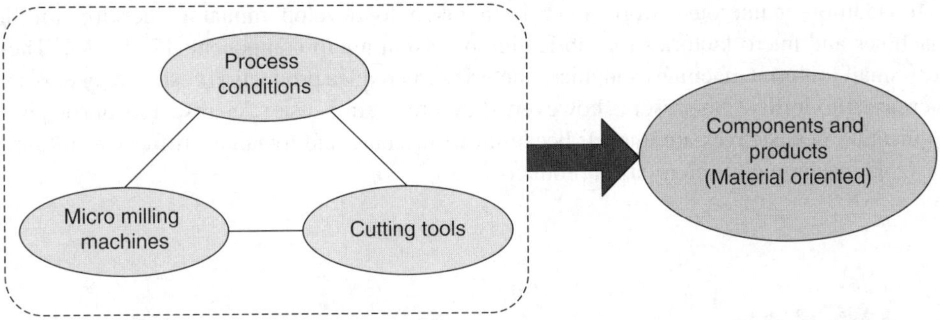

Figure 8.2 The relationship between fundamental elements for micro milling

8.2.1 Micro Milling Machines

Ultra-precision machines are critical for producing high quality parts. To meet the requirements for high dimensional accuracy, good form and surface finish for micro component manufacturing, micro milling machines are required to have high static loop stiffness, good thermal stability, low motion errors and high damping or dynamic stiffness [12]. In addition, a high maximum cutting speed is necessary to achieve optimum cutting performance and high material removal rates over a useful component size range so as to maintain acceptable productivity. Because of small diameter micro cutting tools, micro milling machines are commonly equipped with high speed and low run-out spindles. Furthermore, multi-axis configurations of machines are also important for ensuring process flexibility and forming accuracy in machining freeform components. The advantage is manufacturing without the need for resetting thus greatly increasing productivity and eliminating re-clamping errors. At the early stage of micro manufacturing development micro milling trials were mainly carried out on high speed conventional machines, which were not cost effective, less flexible and tedious in use. Along with increased understanding of the micro machining process and special requirements for micro milling machines, some commercially available precision and ultra-precision machines [13–18], as shown in Figure 8.3 (a–d), have been designed and configured to have micro milling capabilities.

Apart from commercial machines, a five-axis ultra-precision micro milling machine, as illustrated in Figure 8.4, has been built jointly by UPM Ltd and Brunel University [12, 19], featuring a small footprint with good energy and resource efficiency, low production cost and high machining accuracy. The overall footprint including the periphery is only about $1m^2$ and its aerostatic bearing spindle is capable of producing speed up to 200 000 rpm [20]. Surface finishes down to 10nm Ra, surface flatnesses or straightnesses better than 0.1μm over 50 mm of travel, and a repeatability of less than 1μm have been achieved in trials on micro milling non-ferrous metals.

The main specification of the above mentioned ultra-precision machines are summarized in Table 8.1, which provides a comprehensive comparison of characteristics and the state-of-the-art of micro milling machines.

In addition, numerous efforts have been taken to develop miniature desktop milling machines and micro factories for fabricating precision micro components [7, 21–24]. These very small footprint machines and micro factories can reduce production cost, save energy and increase flexibility. At present, however, they are mainly used for research purposes as application perspectives are unclear because of low static and dynamic stiffness resulting in unstable machining of micro-sized products.

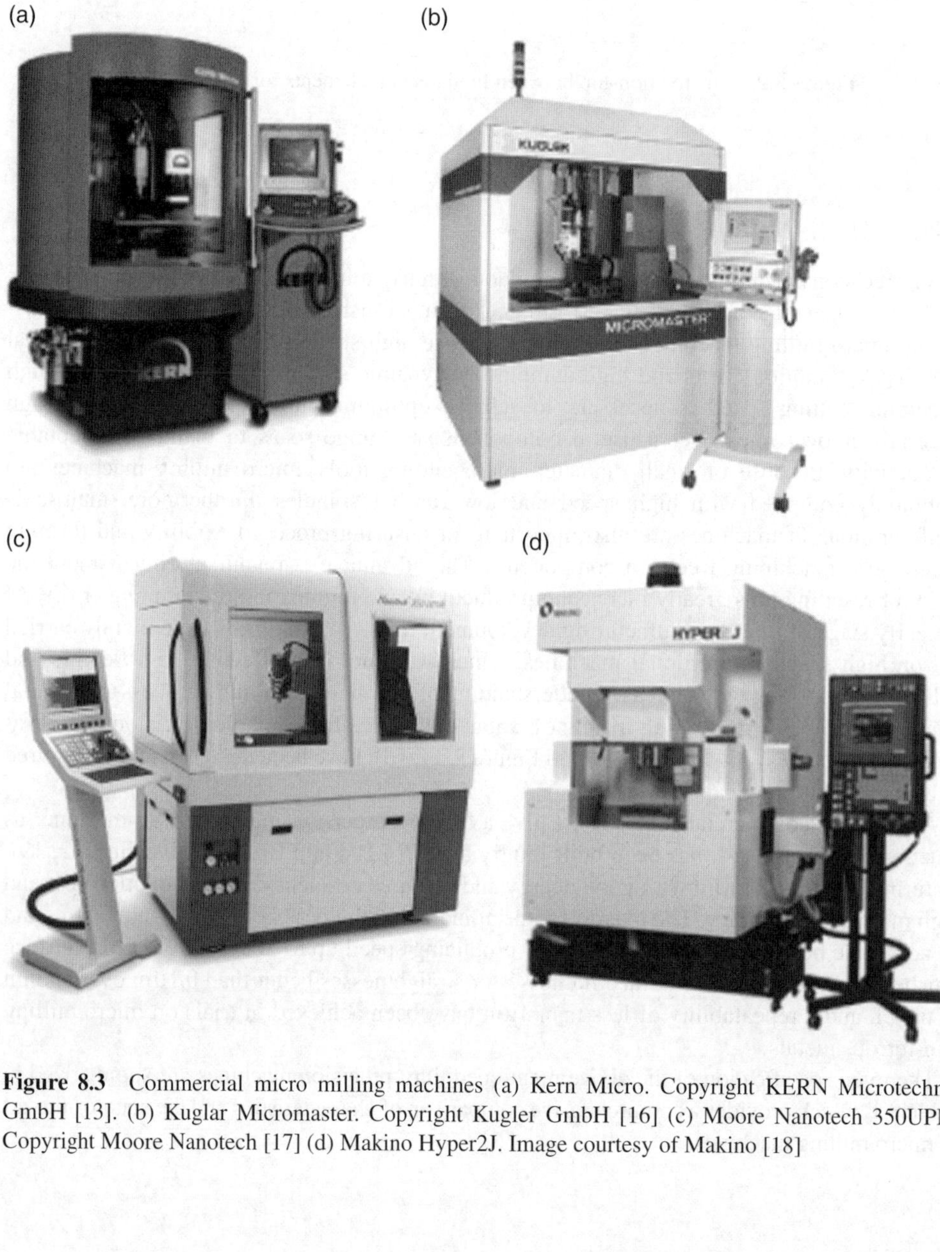

Figure 8.3 Commercial micro milling machines (a) Kern Micro. Copyright KERN Microtechnik GmbH [13]. (b) Kuglar Micromaster. Copyright Kugler GmbH [16] (c) Moore Nanotech 350UPM. Copyright Moore Nanotech [17] (d) Makino Hyper2J. Image courtesy of Makino [18]

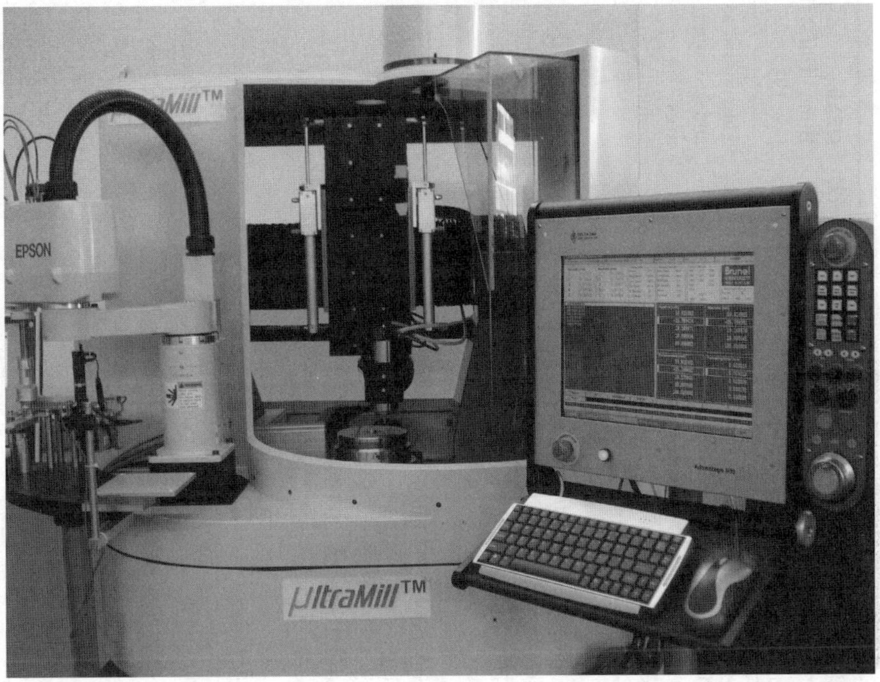

Figure 8.4 Five-axis ultra-precision micro milling machine UltraMill. Reproduced with permission from [20]. Copyright 2010 Sage Publications

8.2.2 Cutting Tools

8.2.2.1 Tool Materials

Tool materials together with tool geometry are the two most important factors determining effective tooling performance. The best tool materials have high hardness combined with adequate toughness and chemical stability at high temperature. Any given material does not generally have outstanding properties in all respects. The materials mainly used for micro milling tools are high speed steel, carbide, ceramics, cubic boron nitride (CBN) and diamond. A comparison of their material properties and applicable work materials are shown in Table 8.2 and Table 8.3 respectively.

Among these tool materials, diamond has some outstanding chemical, physical and mechanical properties, the high hardness enables sharper cutting edges to be achieved, a much lower rate of wear and longer tool life compared to other materials. However, high cost, brittleness and chemical affinity to iron have restricted its applications. For commercial micro tools, the most common and popular material is ultra-fine grain (normally grain size less than 0.5μm) cobalt-bonded tungsten carbide due to comparably low cost, high hardness, strength and adequate wear resistance characteristics. Of the most importance is that it is suitable for machining a wide variety of engineering materials including steels. The performance of tungsten carbide tools is largely dependent upon the composition and grain size. The content

Table 8.1 Main specification of precision and ultra-precision machines

	KERN Micro [13]	FANUC ROBO NANO α-0iB [14]	Sodick AZ250L [15]	Kuglar Micromaster [16]	Moore Nanotech 350UPM [17]	Makino Hyper2J [18]	UPM/ Brunel UltraMill [12, 19]
Configurations	Three linear axes and two rotary axes	Three linear axes and two rotary axes	Three linear axes	Three axes (standard) or five axes (optional)	Three axes (standard) or four axes (optional)	Three linear axes	Three linear axes and two rotary axes
Base structure	Polymer concrete	Cast iron base with concrete	Box-shape structure without overhang	Solid granite	Cast epoxy-granite base with carbon fibre reinforcements	Granite base	Natural granite
Vibration isolation	N/A	Air and oil damper	Counter-balance tables	Passive air dampers	Self levelling air isolation system	N/A	Squeeze film damper
Controller	Heidenhain	Fanuc	Sodick	Multiprocessor path control	Delta Tau	Fanuc	Delta tau
Spindles	Vector-controlled spindle	Static air bearing, air turbine	pneumatic static bearing, air turbine	hydrostatic bearing	Liquid cooled air bearing	N/A	Aerostatic bearing
Speed range	Up to 160,000 rpm	Up to 50,000rpm	Up to 120,000 rpm	Up to 200,000 rpm	200~60,000 rpm	3000~ 40,000 rpm	Up to 200,000 rpm
Motion accuracy	<0.1μm	<0.2μm	N/A	±0.5μm	<0.05μm	±0.3μm	<1μm
Drive system	AC Servo	Linear motors and built-in servo motors	AC coreless linear motor	Linear motor	Brushless DC linear motor	N/A	Brushless drive motors
Drive system resolution	N/A	1nm	3nm	N/A	0.034nm	2nm	5nm
Work volume	250×220 ×200mm	280×150 ×40mm	250×150 ×100mm	300×300 ×200mm	350×350 ×150mm	200×150 ×150 mm	150×150× 80mm

Table 8.2 Comparison of typical tool material properties

	Hot hardness	Toughness	Wear resistance	Thermal shock resistance	Cutting speed	Surface finish	Material costs
High speed steel	Low	High	Low	High	Low	Low	Low
Carbide	↓	↑	↓	↑	↓	↓	↓
Ceramics	↓	↑	↓	↑	↓	↓	↓
CBN	↓	↑	↓	↑	↓	↓	↓
Diamond	High	Low	High	Low	High	High	High

Table 8.3 Typical tool materials and applicable work materials

	Soft non-ferrous	Cast iron	Carbon and low alloy steels	Hardened steels	Nickel-based alloys	Titanium alloys
High speed steel	√	×	O	×	×	×
Carbide	√	√	√	O	√	O
Ceramics	×	√	√	O	√	×
CBN	×	O	×	√	O	O
Diamond	√	×	×	×	×	√

(√: Good, O: Suitable, ×: to be avoided)

of cobalt can affect the tool physical properties. A smaller amount of cobalt will result in higher hardness and brittleness of tools, generally, the finer grain size, the higher hardness, and vice versa.

The micro structure that gives a tool its required hardness or toughness may not give the best wear resistance. Theoretically, tungsten carbide tools can be employed in machining almost all engineering materials, but they suffer from severe wear when cutting hard materials at high speed. In order to prolong tool life and improve tooling performance, hardness, toughness and wear resistance of a tool material must be consistently maintained during machining operations. Coating techniques potentially offer a promising solution to meet this need. They also enable the possibility of dry machining due to a lower friction coefficient between tools and work materials. Coatings should be harder than the substrates in order to resist wear, be more inert to resist chemical wear and must adhere well to the substrates [25]. For coated tools, benefits will be to extend the cost effectiveness and efficiency of the production process in terms of improved surface quality, increased tool life span, reduction in machine downtime and reduction in risk of damage as a result of tool wear or breakages.

Aramacharoen *et al.* [26] reported that AlCrTiN coated tungsten carbide micro end mills provide better performance in machining hardened die steels in terms of reduced

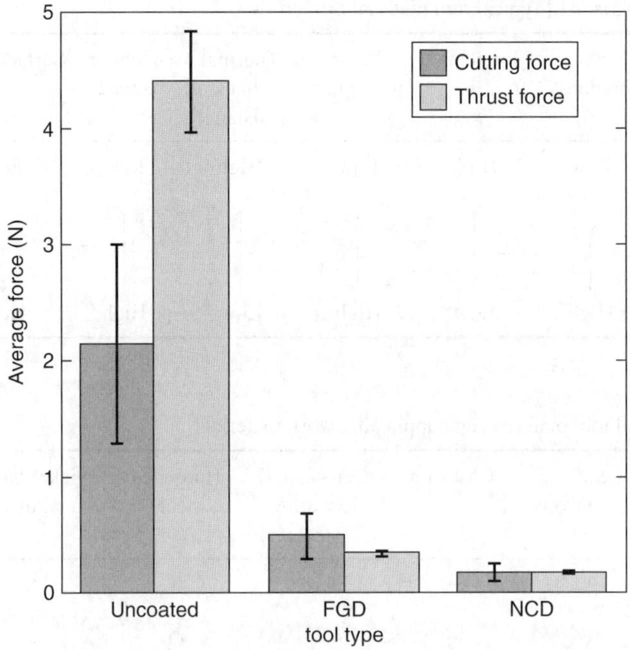

Figure 8.5 Cutting forces using FGD, NCD coated and uncoated micro end mills. Reproduced from [30]. Copyright 2009 Elsevier

tool wear and improved surface quality compared to uncoated ones. In their later research work [27], they compared the performance of TiN, TiCN, TiAlN, CrN and CrTiAlN coatings against uncoated tungsten carbide tools in micro milling of hardened tool steels and demonstrated that these physical vapour depositions (PVD) coatings help reducing cutting edge chipping and the edge radius wear as well as burr size. In addition, it was also reported that TiN coatings deliver the greatest benefits in reducing tool wear and improving machined surface quality.

The most common coatings for commercial micro tools are titanium-based and TiAlN in particular. Although carbide tools with diamond coating layers have recently come on to the market [28], they are not widely employed and are still under research. Diamond coatings for micro mills are advantageous and promising because they are capable of reducing flute clogging by eliminating adhesion of the workpiece material to the tool surface, reducing tool wear due to high hardness and reducing cutting forces due to low friction coefficients against many materials [29]. Therefore, there is strong interest in using diamond coatings to improve micro tool life.

Torres *et al.* [30] successfully coated fine-grained diamond (FGD) and nano-crystalline diamond (NCD) films onto tungsten carbide micro end mills using a hot-filament chemical vapour deposition (CVD) process. The grain size of FGD is between 0.5~1 μm, whereas for NCD is less than 0.1 μm. They studied the performance of both coated tools by dry slot milling of aluminium and reported that there was a dramatic improvement in tool integrity, lower wear rate, no observable adhesion and a significant reduction in the cutting forces compared to the performance of uncoated ones. Figure 8.5 shows a recorded force comparison using

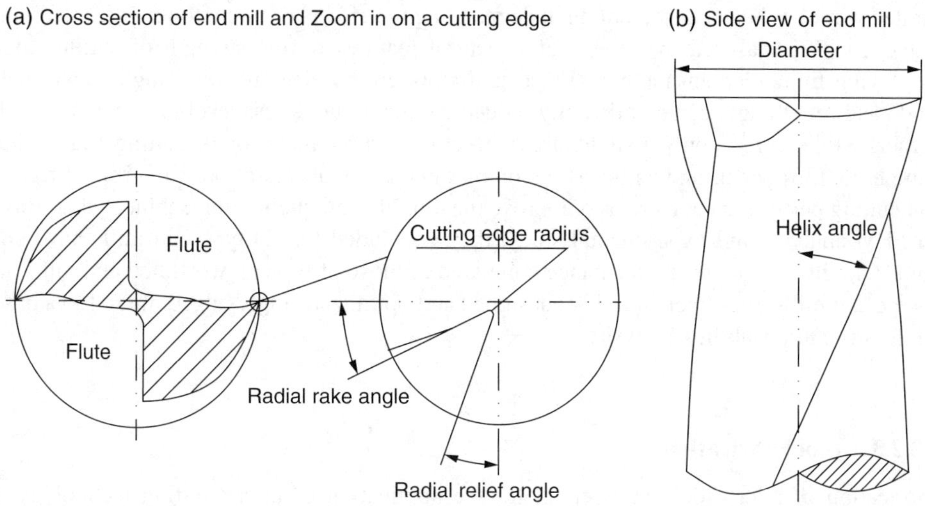

Figure 8.6 Schematic illustration of two-fluted micro end mill geometry

FGD, NCD coated and uncoated tools in micro milling aluminium. Heaney *et al.* [29] also noted similar advantages for FGD coatings. However, in their study, diamond coatings experienced severe delamination due to insufficient bonding strength to the substrate, resulting in catastrophic tool fracture of the cutting edges. The authors suggested further development is indispensable for improving coating adherence upon tungsten carbide so as to increase longevity of diamond coatings.

8.2.2.2 Tooling Shape and Geometry

The typical shape and geometry of micro tools are directly scaled down from those of conventional tools due to analogous cutting process. Micro tools are commonly seen in end-mill or ball-nosed tips. The main geometrical features for a two-fluted micro end mill are diameter, flutes, radial rake angle, radial relief angle, helix angle and cutting edge radius, as schematically illustrated in Figure 8.6.

At present, commercially available micro end mills with Ø25μm diameter are emerging and smaller sizes down to Ø5μm are also available by special order on [31]. With decreasing diameter, stiffness of tools tends to be weakening dramatically, which can easily cause tool deflection, unpredictable tool life and premature tool failures. Increasing radial rake angle and relief angle together with reducing cutting edge radius could lead to sharper but weaker tool tips, which imposes a major limitation on cutting edge strength. Generally, cutting edge radius for tungsten carbide micro tools are between 0.8~5μm due to restraints of the tool fabricating process. Selection of appropriate helix angle is crucial for both tool stiffness and chip evacuation, and 30° helix angle is commonly employed for micro tools design and configurations.

In micro tool design, optimum geometry is essential for maintaining good stiffness and strength, avoiding chip clogging and therefore improving tooling performance. So far, only limited work has been carried out on understanding the effects of micro tooling geometry and shape. Li [32] analyzed the effects of structural features on the strength of cutting edges employing finite element method (FEM) and reported that negative rake angle, lower helix angle and relief angle can significantly reduce the maximum stress level on the cutting edge corners while there is only a slight stress effect caused by alteration of cutting edge radius. Fang *et al.* [33] performed an investigation on three types of micro tools in terms of rigidity and cutting performance in order to improve the tool life and quality of machined parts. Based on FEM analysis and experimental study, they concluded that D-type end mills have both higher rigidity and better performance compared to those of typical two-fluted end mills and Δ-type end mills. Fleischer *et al.* [34] designed and optimized single-edged micro tools based on the structural stability analysis.

8.2.2.3 Tool Fabrication

Fabrication of micro tools is essential for the application of micro cutting technology. In general, grinding is a traditional technique for manufacturing both conventional and micro milling tools. The quality of the tools is largely dependent upon the properties of the grinding wheel, such as abrasive grain size and grinding edge radius. Grinding technology for micro tool fabrication is rarely reported in the literature except for diamond grinding attempted by Schaller *et al.* [35]. With the decrease in tool diameter to below 100μm, such technology becomes challenging due to a reduction in tool rigidity and the requirement for tighter tolerances and sharper cutting edges.

Focused ion beam (FIB), which is cutting force free, is capable of producing micro tools of high quality, but not on a commercial scale. Adams *et al.* [36] fabricated a number of high speed steel and tungsten carbide micro end mills around 25μm diameter and 2, 4, 5 cutting edges by FIB sputtering processes. The cutting edge radius of these tools was less than 0.1μm and Figure 8.7 shows these good quality tools in different shapes. Ali [37] also fabricated micro end mills of 20μm diameter with different aspect ratios by the same process and experiments on Polymethyl-methacrylate (PMMA) indicated surface finishes of 80nm Ra could be achieved. In addition, Friedrich and his colleagues [10, 38] developed some two-fluted and four-fluted micro end mills of 22μm~100μm diameter with a submicron edge radius using an FIB micro machining process. These tools exhibited good forming quality in micro milling high aspect ratio micro features on PMMA as previously shown in Figure 8.1f.

In addition to FIB, electro discharge machining (EDM) is another potential process for fabricating micro tools. Egashira *et al.* [41] developed micro ball end mills with a radius of 10μm using electrical discharge machining (EDM). Fleischer *et al.* [34] showed the feasibility of manufacturing tungsten carbide end mills with helical shapes in 50μm and 100μm diameters and with semi-circular shapes employing wire EDM technique. Cheng *et al.* [39] also fabricated polycrystalline diamond ball end mills of 300μm diameter based on the wire EDM process. End mills of 50μm diameter with non-circular cross-sectional profiles were produced utilizing a similar process by Yan *et al.* [40], and micro grooving tests on electroless nickel plating demonstrated good performance of these tools in terms of wear and feature forming quality. Some of the aforementioned EDM-fabricated micro tools are shown in Figure 8.8.

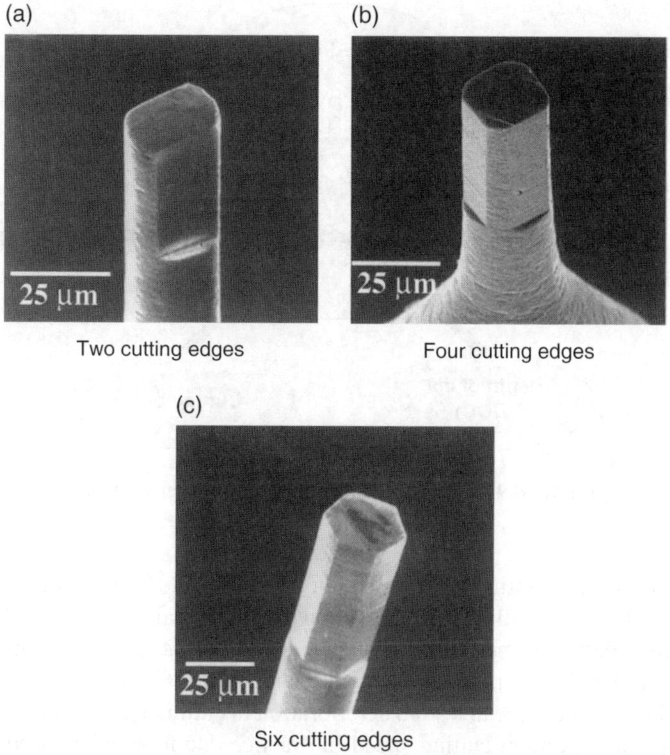

Figure 8.7 Micro end mills fabricated by FIB. Reproduced from [36]. Copyright 2001 Elsevier

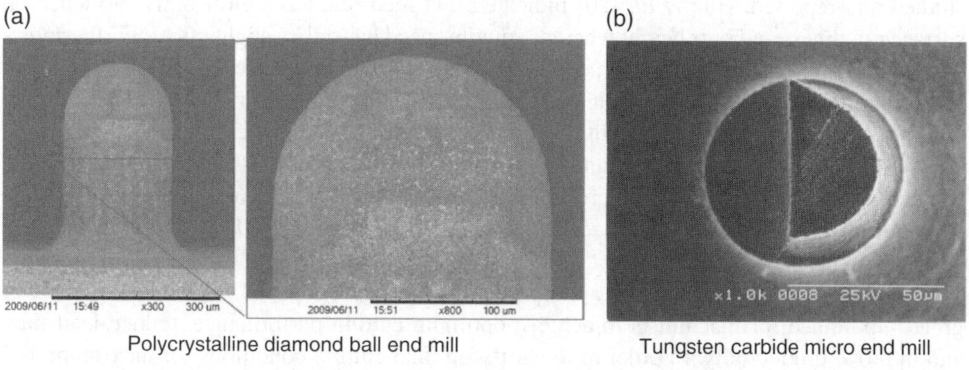

Figure 8.8 Micro tools fabricated by EDM (a) Reproduced with permission from [39]. Copyright 2009 IOP Publishing (b) Reproduced with permission from [40]. Copyright 2009 IOP Publishing

8.2.3 Process Conditions

In micro-scale milling, cutting performance is characterized by a combination of various factors, including part forming accuracy, surface quality, tool wear and life, chip formation, burr formation, cutting forces and power, and so on. To achieve the desired performance,

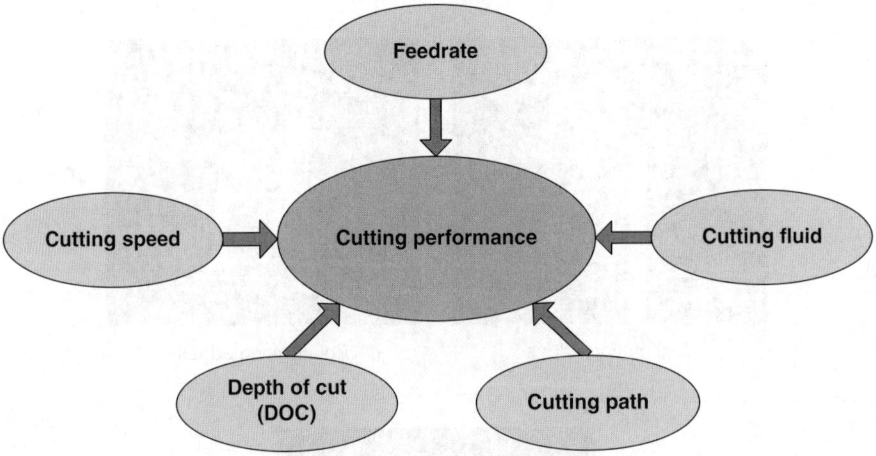

Figure 8.9 Process conditions for cutting performance

influential factors such as cutting speed, feed rate, depth of cut (DOC), cutting path as well as cutting fluid should be carefully selected, as shown schematically in Figure 8.9.

Cutting speed, feed rate and DOC It is well recognized that increasing cutting speed, feed rate and DOC can enhance productivity and in most cases, reducing cutting speed and increasing feed rate and DOC can lead to deterioration in cutting performance. It is not always possible to apply an optimum cutting speed in practice due to spindle speed limitations on available machines.

The effects of cutting speed, feed rate and DOC on micro milling have been experimentally studied and reported. Huo *et al.* [20] indicated that feed rate has significantly influence on surface roughness and burr height whereas spindle speed has only a slight effect. Furthermore, within the limits of his experiments, depth of cut has no significant effect on the surface finish in micro milling of OFHC copper. In micro milling tests on pure copper conducted by Rahman *et al.* [42], the larger the depth of cut, the higher the cutting force. As cutting speed increases, the life of the cutters decrease. Aramcharoen *et al.* [43] reported that increasing the spindle speed leads to smaller burrs in micro milling hardened tool steel. When micro milling steel using CBN tools, Fritz Klocke *et al.* [44] concluded decreasing feed rate tends to increase surface quality.

In process planning, selection of appropriate cutting speed, depth of cut and feed rate are of great importance for machinists to achieve optimum cutting performance, reduce lead time and improve productivity. In order to indentify optimal cutting conditions for maximum tool life and minimum production cost, an optimized generic algorithm was developed and demonstrated by Sreeram *et al.* [45].

Cutting path Cutting paths should be developed accurately and effectively in micro-scale milling. Compared to conventional tool paths, tool paths for micro milling not only require tighter tolerances for higher accuracy but also need to be optimized so as to reduce cutting force, maintain constant chip load and avoid premature tool breakages.

To produce good geometric accuracy and surface quality, micro milling operations require tool paths with higher accuracy than those used in macro-scale machining. Therefore, cutting tools, CAM systems and micro milling machines need to be well matched to produce tighter tolerances. This requires many more interpolated points on small tool paths positioned with greater accuracy than those in conventional machining.

In addition, machining strategies need to be optimized as different cutting paths have a significant impact on cutting performance. According to strategy study conducted by Dimov *et al.* [46], eight types of cutting paths in micro honeycomb structure milling can result in considerably different surface qualities. Litwinski *et al.* [47] reported that micro up milling generally leads to a better surface finish compared to that in down milling, and there is almost no significant difference between zig cutting and contour parallel cutting in terms of surface roughness.

Cutting fluid The use of cutting fluid is generally advantageous with regard to cutting performance and its two main functions are cooling and lubrication. Cutting fluid takes away heat from both the cutting tool and workpiece, facilitates flushing of the chips from the cutting zone and minimizes cutting energy consumption by reducing friction at the tool-work interface. If properly selected and applied, cutting fluid can make a substantial improvement in both tool life and workpiece surface quality.

Aramcharoen [43] concluded that the choice of machining environment is the most significant factor for improving surface finish and reducing tool wear in micro milling hardened tool steel, and the performance under minimum quantity lubrication (MQL) is better than for dry milling or the use of flood coolant. Also, Jun *et al.* [48] observed a cooler cutting zone, lower cutting forces, significantly longer tool life and smaller burr formation when cutting fluid was present in comparison with a corresponding performance under dry machining conditions.

8.2.4 Work Materials

Work materials are dependent on applications, that is, components and products. Different industrial sectors may require specific materials, such as stainless steel and titanium alloys for medical devices, aluminium alloys for aerospace, hardened steel for moulding and silicon for electronics. However, their machinablity, traditionally characterized by some combination of material removal rate, power consumption, chip form, surface finish and tool life varies widely due to their respective mechanical, physical and chemical attributes.

Figure 8.10 illustrates the machinability of common metals compared to that of stainless steel 416 in conventional machining. It is clear that aluminium alloys are the easiest metals to machine, followed by copper alloys, cast iron and steels. The equivalent comparison for micromachining is not established yet. In micro-scale cutting, due to significant size reduction, existing knowledge and experiences in macro-scale machinability of common materials cannot be assumed to be the same and certain influential factors, such as micro structures, defects, heat treatment and chemical reactivity with tool materials should also be considered for indentifying micro-scale machinability.

The machinable spectrum materials by micro milling include metal materials, such as aluminium and its alloys, copper and its alloys, cast iron, steels, nickel based alloys and titanium alloys [2, 3, 9, 36, 42, 50–57], and non-metal materials, such as polymers, silicon and ceramics [10, 58, 59]. For many non-metals, poor thermal conductivity, high brittleness or hardness properties make them difficult to produce components with stringent geometric accuracy and good surface quality as well as achieve acceptable tool life.

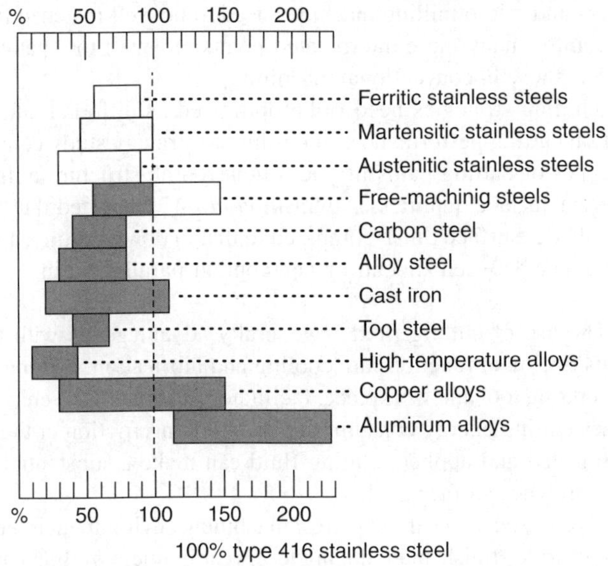

Figure 8.10 Comparative machinability of common metals in conventional machining (Adapted from [49])

8.3 Micro Milling Mechanics

8.3.1 Size Effect in Micro-Scale Cutting

As illustrated in Figure 8.11, in micro-scale cutting, when uncut chip thickness becomes comparable to the cutting edge radius of tools or grain size of work materials, a great number of critical issues, such as cutting edge radius effect, negative rake angle, elastic deformation of machined surface, minimum chip thickness and micro structure, become prominent. These issues may be categorized as size effect, which can influence underlying cutting mechanisms by altering cutting forces, the chip formation process, burr formation, vibration and process stability, energy consumption as well as generation of the machined surface.

In macro-scale cutting, uncut chip thickness is generally larger than the tool cutting edge radius, therefore, edge corners can be neglected and the tools assumed to be sharp. In addition, the effective rake angle is almost equal to the nominal tool rake angle. In this case, material removal is mainly achieved by shearing based on conventional cutting mechanism. When uncut chip thickness decreases to the same order as the cutting edge radius of tools, the tools cannot be considered sharp anymore and the effective rake angle becomes negative. Material removal is then dominated by compressing and ploughing rather than shearing. The schematic of tool-work interaction in macro-scale and micro-scale cutting is shown in Figure 8.12.

When uncut chip thickness reduces to a critical value for a specific cutting edge radius, chips will not be formed and this value is termed as minimum chip thickness (MCT). Once uncut chip thickness is smaller than the minimum chip thickness, not only does a

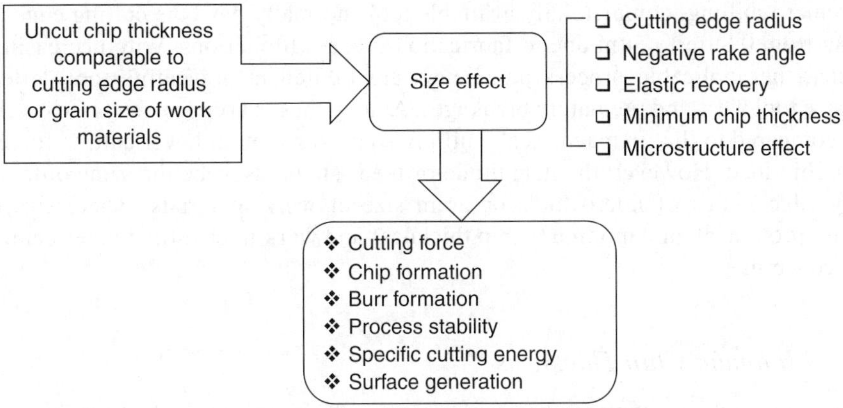

Figure 8.11 Size effect in micro-scale cutting

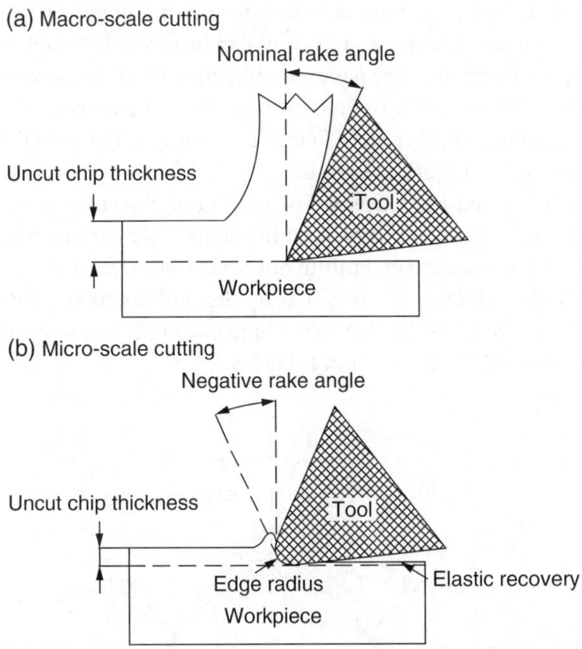

Figure 8.12 Schematic of tool-work interaction

larger negative rake angle appear, but there is no work material removed and elastic recovery resulting from ploughing has a major influence on the machined surface generation. This ploughing phenomenon significantly contributes to increased cutting forces [60] and burr formation [61], larger specific cutting energy [62–64] and poor surface quality [65].

For micro milling, commercially available tools normally have the cutting edge radius ranging from 0.8μm to 5μm due to fabrication process limitations. With decreasing tool diameter, the applicable process parameters are limited in preventing tool deflection, excessive tool wear and premature breakages. Among these variables, feed rate is selected lower compared to that in macro-scale milling so as to maintain lower cutting forces and lighter chip load. However, the magnitude of feed rate tends to be the same order of the cutting edge radius of micro tools or grain size of work materials, where size effect becomes prominent, and minimum chip thickness and work micro structure effect receive major concerns.

8.3.2 Minimum Chip Thickness

Due to an intermittent cutting mechanism for micro milling, uncut chip thickness in the radial direction for each cutting passage varies from zero to the feed per tooth then to zero. It is thus inevitable to encounter ploughing or rubbing because of the MCT effect. In cases where the feed per tooth is larger than MCT, ploughing and shearing mechanisms are present in one cutting passage, as shown in Figure 8.13. When the radial depth of cut is smaller than MCT, the ploughing mechanism becomes dominant, but once radial depth of cut is larger than MCT, shearing of work materials takes place and chips can be formed. Ploughing dominant areas are located at the entrance and exit of the passage while the rest of the zone is dominated by shearing. In cases where feed per tooth is smaller than MCT, there are probably no chips formed during several consecutive revolutions of cutting because of the ploughing dominant mechanism and elastic recovery of work materials.

Weule *et al.* [6] observed the existence of minimum chip thickness in the micro milling process and noticed its significant impact on the achievable surface formation. A saw-tooth-like profile was detected in micro end milling of SAE 1045 steel using around 5μm edge radii tungsten carbide tools, which was explained by the authors as the MCT effect depicted in Figure 8.14. Additionally, Liu *et al.* [66, 67] found cutting force and surface finish variations due to the influence of MCT, as shown in Figure 8.15.

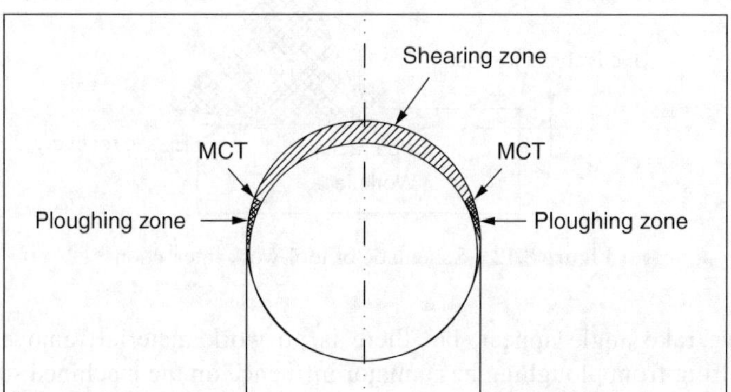

Figure 8.13 Illustration of ploughing and shearing mechanisms for one cutting passage

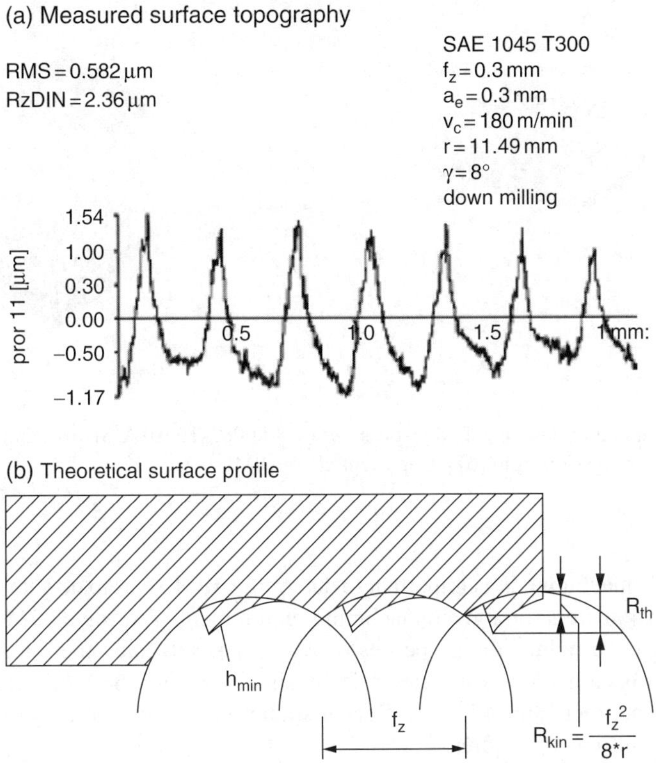

Figure 8.14 The effect of MCT on surface profile. Reproduced from [6]. Copyright 2001 Elsevier

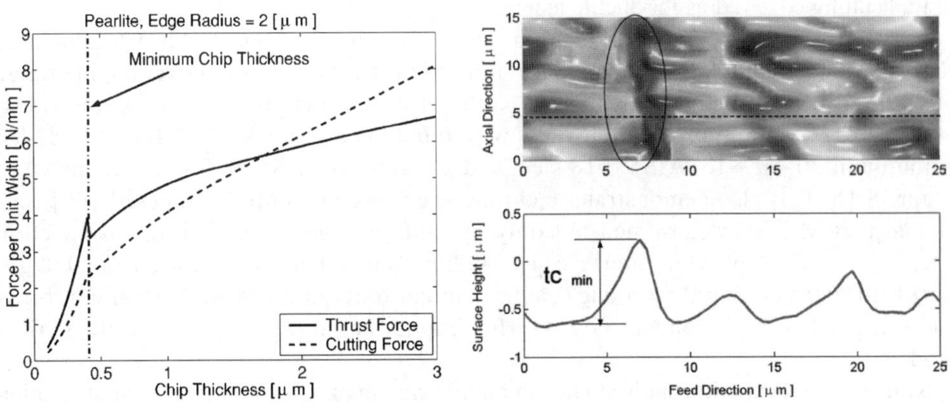

Figure 8.15 The influence of MCT on cutting force (Reproduced with permission from [66]. Copyright 2004 ASME) and surface generation (Reproduced with permission from [67]. Copyright 2006 ASME)

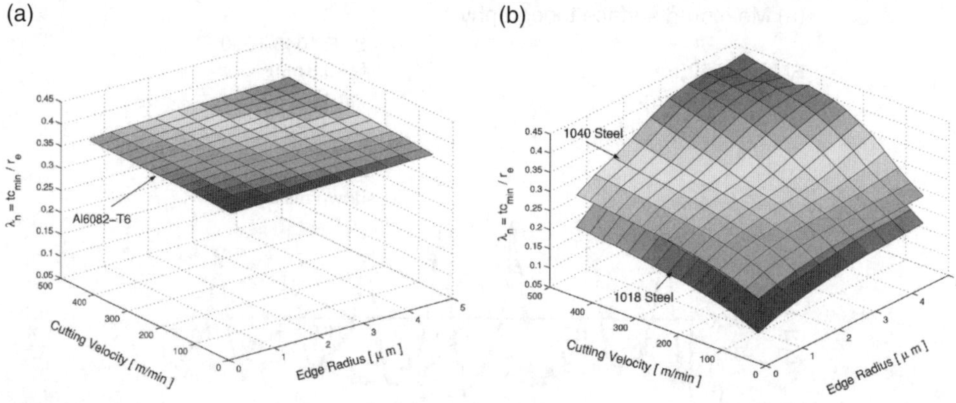

Figure 8.16 Nominal minimum chip thickness for: (a) Al 6082-T6 (b) AISI 1018 and AISI 1040 steel. Reproduced with permission from [67]. Copyright 2006 ASME

In micro-scale machining, it is important to determine MCT so as to be able to select appropriate process parameters for a given machining requirement. The value of MCT depends on the tool cutting edge radius, the properties of work materials and other cutting conditions. Some effort has been given to estimate MCT by simulation [65, 68], theoretical calculation [67, 69], and experimentation [70, 71]. Recent studies show MCT usually lies between 5% and 42% of the cutting edge radius.

Vogler *et al.* [65] studied MCT in micromachining of pearlite and ferrite materials based on a micro structure-level FEA model developed by Chuzoy *et al.* [72]. They pointed out that the ratios of MCT to cutting edge radius were theoretically 0.2 for pearlite and 0.35 for ferrite. Molecular dynamics simulation was adopted by Shimada *et al.* [68] to determine MCT in diamond cutting of aluminium and copper. They estimated that MCT could be as small as 1/20 of tool cutting edge radius for ductile metals.

Liu *et al.* [67] termed normalized minimum chip thickness (λn) as MCT to tool edge radius ratio and calculated the value for aluminium and carbon steel over a wide range of cutting speed and edge radius based on the proposed analytical model. In the study, the normalized minimum chip thickness was found to be about 35–40% for AL6082-T6 aluminium, 20–30% for AISI 1018 steel and 20–42% for AISI 1040 steel, as shown in Figure 8.16. This also demonstrates a cutting speed effect on MCT. Son *et al.* [69] concluded that MCT is determined not only by cutting edge radius but also by friction coefficient at the tool-work interface. The authors found that MCT ranged from 0.09μm to 0.12μm when using 0.5μm edge radii diamond tools in grooving aluminium, brass and oxygen-free high-conductivity (OFHC) copper based on the proposed theoretical model.

Kim *et al.* [70] experimentally studied minimum chip thickness by cutting force fluctuation analysis, and the ratio of uncut chip thickness to the cutting edge radius was estimated to be between 22–25% for brass. Yuan *et al.* [71] studied MCT by micro cutting of aluminium alloys with different edge radius tools, and concluded that tool sharpness has a considerable impact on MCT and surface integrity. The obtainable MCT was reported to be 0.05~0.2μm when using tools with an edge radius between 0.2~0.6μm.

8.3.3 Work Micro Structure Effect

In conventional cutting, work materials are considered to be homogeneous and isotropic, and chips are formed by shearing a bulk piece of material. In micro-scale cutting, however, uncut chip thickness becomes the same order as the average grain size of work materials, where cutting is performed by fracturing an individual grain. In this case, the micro structure of work material should be treated as non-homogenous and anisotropic, which imposes a significant impact on the cutting mechanism. Different material micro structures could lead to different machining responses. Researchers have found that altering the micro structure of the work material, such as its grain size and crystallographic orientation, can affect the cutting force [73], chip formation [74–76], surface roughness [73, 77] and burr formation [78] in micro-scale machining. Studies have been conducted to understand the effects of the work material properties on cutting force, part forming quality and surface integrity in micro milling.

Vogler *et al.* [3] studied the multiphase micro structure effects on the cutting force variation in micro end milling of both single-phase and multi-phase work materials. They observed the presence of higher frequency components of cutting force in the ductile iron experiments but not in the ferrite and pearlite machining. This was attributed to grain size and spacing of secondary phase in the material micro structure. In comparison with single-phase and multi-phase surface generation in [65], they observed micro-milled surfaces on multi-phase ductile iron are much rougher than those on single-phase ferrite and pearlitic workpieces, as shown in Figure 8.17. The cause of increased surface finish was explained as the result of phase boundary effects leading to a discontinuous chip formation process when micro milling multi-phase materials. Figure 8.18 illustrates corresponding chips in micro milling single phase ferrite and pearlite, multi-phase ferritic and pearlitic DI materials.

Schmidt *et al.* [79] investigated the influence of material state on the surface quality in micro milling steels and suggested the material should be heat treated in finest and equidistant micro structure prior to cutting so as to achieve constant cutting conditions, resulting in reasonable surface quality. Popov *et al.* [80] modified micro structures of Al 5083 workpieces by equal channel angular pressing (ECAP) to achieve more homogeneous material properties. Figure 8.19 shows micro structures of Al 5083 before processing, after conventional processing (CP) and ECAP processing. The size of ECAP processed grains is observed to be five times smaller than those of the CP material. They also assessed the influence of micro structures on part quality and surface integrity by micro milling rib features and reported that surface roughness can be greatly improved by reduction in grain size and material anisotropy of aluminium alloys. The similar material processing was performed by Pham *et al.* [81] on copper samples, and through experiments, it was found that surface roughness and surface defects reduced significantly as a result of smaller grain size and refinement in mechanical and metallurgical properties at boundaries between the individual grains.

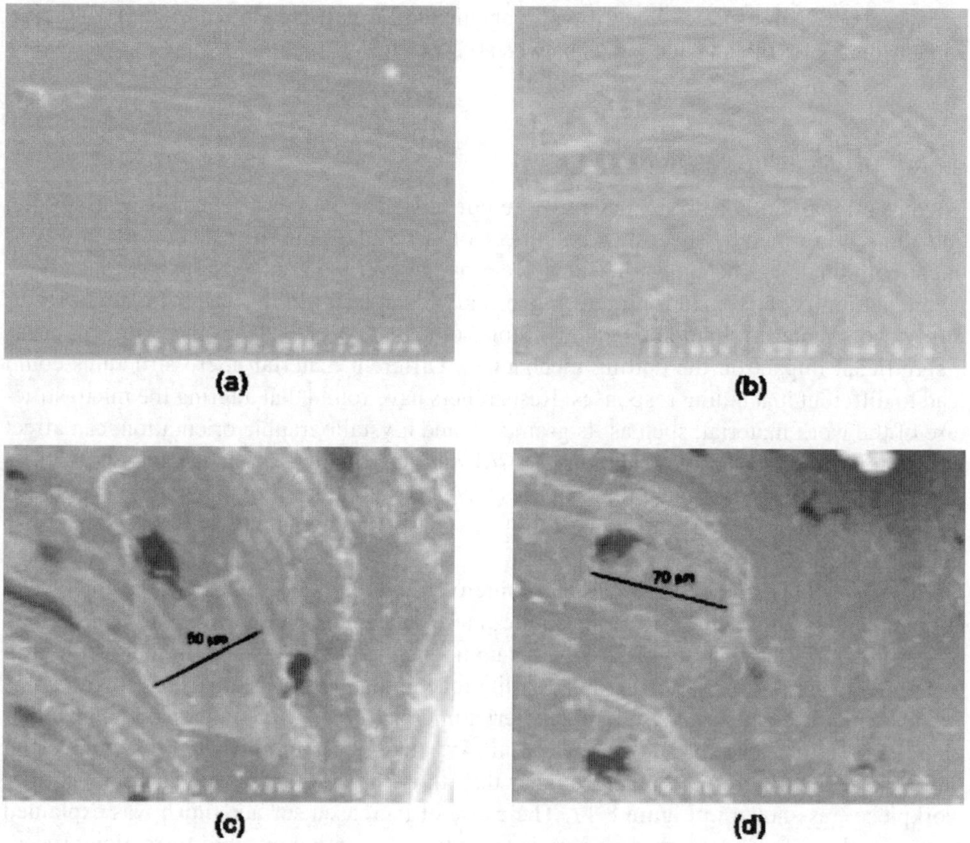

Figure 8.17 Micro-milled slot floor surfaces for (a) pearlite (b) ferrite (c) ferritic DI (d) pearlitic DI. Reproduced with permission from [65]. Copyright 2004 ASME

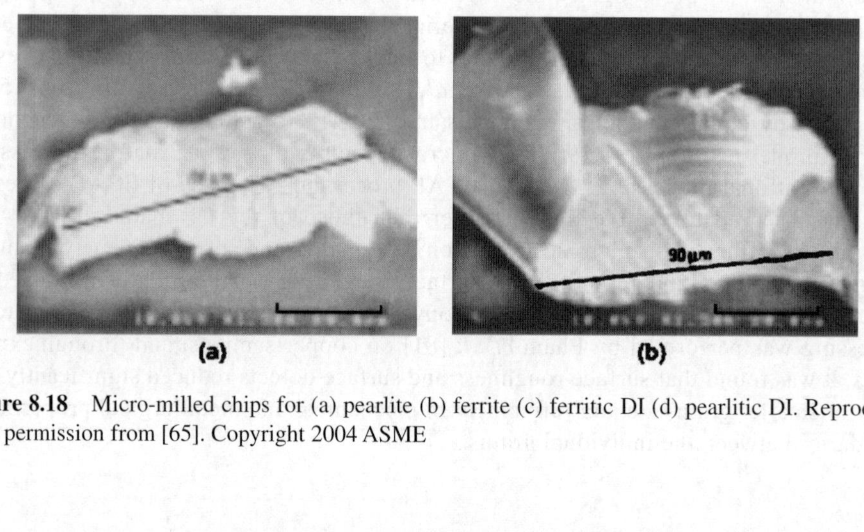

Figure 8.18 Micro-milled chips for (a) pearlite (b) ferrite (c) ferritic DI (d) pearlitic DI. Reproduced with permission from [65]. Copyright 2004 ASME

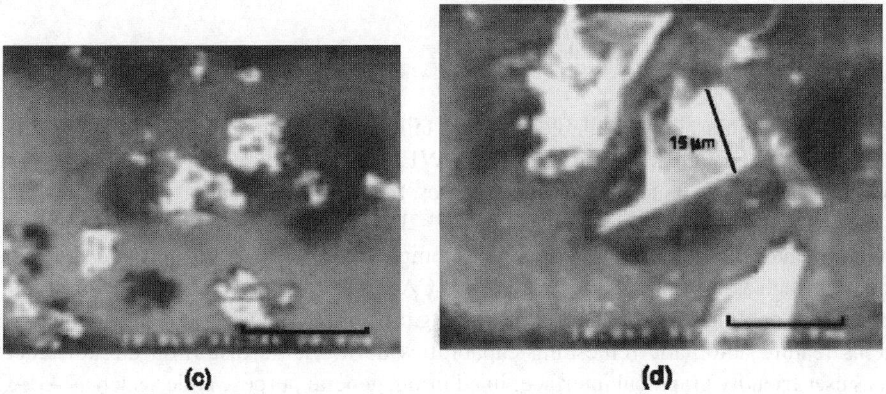

Figure 8.18 (Continued)

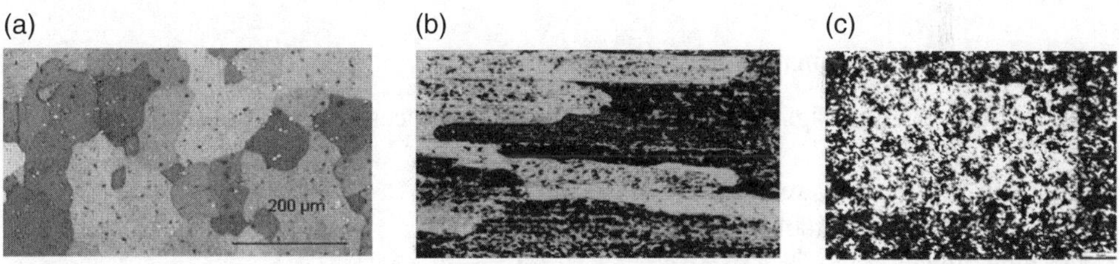

Figure 8.19 Micro structures of Al 5083: (a) before processing (b) after conventional processing (c) after ECAP processing
Reproduced with permission from [80]. Copyright 2006 Sage Publications

8.4 Modelling of the Micro Milling Process

In micro-cutting process characterization, molecular dynamic (MD) simulation, finite element method (FEM), multi-scale modelling and mechanistic process modelling are popularly used for assisting scientific understanding of the cutting physics, chip formation mechanisms, size effects and the machined surface integrity. The MD simulation is based on interatomic force analysis and it is more suitable for nanometric cutting. FEM is based on continuum mechanics principles, in which the materials are assumed as continuous and homogeneous micro structure. This powerful numerical technique has been mostly used for orthogonal cutting simulations. The newly emerging multi-scale modelling techniques combine the strengths of both MD and FEM for extended length scale simulations, and the mechanistic process modelling refers to developing an analytical model relating the process outputs to various input conditions, such as tooling geometries, work material properties and process conditions. Among the above approaches, FEM is an attractive computer-based means for understanding the mechanics of the micro milling process, and mechanistic modelling are commonly applied for micro cutting forces characterization.

8.4.1 Finite Element Modelling

8.4.1.1 FEM software

A number of commercially available tools are tailored for simulating metal cutting process, including SFTC/DEFORM™ and THIRD WAVE/AdvantEdge™. DEFORM™ is originally developed for analyzing metal forming processes, then extended to heat treatment, machining and mechanical joining processes. DEFORM™-2D is widely adopted for orthogonal cutting simulation and DEFORM™-3D is for more complex processes requiring three-dimensional modelling, such as drilling. AdvantEdge™ is a CAE software solution specialized for cutting simulations, enabling users to analyze machining processes in 2D and 3D environments. Both systems feature automatic remeshing capabilities during large distortion circumstances as well as user friendly graphical interface. In addition, general purpose codes, such as ABAQUS and ANSYS LS-DYNA, can also be used for cutting process modelling. However, they are not universally adopted due to a large amount of knowledge and time required for setting up models at pre-processor stage.

8.4.1.2 FEM Developments

The general steps for designing and analyzing a cutting process in FEM are:

- Selecting a FEM software.
- Creating or importing tool and work models.
- Meshing models.
- Applying material properties.
- Inputting cutting conditions.
- Defining boundary conditions and inter-object relationships.
- Running simulations.
- Acquiring simulation results in the post processor.

In order to speed up calculations without compromising the accuracy of simulation results, it is necessary to take appropriate simplifications or assumptions on tool and work models as well as object types for specific applications. Generally, three-dimensional analysis is more accurate but time-consuming than the two-dimensional one, and elastic-plastic to plastic objects as well.

Work material flow stress model and tool-work interface friction characteristics play an extremely important role in adequate and effective modelling of the cutting process. The flow stress, which is dominated by temperature, strain and strain rate, should be determined experimentally under a wide range of temperature and large deformation conditions. The main flow stress constitutive models used in FEM are summarized in Table 8.4, among which the Johnson-Cook model is frequently implemented and tabular data from Split-Hopkinson bar tests is also a promising one due to its ability to follow the true behaviour of a material.

The friction schemes at the tool-work interface are very complicated and difficult to characterize. The contact regions and the friction coefficients between the tool and the chip are affected by various factors, such as material properties, cutting conditions, tooling

Table 8.4 Summary of main flow stress constitutive models

Number	Material model	Mathematical representations	Material constants
1	Tabular format	$\bar{\sigma}=\bar{\sigma}(\bar{\varepsilon},\dot{\bar{\varepsilon}},\mathrm{T})$	N/A
2	Oxley [82]	$\bar{\sigma}=\sigma_1\left(\bar{\varepsilon}\right)^{\mathrm{n}}$	σ_1, n
3	Johnson and Cook [83]	$\bar{\sigma}=[\mathrm{A}+\mathrm{B}(\bar{\varepsilon})^{\mathrm{n}}]\left[1+\mathrm{C}\ln\left(\frac{\dot{\bar{\varepsilon}}}{\dot{\bar{\varepsilon}}_0}\right)\right]\left[1-\left(\frac{\mathrm{T}-\mathrm{T}_{\mathrm{room}}}{\mathrm{T}_{\mathrm{melt}}-\mathrm{T}_{\mathrm{room}}}\right)^{\mathrm{m}}\right]$	A, B, C, m, n
4	Shirakashi and Usui [84]	$\bar{\sigma}=\mathrm{A}\cdot(\bar{\varepsilon})^{\mathrm{n}}\cdot(\dot{\bar{\varepsilon}})^{\mathrm{m}}\cdot[-\lambda(\mathrm{T}-\mathrm{T}_0)]$	A, m, n, λ
5	Zerilli and Armstrong [85]	$\bar{\sigma}=\mathrm{C}_0+\mathrm{C}_1\cdot\exp\left[-\mathrm{C}_2\mathrm{T}+\mathrm{C}_3\mathrm{T}\ln\left(\dot{\bar{\varepsilon}}\right)\right]+\mathrm{C}_4(\bar{\varepsilon})^{1/2}\cdot\exp\left[-\mathrm{C}_5\mathrm{T}+\mathrm{C}_6\mathrm{T}\ln\left(\dot{\bar{\varepsilon}}\right)\right]$	C_0, C_1, C_2, C_3, C_4, C_5

geometries, and so on. Whether sliding, shearing or combination of two behaviours dominates machining needs further investigation for specific applications. Özel [86] analyzed the effects of friction models on the cutting forces, tooling stresses and too-chip temperatures using orthogonal FE simulations and found that experimental measurement based variable friction models are more suitable for prediction in terms of accuracy. The friction models for FE programmes include constant sliding, constant shearing, variable shearing, variable sliding, combination of constant shearing and sliding, and combination of variable shearing and sliding, among which pure sliding is considered as appropriate contact scheme at low cutting speed [86, 87]. Although spindle speed is high in micro milling, cutting speed is relatively low compared to conventional cutting due to small tool diameter. Therefore, pure sliding pattern is assumed reasonable and applicable, and the friction coefficients can be derived from pin-on-disk tests.

8.4.1.3 FEM for Micro Milling

Dhanorker and Özel [50, 88] developed a rigid-plastic plane strain model to predict cutting forces, temperature distributions and chip formation in the meso/micro scale milling process. The authors found that predicted temperatures are much lower whereas the specific cutting forces are larger when compared to those in conventional milling. Wang *et al.* [89] studied micro chip formation mechanism in micro milling Al6061-T6 alloy adopting similar approach using DEFORM™-2D. The simulation was experimental validated by comparisons of

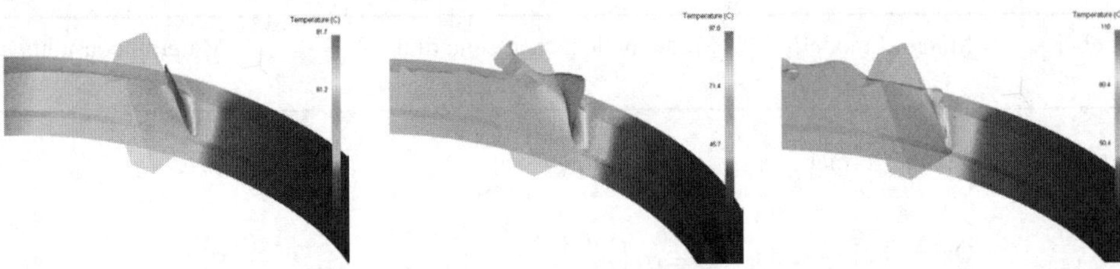

Figure 8.20 Predicted chip formation at different rotation angles

predicted cutting forces and chip formation with measured ones. Chip formation calculations and analyses were undertaken based on the proposed FE model.

Afazov *et al.* [90] performed a number of orthogonal cutting simulations for investigating micro milling forces in an indirect manner. In the study, cutting forces were determined based on a combination of simulated results from orthogonal cutting and uncut chip thickness mathematical model. The predicted forces are shown in good agreement with measured micro milling forces under different cutting conditions.

A 3D rigid-plastic finite element model was introduced by Wu *et al.* [91] for cutting forces, tool temperature and chip formation prediction in micro end milling. The rake angle, clearance angle, cutting edge radius, diameter and helix angle were taken into account for the tool modelling, and the tool and workpiece models were created by Pro/Engineer then exported to the FE code DEDORM™-3D. Predicted chip formation at different rotation angles are shown in Figure 8.20. Cutting forces and chip formation from simulations reasonably match experimental results from dry slots milling of AISI 1045 steel, and the proposed model was realized to be capable of scientific understanding and characterizing micro milling process with acceptable accuracy.

8.4.2 Mechanistic Modelling

8.4.2.1 Cutting Forces

Bao and Tansel [92] developed an analytical model for micro milling forces prediction based on Tlusty and Macneil's conventional model [93]. Process conditions, tool geometry and work material were incorporated into the model, in which the chip thickness was calculated by considering the trajectory of cutting tool tip. Compared to experimental forces data, the model was reported to have an average deviation of around 10%. In addition, the authors compared the differences of the proposed model with the conventional model, and concluded that the former estimates the cutting forces more accurately with an aggressive ratio of feed per tooth to tool radius, and the latter is only a special case of the former when the ratio becomes zero. Later, the authors modified their modelling work by taking the effects of tool run-out [94] and tool wear [95] into account. Based on the same conventional force model, Kang *et al.* [96] presented a cutting forces model of micro end milling considering the tool edge radius effects.

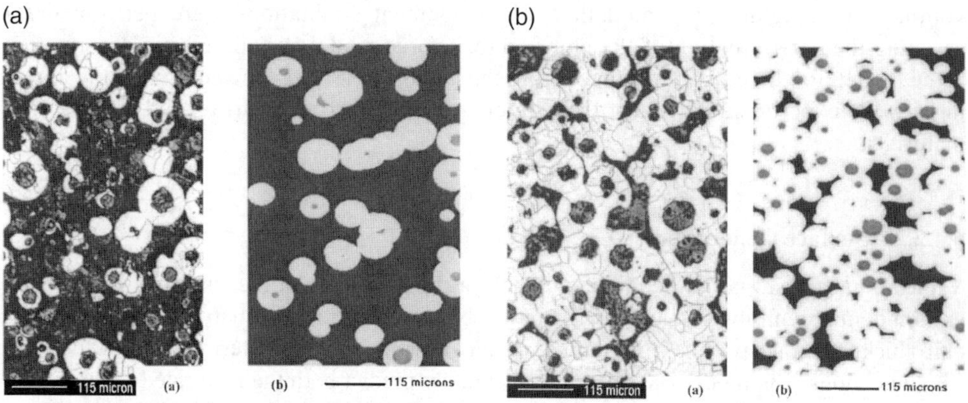

Actual and simulated pearlitic ductile iron

Actual and simulated ferritic ductile iron

Figure 8.21 Comparisons of actual and simulated micro structure of ductile iron

Zaman *et al.* [97] introduced a first three-dimensional cutting force model for micro end milling operations. The theoretical chip area with the variation at tool rotation angle rather than uncut chip thickness was integrated and the predicted maximum cutting forces values are found well matched to the experimental results. Li *et al.* [98] developed a three-dimensional force model considering the combination of tool tip trochoidal trajectory, tool run-out and minimum chip thickness effects, and instantaneous uncut chip thickness algorithms were used for cutting forces calculations.

Jun *et al.* [99, 100] developed a dynamic cutting force and vibration model considering both chip removal and ploughing cutting mechanisms in the micro milling process. Influences of the stable built-up edge and the effects of minimum chip thickness, elastic recovery and the elastic-plastic nature were taken into account in the chip thickness model. In the study, the authors noticed uneven variations for thrust force and attributed to the minimum chip thickness effect. In addition, the rate of increase for the cutting forces was found to be much higher when the chip thickness is smaller than the minimum chip thickness compared to that when the chip thickness is larger than the minimum chip thickness, as clearly shown in Figure 8.15.

Homogeneous micro structures of work material were assumed in the above models. In an effort to better understand cutting forces in micro milling heterogeneous materials, Vogler *et al.* [3] developed a mechanistic model by using micro-structural mapping technique. A mapping of multiple phases was created to represent the actual work micro structure. Comparisons of actual and simulated micro structure of pearlitic and ferritic ductile iron are illustrated in Figure 8.21. This model was capable of predicting higher frequency variation of cutting forces when micro-milling ductile iron. Through simulations, the authors attributed the frequency of the force variation to the spacing of the secondary phase, and the magnitude of this variation to the size of the secondary phase grains.

In their later work, Vogler *et al.* [60] enhanced the above force model by incorporating cutting edge radius effects. The minimum chip thickness resulting in two separate force generation mechanisms including slip-line plasticity model for shearing and interference volume for ploughing was modelled. Finite element simulations were performed for assisting the determination of the chip removal force model and the elastic deformation force model. Predicted cutting forces from the enhanced model showed favourable agreement with experimental results from micro milling ferrite, pearlite and ductile iron materials.

8.4.2.2 Surface Generation

Vogler *et al.* [65] introduced a process model for the prediction of the slot floor surface generation in micro milling single-phase materials based on the tool geometry and the minimum chip thickness concept. The minimum chip thickness ratios were determined by orthogonal FE simulations. The model was shown to accurately predict both the feed rate trends and the magnitude of the surface roughness for ferrite and pearlite materials, and indicated that there exists an optimum feed rate for the smallest surface roughness due to trade-off between conventional feed effect and minimum chip thickness effect.

X. Liu *et al.* [101, 102] developed surface generation models being capable of accounting for deterministic and stochastic surface roughness components for both sidewall and floor surfaces in micro end milling. The effects of the process kinematics, dynamics, minimum chip thickness and elastic recovery, and cutting edge serration as well as ploughing due to the significant tool edge radius effect were considered for the side wall surface model. For the floor surface model, the effects of the minimum chip thickness, elastic recovery and the transverse vibrations were taken into account and the extent of ploughing at end cutting edge as well. Examples of the predicted deterministic surface topographies for side walls in upmilling are shown in Figure 8. 22. The side wall and floor models were shown to predict surface roughness within 10% and 15% respectively over a wide range of process conditions.

8.4.2.3 Chip Formation

X. Liu *et al.* [67] developed an analytical model to predict the minimum chip thickness values based on molecular-mechanical theory of friction. The workpiece and the tool thermo-mechanical properties were incorporated into the model, which can account for the influence of cutting velocity and tool cutting edge radius on the minimum chip thickness. The authors found that the higher the carbon content, the larger the normalized minimum chip thickness for carbon steels, whereas for Al6082-T6, the normalized minimum chip thickness almost remains constant due to both the thermal softening effect and strain hardening effect.

Kim *et al.* [70] developed a static model of chip formation in micro milling, which is capable of accounting for the coupled minimum chip thickness and edge radius effects. The model was validated by verifying the level of periodicity present in the cutting forces at various feed per tooth. It was realized that the tool may rotate without removing any materials when the feed per tooth is smaller than the minimum chip thickness and that the periodicity of cutting forces is a function of the minimum chip thickness, feed per tooth and position angle. The authors suggested the appropriate feed per tooth for a given tool diameter can be determined by an upper and lower limit, as shown in Figure 8.23.

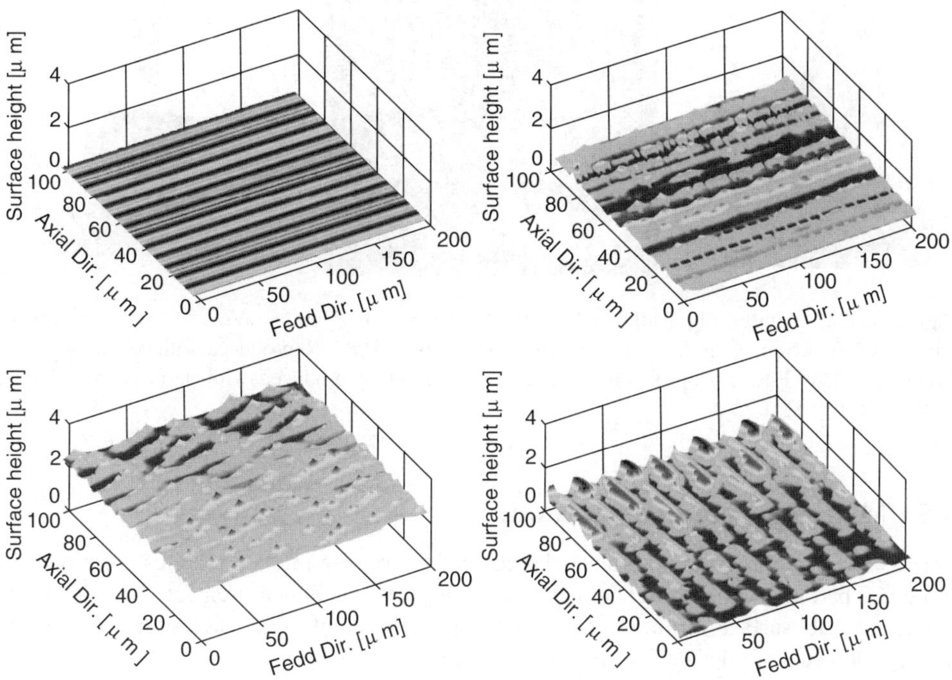

Figure 8.22 Predicted 3D surface topographies for side walls under different process conditions

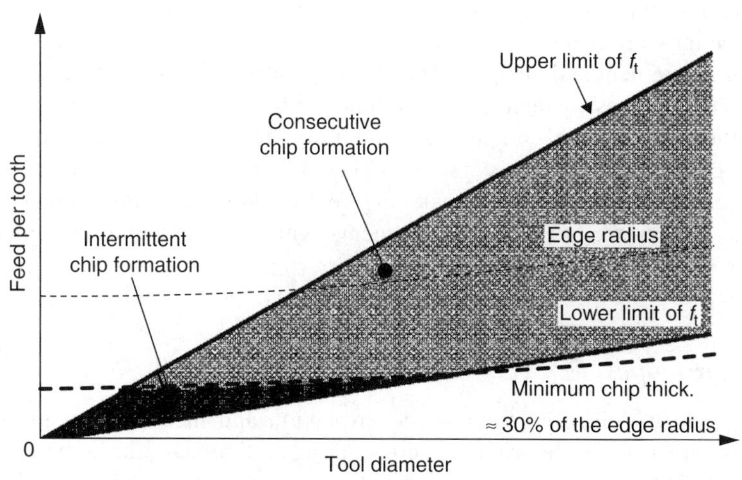

Figure 8.23 Feed per tooth and the minimum chip thickness

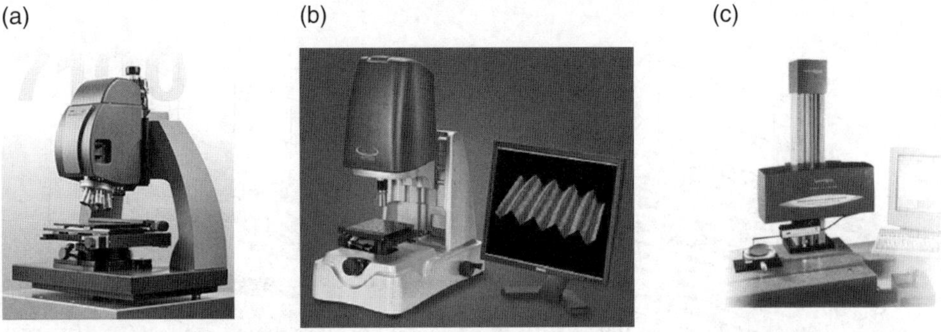

Figure 8.24 Commercially available 3D surface profilers: (a) Zygo NewView™ 7100. Reproduced with permission from Zygo Corporation [103] (b) Wyko NT9080. Reproduced with permission from Bruker [104] (c) Taylor Hobson Form Talysurf PGI 1240. Reproduced with permission of Taylor Hobson Ltd. [105]

8.5 Metrology and Instrumentation

The characterization of micro tools and machined micro or miniature features constitutes an important part in micro milling. For performing inspection and measurement at the micro and nano scale, 3D surface profilers, microscopes, process monitoring sensors and systems are widely adopted, the relevant features of which are given below.

8.5.1 3D Surface Profilers

A 3D surface profiler is an instrument for making quantitative measurements of surface texture. The characterization of a surface typically includes surface roughness and surface shape measurements. In most cases, surface profilers can be used for 3D surface roughness measurement and analysis. Surface roughness reflects characteristic marks left by the machining process, which is a direct indicator for the quality of a machined surface, and it can affect a mechanical component's suitability for its function. Therefore, measuring surface topography and texture accurately in 3D is a crucial concern for industrial applications, particularly at micro and nano scale. Figure 8.24 shows the latest typical non-contact and contact 3D surface metrology instruments which are essential for micro milling applications.

8.5.2 Microscopes

Microscopes are widely used for off-line inspection and measurement in engineering applications. Basically, electronic microscopes, particularly the scanning electron microscope (SEM), are commonly applied in micro-scale milling to observe chip formation, the state of micro tools and quality of micro features, as illustrated in Figure 8.1, Figures 8.7–8.8 and Figures 8.17–8.19. SEM is an instrument that delivers

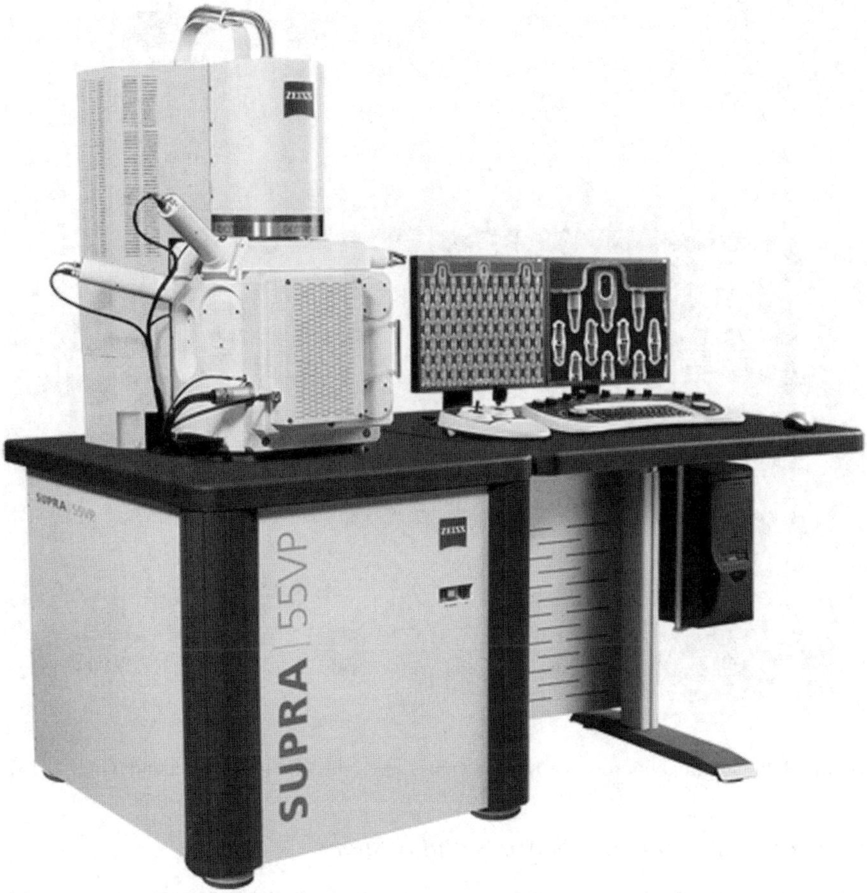

Figure 8.25 Scanning electron microscope ZEISS SUPRA 55VP Image courtesy Carl Zeiss Microscopy [106]

high resolution images by using a focused beam of high-energy electrons to generate a variety of signals at the surface of samples. These electron-sample interactive signals contain information including surface topography, chemical composition, crystalline structure and orientation of materials. Figure 8.25 shows a typical scanning electron microscope ZEISS SUPRA 55VP, which has a resolution of 0.8nm and magnification of up to 1,000,000 times.

In some cases, high magnifications microscopes are not always necessary and a low magnification optical microscope is sufficient for laboratories and shop floor applications. Generally, an optical microscope is equipped with several objectives to provide different magnifications, which can reach as high as 1000 times. Figure 8.26 illustrates a specialized optical microscope KERN μ-view with 450 times magnification for micro tooling checks. In addition to off-line inspection and measurement microscopes, some modern microscopes are equipped with CCD cameras, by which the machining process can be observed in real time.

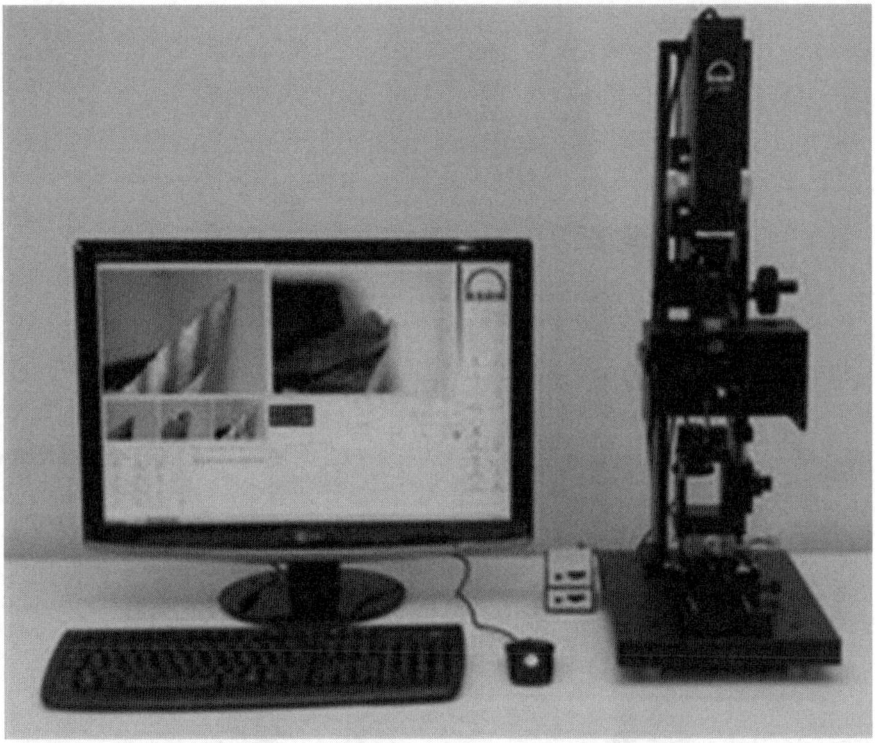

Figure 8.26 Optical microscope Kern μ-view Copyright KERN Microtechnik GmbH [13]

8.5.3 Process Monitoring Sensors and Systems

Due to fragility of micro tools, excessive cutting forces and vibrations can significantly affect the overall quality of machined parts and longevity of tools. Therefore, on-line monitoring of operations becomes indispensable for evaluating, controlling and improving production processes so as to meet increasing precision and quality demands. Traditionally, two main monitoring approaches can be employed. One is direct monitoring, such as the aforementioned optical microscope with CCD cameras, by which valuable cutting information can be observed or recorded directly. The other is an indirect method in which sensors like force sensors, acoustic emission (AE) sensors and accelerometers gather specific data correlated to cutting operations.

Among various indirect sensors, the force dynamometer is the most common and useful tool. The forces generated in machining operations, reflecting interaction between cutting tools and work materials, have a direct impact on cutting performance such as tool wear or failure, quality of machined surface and accuracy of components. Therefore, it is necessary to characterize these forces quantitatively and precisely so as to examine, evaluate and optimize cutting processes. A force dynamometer is capable of performing the function accurately both in magnitude and direction by creating electrical signals proportional to the forces in three orthogonal directions. Ideally, a dynamometer should be developed and configured with high sensitivity, high rigidity, high natural frequency response, high linearity, low drift and low measuring threshold in order to achieve better static and dynamic characteristics.

Figure 8.27 Kistler MiniDyn 9256C1. Image courtesy of Kistler [107]

The latest and most popular commercial miniature force dynamometers are Kistler MiniDyn series [107]. Figure 8.27 shows one of the multi-component dynamometers. There are four 3-component piezoelectric force sensors mounted under high preload between the cover plate and the two lateral base plates inside the dynamometer. Its high sensitivity and rigidity as well as a 2mN threshold make it applicable to measuring extremely small forces for ultra-precision applications. The instrument has been extensively used by machining researchers and technologists for micro forces characterization and process optimization [59, 108–111].

Each sensor has its limitations. For example, it is difficult to capture and characterize the real-time state of tools by direct monitoring due to visually undetectable tool-work interface in micro-scale cutting. Some indirect sensors may not have the necessary signal-to-noise (S/N) ratio and sensitivity required to adequately and reliably characterize surface finish, subsurface damage, and cutting forces at the ultra-precision scale due to extremely low cutting forces and low power consumption present [112]. Figure 8.28 shows main application fields, level of precision and control parameters for some typical indirect sensors. To overcome their respective limits, sensor fusion is regarded as an effective, flexible and reliable approach as the fusion and filtering of various process monitoring signals provide solutions for micro-control to enhance robustness and reconcile the limited frequency bandwidth of currently available sensors [1].

Malekian *et al.* [114] presented an integrated monitoring method consisting of AE sensor, force sensor and accelerometer to examine tool condition during micro milling operations, as shown in Figure 8.29. In the study, a vision system with CCD camera was utilized to observe actual tool state and measure edge corners of micro tools. Different bandwidth signals derived from various sensors were fused together through an adaptive neuro-fuzzy algorithm for tool wear determination. It was demonstrated that the proposed monitoring approach can predict tool wear effectively and reliably.

Gandarias *et al.* [115] proposed three indirect methods including reflective single-beam laser system (RSBLS), tool-workpiece voltage monitoring system (TWVMS) and off-line

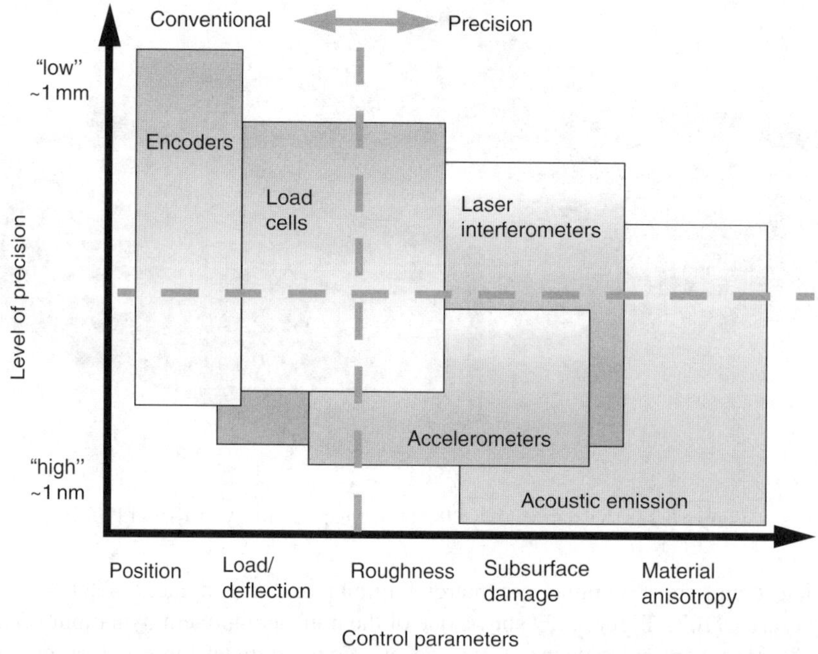

Figure 8.28 Sensor application vs. level of precision and control parameters. Reproduced from [113]. Copyright 2006 Elsevier

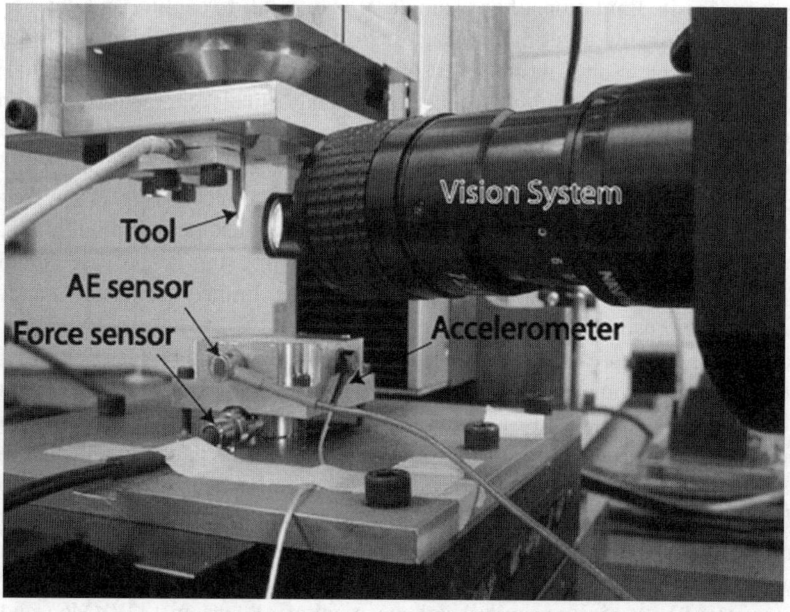

Figure 8.29 Setup of tool monitoring system. Reproduced from [114]. Copyright 2009 Elsevier

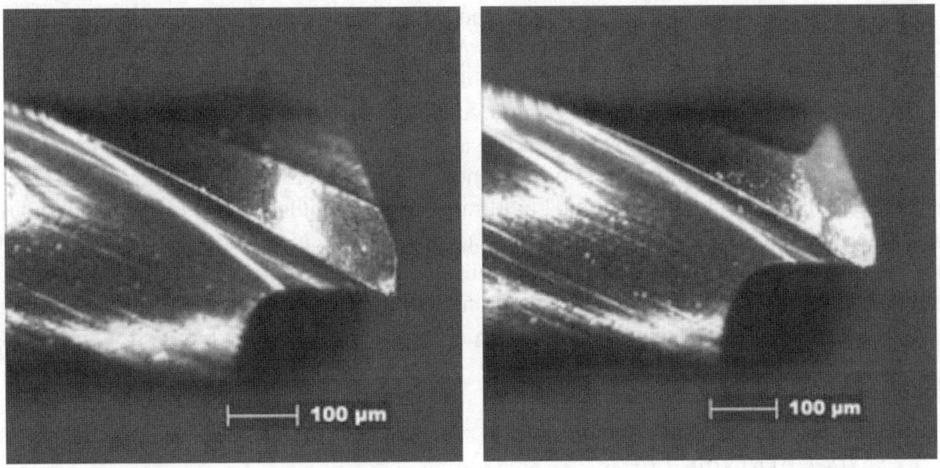

(a) Wear on the 1st cutting edge (b) Wear on the 2nd cutting edge

Figure 8.30 Uneven cutting edge wear caused by tool run-out (Reproduced from [32])

laser system (OLS) for micro tool failure detection. Due to the fact that each system has its limitations, such as precise laser focusing or work material property requirements, the authors suggested an integrated system combining the advantages of RSBLS and TWVMS should be promising for further development of a tool monitoring system.

8.6 Scientific and Technological Challenges

Micro milling represents an emerging fabricating technology with a promising future for the mass manufacturing of miniature and micro components and products. Nevertheless, this fast-growing field presents numerous challenges if practical application is to be increased, Examples are tool run-out, tool wear and life, micro-burr formation and process optimization. These critical issues constitute main limitations for micro milling technology reaching its full industrial potential.

8.6.1 Tool Run-out

Tool run-out is characterized as misalignment and/or eccentricity between the axis of rotation of the spindle and the axis of symmetry of the tool. A small run-out may cause negligible effect in macro-scale milling, while in micro-scale milling, it can result in unsTableperiodic variations in chip load and forces. As a result of the tool run-out, it is common that one cutting edge performs machining while the other does not, leading to uneven wear on tool cutting edges as an example shown in Figure 8.30. It is clearly seen that one cutting edge wears more than another one for a two-fluted cutter. In this case, cutters become blunt quickly, leading to significant force variations, chatter and rougher surface profile as well as increasing the possibility of tool failures.

The causes of tool run-out include asymmetric tool geometry, imperfect tool alignment in tool holder, mismatch between the tool holder and machining spindle, mass imbalance, eccentricity of spindle bearings, and vibration of tools during machining, and so on [116]. Run-out of micro tool tip is a combination of error sources and it is inevitable. Controlling and reducing tool run-out is critical for improving tool life and machining accuracy. In order to minimize dynamic tool run-out in micro-scale milling, not only precision tool, tool holder and spindle but also their precision fit are required. For ultra-precision spindles, run-out inaccuracy should be within 1μm so as to ensure high geometric accuracy of machined parts. According to existing experience and instruments, dynamic tool run-out is extremely difficult to determine or predict due mostly to the very small dimensions of the micro tools. Consequently, characterizing, controlling and reducing micro tool run-out remains a major barrier for improving micro cutting performance.

8.6.2 Tool Wear and Life

In conventional milling, the tool wears gradually and tool life is assessed based on machining precision and quality. For micro tools in normal condition, their low rigidity leads to rapid wear, causing large force variations and deteriorating geometric accuracy and part quality. Figure 8.31 shows cutting forces and burr formation variations using a new and worn tool in micro milling steel. When this phenomenon continues for an extended period of time, the increasing forces and stresses can lead to fatigue-related fracture of micro tools.

While in extreme conditions, premature tool failures take place. There might be two reasons associated with premature tool breakages: One is because process parameters are not properly selected, leading to cutting forces beyond the strength of the tool in the first cut. The other is due to chip clogging in the tool flutes, resulting in higher temperature, greater cutting forces and stresses on the tool tip.

Occasionally, partial damage of cutting edge, machining vibrations due to impurities of work materials and tool rotational deviations also give rise to the probability of tool breakages. As a result, unpredictable tool life and premature failures are major problems for the application of micro milling technology and prolongation of tool life presents a key topic for researchers and practitioners.

8.6.3 Micro-Burr Formation

For micro mechanical machining, burr formation is inevitable. The presence of burrs not only reduces part accuracy and quality but also affects its assembly and functionality. In conventional machining, deburring is routinely performed to meet specified tolerance and accuracy requirements. However, it cannot always be applied to micro parts because of tiny part features as well as high surface quality requirements. Furthermore, burr removal is often a major obstacle in terms of time, effort or automation. The performance of micro products cannot be maintained as required due to alteration of inherent material properties and functionality across the deburring zone. Consequently, micro-burr formation should be controlled and minimized as far as possible during the machining process.

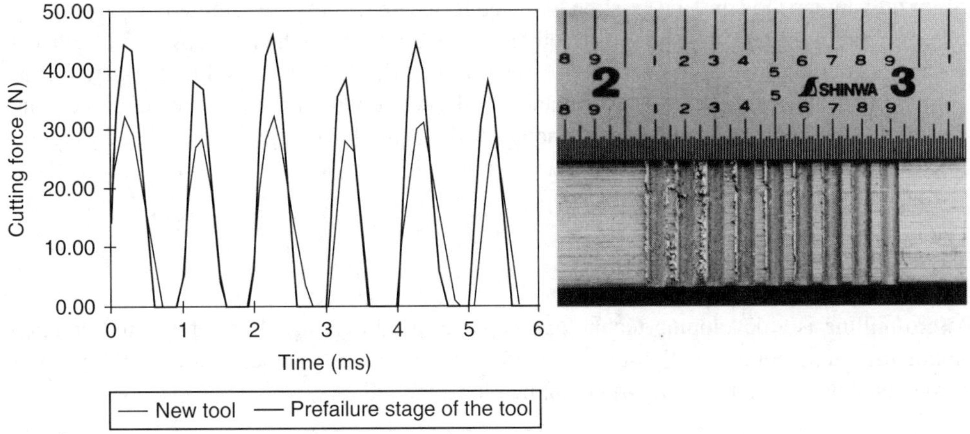

Figure 8.31 Comparison of cutting forces and burr formation using a new and worn tool. Reproduced from [95]. Copyright 2000 Elsevier

A variety of factors, including work material, tooling properties and process conditions, can affect micro-burr formation more or less during machining operations. Aurich *et al.* [117] summarized potential burr control strategies including:

- Tool and tooling: tool geometry alteration, proper tool material, selection corresponding to the work material, coating technology, tool size.
- Coolant: application method, coolant media, application location.
- Process parameters: proper combination of cutting speed, feed, and so on.
- Work material: replacing work material for less burr or preferable burr type.
- Work geometry: design change.
- Process sequencing: order of processes.
- Tool path and machining strategies: better tool path planning and effective machining strategies.

As burr formation is an integrated result of various influential factors, it may not be sufficient to achieve smaller burrs by changing one or two factors. Burr minimization should be considered at all stages, from part design to the entire manufacturing process. So far, optimization and optimal combinations of these influential factors so as to control and minimize micro burrs are largely dependent on specific machined components.

8.6.4 Process Conditions Optimization

For industrial production, there is always a trade off between performance and productivity, and improving both together is a challenge. The holistic considerations for optimization might consist of tool life, work material machinability, geometric accuracy and machining quality of machined parts, material removal rate, and so on. For conventional milling, the appropriate process conditions are mainly adopted by recommendations provided by tool suppliers and experiences of machine operators. However, for micro-scale milling, the emerging technology is not fully understood and the existing knowledge and experiences are limited. As a result, the process conditions used in practice are quite conservative due to the high cost and fragility of micro tools. Effective optimization of process conditions so as to achieve a satisfactory compromise between cutting performance and productivity needs to be addressed and developed before this technology can reach its full potential in cost-effective, high throughput micro manufacturing.

8.7 Application Perspectives

Micro milling is a developing technology capable of fabricating 3D freeform and complex miniaturized components with high geometric accuracy directly and cost-effectively. So far, it has found a wide variety of important applications, including:

- Medical: Dental implants, bridges, hip prosthesis, bone plates and screws, spinal hook, surgical forceps and scissors, lab-on-a-chip, micro pumping mechanisms, and so on.
- Watch-making: Precision watch cases, wristwatch parts, movements, electrodes, dial, and so on.
- Electronics: Micro moulds for inserted connectors, miniature electronic devices, micro sensors, radio-frequency identification chips, and so on.
- Optics: Micro lens arrays, optical devices, and so on.
- Aerospace: Mould for planetary gear wheels, miniature devices for rockets, micro gyroscope components, micro valves and servo devices, and so on.
- Others: Moulds for rings and pendants, components for measuring devices, injection nozzles, electrodes for cutting inserts, and so on.

Some specific micro-milled components for the industrial applications above are illustrated in Figure 8.32 [118–123].

8.8 Concluding Remarks

The chapter has provided a critical and systematic review of micro milling research and applications. Advantageous process capabilities, such as flexibility, cost efficiency and high form quality, gives the emerging approach considerable promise for industrial production, however, existing knowledge and experiences are limited and the present capability of manufacturing techniques need to be continually developed to meet current and future production demands. Based on the review of the state-of-the-art technology, the following aspects are of great interest for future research:

- Development of ultra-precision spindles or machines with higher rotational speeds and lower tool run-out.
- Optimal design and fabrication of micro tools for improving tool life: including geometry optimization, appropriate fabricating and coating technology, and so on.
- Increasing the knowledge related to micro-machinability and optimized process conditions for a range of different engineering materials.

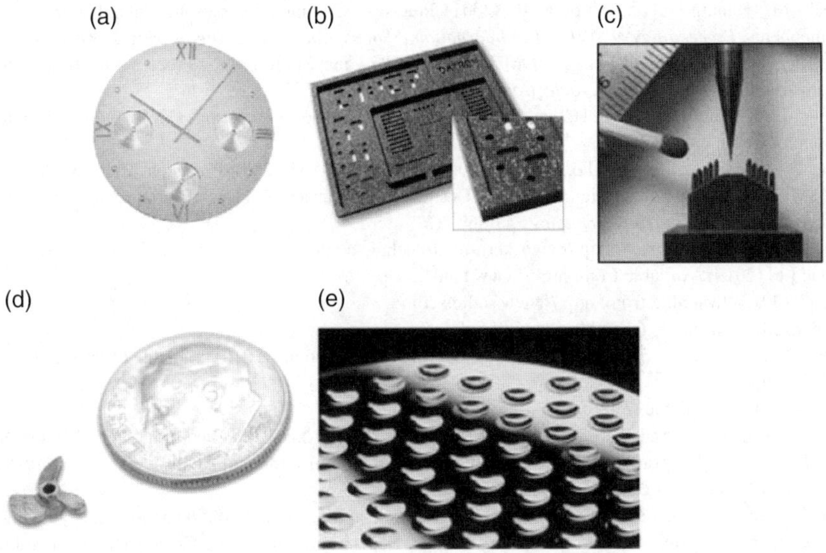

Figure 8.32 Specific micro-milled components: (a) Precision watch dial. Copyright Willemin Macodel SA [118] (b) Soldering frame. Images courtesy of DATRON AG, Germany [119] (c) Micro electrode. Image courtesy of Cimatron Group [120] (d) Micro propeller. Copyright Inventables [121] (e) Wafer-level optics. Copyright Kaleido Technology ApS [122]

- Development of specific CAD/CAM modules for optimized cutting strategies to achieve tighter tolerances and maintain constant chip load.
- Investigation of practical and reliable sensing methods for monitoring and controlling micro milling processes.

References

[1] Chae, J., Park, S.S., Freiheit, T. (2006) Investigation of Micro-Cutting Operations. *International Journal of Machine Tools and Manufacture* 46:313–332.

[2] Liu, X., DeVor, R.E., Kapoor, S.G., Ehmann, K.F. (2004) The Mechanics of Machining at the Microscale: Assessment of the Current State of the Science. *Journal of Manufacturing Science & Engineering* 126:666–678.

[3] Vogler, M.P., DeVor, R.E., Kapoor, S.G. (2003) Microstructure-Level Force Prediction Model for Micro-Milling of Multi-Phase Materials. *Journal of Manufacturing Science and Engineering*, Transactions of the ASME 125:202–209.

[4] Yole Developpement (2010) Available from: http://www.yole.fr/.

[5] Brecher, C., Wenzel, C., and Klar, R. (2008) Characterization and Optimization of the Dynamic Tool Path of a Highly Dynamic Micromilling Machine. *CIRP Journal of Manufacturing Science and Technology* 1:86–91.

[6] Weule, H., Hüntrup, V., and Tritschle, H. (2001) Micro-Cutting of Steel to Meet New Requirements in Miniaturization. *CIRP Annals – Manufacturing Technology* 50:61–64.

[7] Bang, Y.B., Lee, K.M., and Oh, S. (2005) 5-Axis Micro Milling Machine for Machining Micro Parts. *International Journal of Advanced Manufacturing Technology* 25:888–894.

[8] Takeuchi, Y., Suzukawa, H., Kawai, T., and Sakaida Y. (2006) Creation of Ultra-Precision Microstructures with High Aspect Ratios. *CIRP Annals – Manufacturing Technology* 55:107–110.

[9] Weck, M., Hennig, J., and Hilbing, R. (2001) Precision Cutting Processes for Manufacturing of Optical Components. *Proceedings of SPIE – The International Society for Optical Engineering* 4440: 145–151.

[10] Friedrich, C.R., and Vasile, M.J. (1996) Development of the Micromilling Process for High-Aspect-Ratio Microstructures. *Journal of Microelectromechanical Systems* 5:33–38.

[11] Alting, L., Kimura, F., Hansen, H., and Bissacco, G. (2003) Micro Engineering. *CIRP Annals – Manufacturing Technology* 52:635–657.

[12] Huo, D., Cheng, K., and Wardle, F. (2009) Design of a Five-Axis Ultra-Precision Micro-Milling Machine-UltraMill. Part 1: Holistic Design Approach, Design Considerations and Specifications. *International Journal of Advanced Manufacturing Technology* 47:867–877.

[13] Kern (2010) Available from: http://www.kern-microtechnic.com/.

[14] Fanuc Ltd (2010) Available from: http://www.fanuc.co.jp/.

[15] Sodick (2010) Available from: http://www.sodick.com/.

[16] Kugler (2010) Available from: http://www.kuglerofamerica.com/.

[17] Moore Nanotechnology Systems: Ultra-Precision Machining Systems (2010) Available from: http://www.nanotechsys.com/.

[18] Makino (2010) Available from: http://www.makino.com/.

[19] Huo, D., Cheng, K., and Wardle, F. (2009) Design of a Five-Axis Ultra-Precision Micro-Milling Machine-UltraMill. Part 2: Integrated Dynamic Modelling, Design Optimisation and Analysis. *International Journal of Advanced Manufacturing Technology* 47:879–890.

[20] Huo, D., and Cheng, K. (2010) Experimental Investigation on Micromilling of Oxygen-Free, High-Conductivity Copper using Tungsten Carbide, Chemistry Vapour Deposition, and Single-Crystal Diamond Micro Tools. Proceedings of the Institution of Mechanical Engineers, Part B: *Journal of Engineering Manufacture* 224:995–1003.

[21] Tanaka, M. (2001) Development of Desktop Machining Microfactory. *Riken Review* 34:46–49.

[22] Okazaki, Y., Mori, T., and Morita, N. (2001) *Desk-Top NC Milling Machine with 200 Krpm Spindle*. Proceedings of 2001 ASPE Annual Meeting, pp. 192–195.

[23] Lee, S.W., Mayor, R., and Ni, J. (2006) Dynamic Analysis of a Mesoscale Machine Tool. *Journal of Manufacturing Science and Engineering*, Transactions of the ASME 128:194–203.

[24] Li, H., Lai, X., Li, C., Lin, Z., Miao, J., and Ni, J. (2008) Development of Meso-Scale Milling Machine Tool and its Performance Analysis. *Frontiers of Mechanical Engineering in China* 3:59–65.

[25] Childs, T.H.C., Maekawa, K., Obikawa, T., and Yamane, Y. (2000) *Metal Machining: Theory and Applications*. Arnold, London.

[26] Aramcharoen, A., Mativenga, P.T., and Yang, S. (2007) *The Effect of AlCrTiN Coatings on Product Quality in Micro-Milling of 45 HRC Hardened H13 Die Steel*. Proceedings of the 35th International MATADOR Conference, pp. 203–206.

[27] Aramcharoen, A., Mativenga, P.T., Yang, S., Cooke, K.E., and Teer, D.G. (2008) Evaluation and Selection of Hard Coatings for Micro Milling of Hardened Tool Steel. *International Journal of Machine Tools and Manufacture* 48:1578–1584.

[28] Union Tool (2010) Available from: http://www.uniontool.co.jp/.

[29] Heaney, P.J., Sumant, A.V., Torres, C.D., Carpick, R.W., and Pfefferkorn, F.E. (2008) Diamond Coatings for Micro End Mills: Enabling the Dry Machining of Aluminum at the Micro-Scale. *Diamond and Related Materials* 17:223–233.

[30] Torres, C.D., Heaney, P.J., Sumant, A.V., Hamilton, M.A., Carpick, R.W., and Pfefferkorn, F.E. (2009) Analyzing the Performance of Diamond-Coated Micro End Mills. *International Journal of Machine Tools and Manufacture* 49:599–612.

[31] Performance Micro Tool (2010) Available from: http://www.pmtnow.com/.

[32] Li, P. (2009) *Micromilling of hardened tool steels*. Dissertation or Thesis, Delft University of Technology.

[33] Fang, F.Z., Wu, H., Liu, X.D., Liu, Y.C., and Ng, S.T. (2003) Tool Geometry Study in Micromachining. *Journal of Micromechanics and Microengineering* 13:726–731.

[34] Fleischer, J., Deuchert, M., Ruhs, C., Kühlewein, C., Halvadjiysky, G., and Schmidt, C. (2008) Design and Manufacturing of Micro Milling Tools. *Microsystem Technologies* 14:1771–1775.

[35] Schaller, T., Bohn, L., Mayer, J., and Schubert, K. (1999) Microstructure Grooves with a Width of Less than 50 Mm Cut with Ground Hard Metal Micro End Mills. *Precision Engineering* 23:229–235.

[36] Adams, D.P., Vasile, M.J., Benavides, G., and Campbell, A.N. (2001) Micromilling of Metal Alloys with Focused Ion beam–fabricated Tools. *Precision Engineering* 25:107–113.

[37] Ali, M.Y., and Ong, A.S. (2006) Fabricating Micromilling Tool using Wire Electrodischarge Grinding and Focused Ion Beam Sputtering. *International Journal of Advanced Manufacturing Technology* 31:501–508.

[38] Vasile, M.J., Friedrich, C.R., Kikkeri, B., and McElhannon, R. (1996) Micrometer-Scale Machining: Tool Fabrication and Initial Results. *Precision Engineering* 19:180–186.

[39] Cheng, X., Wang, Z.G., Nakamoto, K., and Yamazaki, K. (2009) Design and Development of a Micro Polycrystalline Diamond Ball End Mill for micro/nano Freeform Machining of Hard and Brittle Materials. *Journal of Micromechanics and Microengineering* 19(115022).

[40] Yan, J., Uchida, K., Yoshihara, N., and Kuriyagawa, T. (2009) Fabrication of Micro End Mills by Wire EDM and some Micro Cutting Tests. *Journal of Micromechanics and Microengineering* 19(025004).

[41] Egashira, K., and Mizutani, K. (2003) Milling using Ultra-Small Diameter Ball End Mills Fabricated by Electrical Discharge Machining. *Seimitsu Kogaku Kaishi/Journal of the Japan Society for Precision Engineering* 69:1449–1453.

[42] Rahman, M., Senthil Kumar, A., and Prakash, J.R.S. (2001) Micro Milling of Pure Copper. *Journal of Materials Processing Technology* 116:39–43.

[43] Aramcharoen, A., and Mativenga, P.T. (2009) Evaluation of Critical Parameters in Micro Machining of Hardened Tool Steel. *International Journal of Nanomanufacturing* 3:100–111.

[44] Klocke, F., Quito, F., Arntz, K., and Souza, A. (2009) *A Study of the Influence of Cutting Parameters on Micro Milling of Steel with Cubic Boron Nitride (CBN) Tools*. Proceedings of SPIE – The International Society for Optical Engineering No.7204.

[45] Sreeram, S., Kumar, A.S., Rahman, M., and Zaman, M.T. (2006) Optimization of Cutting Parameters in Micro End Milling Operations in Dry Cutting Condition using Genetic Algorithms. *International Journal of Advanced Manufacturing Technology* 30:1030–1039.

[46] Dimov, S., Pham, D.T., Ivanov, A., Popov, K., and Fansen, K. (2004) Micromilling Strategies: Optimization Issues. Proceedings of the Institution of Mechanical Engineers, Part B: *Journal of Engineering Manufacture* 218:731–736.

[47] Litwinski, K.M., Min, S., Lee, D.E., Dornfeld, D.A., and Lee, N. (2006) *Scalability of Tool Path Planning to Micro Machining*. Proceedings of the 1st International Conference on Micromanufacturing – ICOMM pp. 174–179.

[48] Jun, M.B.G., Joshi, S.S., DeVor, R.E., and Kapoor, S.G. (2008) An Experimental Evaluation of an Atomization-Based Cutting Fluid Application System for Micromachining. *Journal of Manufacturing Science and Engineering*, Transactions of the ASME 130 pp:0311181-0311188.

[49] PMPA: Precision Machined Products Assocation (2010) Available from: http://www.pmpa.org/.

[50] Özel, T., Liu, X., and Dhanorker, A. (2007) *Modelling and Simulation of Micro-Milling Process*. 4th International Conference and Exhibition on Design and Production of Machines and Dies/Molds pp: 21–23.

[51] Weinert, K., and Petzoldt, V. (2008) *Machining NiTi Micro-Parts by Micro-Milling*. Materials Science and Engineering: A 481–482,:672–675.

[52] Uhlmann, E., Piltz, S., and Schauer, K. (2005) Micro Milling of Sintered tungsten–copper Composite Materials. *Journal of Materials Processing Technology* 167:402–407.

[53] Bissacco, G., Hansen, H.N., and De Chiffre, L. (2005) Micromilling of Hardened Tool Steel for Mould Making Applications. *Journal of Materials Processing Technology* 167:201–207.

[54] Shelton, J.A., and Shin, Y.C. (2010) Comparative Evaluation of Laser-Assisted Micro-Milling for AISI 316, AISI 422, TI-6AL-4V and Inconel 718 in a Side-Cutting Configuration. *Journal of Micromechanics and Microengineering* 20, No.075012.

[55] Shelton, J.A., and Shin, Y.C. (2010) Experimental Evaluation of Laser-Assisted Micromilling in a Slotting Configuration. *Journal of Manufacturing Science and Engineering*, Transactions of the ASME 132 No.021008.

[56] Schueler, G.M., Engmann, J., Marx, T., Haberland, R., and Aurich, J.C. (2010) *Burr Formation and Surface Characteristics in Micro-End Milling of Titanium Alloys*. Burrs – Analysis, Control and Removal pp. 129–138.

[57] Gietzelt, T., Eichhorn, L., and Schubert, K. (2008) Manufacturing of Microstructures with High Aspect Ratio by Micromachining. *Microsystem Technologies* 14:1525–1529.

[58] Rusnaldy, X., Ko, T.J., and Kim, H.S. (2007) Micro-End-Milling of Single-Crystal Silicon. *International Journal of Machine Tools and Manufacture* 47:2111–2119.

[59] Rusnaldy, Ko, T.J., and Kim, H.S. (2008) An Experimental Study on Microcutting of Silicon using a Micromilling Machine. *International Journal of Advanced Manufacturing Technology* 39:85–91.

[60] Vogler, M.P., Kapoor, S.G., and DeVor, R.E. (2004) On the Modeling and Analysis of Machining Performance in Micro-Endmilling, Part II: Cutting Force Prediction. *Journal of Manufacturing Science and Engineering*, Transactions of the ASME 126:695–705.

[61] Lee, K., and Dornfeld, D.A. (2002) An Experimental Study on Burr Formation in Micro Milling Aluminum and Copper. *Trans. North American Manufacturing Research Institute* 30: 255–262.

[62] Lucca, D.A., Rhorer, R.L., and Komanduri, R. (1991) Energy Dissipation in the Ultraprecision Machining of Copper. *CIRP Annals – Manufacturing Technology* 40:69–72.

[63] Ng, C.K., Melkote, S.N., Rahman, M., and Senthil, K.A. (2006) Experimental Study of Micro- and Nano-Scale Cutting of Aluminum 7075-T6. *International Journal of Machine Tools and Manufacture* 46:929–936.

[64] Liu, K., and Melkote, S.N. (2007) Finite Element Analysis of the Influence of Tool Edge Radius on Size Effect in Orthogonal Micro-Cutting Process. *International Journal of Mechanical Sciences* 49 pp:650–660.

[65] Vogler M.P., DeVor R.E., Kapoor S.G. (2004) *On the Modeling and Analysis of Machining Performance in Micro-End Milling, Part I: Surface Generation.* Journal of Manufacturing Science and Engineering 126:685–694.

[66] Liu, X., Jun, M.B.G., DeVor, R.E., and Kappor, S.G. (2004) *Cutting Mechanisms and their Influence on Dynamic Forces, Vibrations and Stability in Micro-Endmilling*. ASME Conference Proceedings pp.583–592.

[67] Liu, X., DeVor, R.E., and Kapoor, S.G. (2006) An Analytical Model for the Prediction of Minimum Chip Thickness in Micromachining. *Journal of Manufacturing Science and Engineering* 128:474–481.

[68] Shimada, S., Ikawa, N., Tanaka, H., Ohmori, G., Uchikoshi, J., and Yoshinaga, H. (1993) Feasibility Study on Ultimate Accuracy in Microcutting using Molecular Dynamics Simulation. *CIRP Annals – Manufacturing Technology* 42:91–94.

[69] Son, S.M., Lim, H.S., and Ahn, J.H. (2005) Effects of the Friction Coefficient on the Minimum Cutting Thickness in Micro Cutting. *International Journal of Machine Tools and Manufacture* 45:529–535.

[70] Kim, C.J., Mayor, J.R., and Ni, J. (2004) A Static Model of Chip Formation in Microscale Milling. *ASME Journal of Manufacturing Science and Engineering* 126:710–718.

[71] Yuan, Z.J., Zhou, M., and Dong, S. (1996) Effect of Diamond Tool Sharpness on Minimum Cutting Thickness and Cutting Surface Integrity in Ultraprecision Machining. *Journal of Materials Processing Technology* 62:327–330.

[72] Chuzhoy, L., Devor, R.E., Kapoor, S.G., and Bammann, D.J. (2002) Microstructure-Level Modeling of Ductile Iron Machining. *Journal of Manufacturing Science and Engineering* 124:162–169.

[73] Yuan, Z.J., Lee, W.B., Yao, Y.X., and Zhou, M. (1994) Effect of Crystallographic Orientation on Cutting Forces and Surface Quality in Diamond Cutting of Single Crystal. *CIRP Annals – Manufacturing Technology* 43:39–42.

[74] Simoneau, A., Ng, E., and Elbestawi, M.A. (2006) Chip Formation during Microscale Cutting of a Medium Carbon Steel. *International Journal of Machine Tools and Manufacture* 46:467-481.

[75] Simoneau, A., Ng, E., and Elbestawi, M.A. (2006) The Effect of Microstructure on Chip Formation and Surface Defects in Microscale, Mesoscale, and Macroscale Cutting of Steel. *CIRP Annals – Manufacturing Technology* 55:97–102.

[76] Simoneau, A., Ng, E., and Elbestawi, M.A. (2007) Grain Size and Orientation Effects when Microcutting AISI 1045 Steel. *CIRP Annals – Manufacturing Technology* 56:57–60.

[77] Cheung, C., To, S., and Lee, W. (2002) Anisotropy of Surface Roughness in Diamond Turning of Brittle Single Crystals. *Materials and Manufacturing Processes* 17:251–267.

[78] Min, S., Dornfeld, D., Inasaki, I., Ohmori, H., Lee, D., Deichmueller, M., Yasuda, T., and Niwa, K. (2006) Variation in Machinability of Single Crystal Materials in Micromachining. *CIRP Annals – Manufacturing Technology* 55:103–106.

[79] Schmidt, J., Spath, D., Elsner, J., Hüntrup, V., and Tritschler, H. (2002) Requirements of an Industrially Applicable Microcutting Process for Steel Micro Structures. *Microsystem Technologies* pp.402–408.

[80] Popov, K.B., Dimov, S.S., Pham, D.T., Minev, R.M., Rosochowski, A., and Olejnik, L. (2006) Micromilling: Material Microstructure Effects. Proceedings of the Institution of Mechanical Engineers, Part B: *Journal of Engineering Manufacture* 220:1807–1813.

[81] Pham, D., Elkaseer, A., Popov, K., Dimov, S., Olejnik, L., and Rosochowski, A. (2009) *Micromilling of Coarse-Grained and Ultrafine-Grained Cu99.9E: Effects of Material Microstructure on Machining Conditions and Surface Quality*. 4M/ICOMM 2009 – The Global Conference on Micro Manufacture pp: 241–244.

[82] Oxley, P.L.B. (1989) *Mechanics of Machining: an Analytical Approach to Assessing Machinability*. Ellis Horwood, Chichester,West Sussex, London.

[83] Johnson, G.R., and Cook, W.H. (1983) *A Constitutive Model and Data for Metals Subjected to Large Strains, High Strain Rates, and High Temperatures*. Proceedings of the 7th International Symposium on Ballistics pp: 541–547.

[84] Shirakashi, T., and Usui, E. (1970) *Effect of Temperature and Strain Rate upon Flow Stress of Metal in Compression*. Bulletin of the Japan Society of Precision Engineering 1–4 pp.91.

[85] Zerilli, F.J., and Armstrong, R.W. (1995) Constitutive Equation for HCP Metals and High Strength Alloy Steels. *High Strain Rate Effects on Polymer, Metal and Ceramic Matrix Composites and other Advanced Materials* 48:121–126.

[86] Özel, T. (2006) The Influence of Friction Models on Finite Element Simulations of Machining. *International Journal of Machine Tools and Manufacture* 46:518–530.

[87] Sheikh, M.A., Mativenga, P.T., and Iqbal, S.A. (2007) *Characterization of Machining of AISI 1045 Steel Over a Wide Range of Cutting Speeds. Part 2: Evaluation of Flow Stress Models and Interface Friction Distribution Schemes*. Proceedings of the Institution of Mechanical Engineers, Part B: Journal of Engineering Manufacture 221:917–926.

[88] Dhanorker, A., and Özel, T. (2006) *An Experimental and Modeling Study on meso/micro End Milling Process*. Proceedings of 2006 ASME Conference on Manufacturing Science and Engineering Paper No. 21127.

[89] Wang, J.S., Gong, Y.D., Abba, G., Antoine, J.F., and Shi, J.S. (2009) Chip Formation Analysis in Micromilling Operation. *The International Journal of Advanced Manufacturing Technology* 45:430–447.

[90] Afazov, S.M., Ratchev, S.M., and Segal, J. (2010) Modelling and Simulation of Micro-Milling Cutting Forces. *Journal of Materials Processing Technology* 210:2154–2162.

[91] Wu, T., and Cheng, K. (2011) *3D FE-Based Modeling and Simulation of the Micro Milling Process*. 4th International Conference of Asian Society for Precision Engineering and Nanotechnology No.0241.

[92] Bao, W.Y., and Tansel, I.N. (2000) Modeling Micro-End-Milling Operations. Part I: Analytical Cutting Force Model. *International Journal of Machine Tools and Manufacture* 40:2155–2173.

[93] Tlusty, J., and Macneil, P. (1975) Dynamics of Cutting Forces in End Milling. *Annals of CIRP* 24: 21–25.

[94] Bao, W.Y., and Tansel, I.N. (2000) Modeling Micro-End-Milling Operations. Part II: Tool Run-Out. *International Journal of Machine Tools and Manufacture* 40:2175–2192.

[95] Bao, W.Y., and Tansel, I.N. (2000) Modeling Micro-End-Milling Operations. Part III: Influence of Tool Wear. *International Journal of Machine Tools and Manufacture* 40:2193–2211.

[96] Kang, I.S., Kim, J.S., Kim, J.H., Kang, M.C., and Seo, Y.W. (2007) A Mechanistic Model of Cutting Force in the Micro End Milling Process. *Journal of Materials Processing Technology* 187–188:250–255.

[97] Zaman, M.T., Kumar, A.S., Rahman, M., and Sreeram, S. (2006) A Three-Dimensional Analytical Cutting Force Model for Micro End Milling Operation. *International Journal of Machine Tools and Manufacture* 46:353–366.

[98] Li, C., Lai, X., Peng, L., Li, H., and Ni, J. (2007) Modeling of Three-Dimensional Cutting Forces in Micro-End-Milling. *Journal of Micromechanics and Microengineering* 17: 671.

[99] Jun, M.B.G., Liu, X., DeVor, R.E., and Kapoor, S.G. (2006) Investigation of the Dynamics of Microend Milling-Part I: Model Development. *Journal of Manufacturing Science and Engineering* 128:893–900.

[100] Jun, M.B.G., DeVor, R.E., and Kapoor, S.G. (2006) Investigation of the Dynamics of Microend Milling-Part II: Model Validation and Interpretation. *Journal of Manufacturing Science and Engineering* 128:901–912.

[101] Liu, X., DeVor, R.E., and Kapoor, S.G. (2007) Model-Based Analysis of the Surface Generation in Microendmilling---Part I: Model Development. *Journal of Manufacturing Science and Engineering* 129:453–460.

[102] Liu, X., DeVor, R.E., and Kapoor, S.G. (2007) Model-Based Analysis of the Surface Generation in Microendmilling---Part II: Experimental Validation and Analysis. *Journal of Manufacturing Science and Engineering* 129:461–469.

[103] Zygo Corporation (2010) Available from: http://www.zygo.com/.

[104] Veeco (2010) Available from: http://www.veeco.com.

[105] Taylor Hobson (2010) Available from: http://www.taylor-hobson.com.

[106] Zeiss (2010) Available from: http://www.zeiss.com.

[107] Kistler (2010) Available from: http://www.kistler.com/.

[108] Biermann, D., and Kahnis, P. (2010) Analysis and Simulation of Size Effects in Micromilling. *Production Engineering* 4:25–34.

[109] Lai, X., Li, H., Li, C., Lin, Z., and Ni. J. (2008) Modelling and Analysis of Micro Scale Milling Considering Size Effect, Micro Cutter Edge Radius and Minimum Chip Thickness. *International Journal of Machine Tools and Manufacture* 48:1–14.

[110] Filiz, S., Xie, L., Weiss, L.E., and Ozdoganlar, O.B. (2008) Micromilling of Microbarbs for Medical Implants. *International Journal of Machine Tools and Manufacture* 48:459–472.

[111] Park, J.B., Wie, K.H., Park, J.S., and Ahn, S.H. (2009) Evaluation of Machinability in the Micro End Milling of Printed Circuit Boards. Proceedings of the Institution of Mechanical Engineers, Part B: *Journal of Engineering Manufacture* 223:1465–1474.

[112] Dornfeld, D., Min, S., and Takeuchi, Y. (2006) Recent Advances in Mechanical Micromachining. CIRP Annals – *Manufacturing Technology* 55:745–768.

[113] Lee, D.E., Hwang, I., Valente, C.M.O., Oliveira, J.F.G., and Dornfeld, D.A. (2006) Precision Manufacturing Process Monitoring with Acoustic Emission. *International Journal of Machine Tools and Manufacture* 46:176–188.

[114] Malekian, M., Park, S.S., Jun, M.B.G., and Jun, M.B.G. (2009) Tool Wear Monitoring of Micro-Milling Operations. *Journal of Materials Processing Technology* 209:4903–4914.

[115] Gandarias, E., Dimov, S., Pham, D., Ivanov, A., Popov, K., Lizarralde, R., and Arrazola, P. (2006) New Methods for Tool Failure Detection in Micromilling. Proceedings of the Institution of Mechanical Engineers, Part B: *Journal of Engineering Manufacture* 220:137–144.

[116] Lee, K., and Dornfeld, D. (2005) Micro-Burr Formation and Minimization through Process Control. *Precision Engineering* 29:246–252.

[117] Aurich, J.C., Dornfeld, D., Arrazola, P.J., Franke, V., Leitz, L., and Min, S. (2009) Burrs—Analysis, Control and Removal. CIRP Annals – *Manufacturing Technology* 58:519–542.

[118] Willemin Macodel (2010) Available from: http://www.willemin-macodel.com/.

[119] Datron Dynamics (2010) Available from: http://www.datron.com/.

[120] Cimatron Group (2010) Available from: http://www.cimatron.com/.

[121] Inventables (2010) Available from: http://www.inventables.com/.

[122] Kaleido Technology (2010) Available from: http://www.kaleido-technology.com/.

[123] Design News (2010) Available from: http://www.designnews.com/.

9

Micro Drilling Applications

M. J. Jackson, T. Novakov and K. Mosiman
Centre for Advanced Manufacturing, MET, College of Technology, Purdue University, United States of America

9.1 Chapter Overview

The advent of drilling at the micro and mesoscales has been accompanied by the wide array of studies concerned with drilling at high speeds with very small metal removal rates. This has led to an understanding of how material deformation mechanisms affect drilling operations at these scales and has spawned a rich area of academic investigation. This has been precipitated by the need to produce burr-free drilling especially in the aerospace and medical device industrial sectors. Both sectors use both fibre-reinforced polymeric materials and plain polymeric materials where the occurrence of burr formation can lead to premature fracture of the material, or by the incorporation of these materials into the human body. Both situations are deleterious to the function of engineered products in both of these industrial sectors. This chapter focuses on understanding the phenomenon of chatter and burr formation during drilling operations and the remedies associated with their elimination using case studies. This chapter was originally published as a paper in the *International Journal of Advanced Manufacturing Technology* entitled, 'Chatter Problems in Micro and Macro-Cutting Operations, Existing Models and Influential Parameters – A Review', 2010, 47: 597–620, Reprinted with kind permission of Springer Science + Business Media B.V.

9.2 Investigation of Chatter in Mesoscale Drilling

Chatter is considered to be a self excited vibration between the machine tool and the workpiece [1, 2]. Regenerative chatter is the most important type of chatter and it is a consequence of both the previously cut surface and current chip formation. As the tool is creating a new cut surface by removing material in a form of a chip, it leaves waviness in the profile that consequently affects the next cut. The new wavy surface induces vibrations in the tool that

Micro-Cutting: Fundamentals and Applications, First Edition. Edited by Kai Cheng and Dehong Huo.

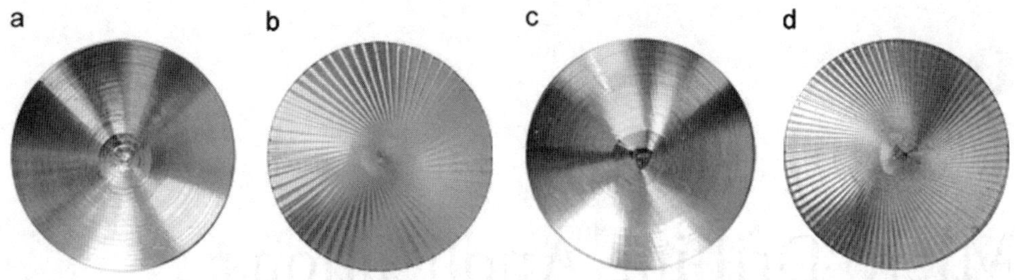

Figure 9.1 Hole shapes: (a) stable cut, no visible vibrations; (b) sunray pattern due to unstable torsional–axial chatter vibration; (c) trigon caused by whirling vibrations and (d) surface resulting from combined torsional–axial chatter and whirling vibration. Reproduced from [7]. Copyright 2007 Elsevier

then affects the surface (Figure 9.1). Chatter is a great problem in cutting due to the fact that it causes decrease in tool life, limited productivity, bad surface finish, poor tolerances as well as unacceptable noise. Chatter problems in turning and milling have been covered extensively throughout the years and can easily be found in literature [1–4].

There are many papers dealing with the analytical models for prediction of maximum stable depths of cut as a function of spindle speeds. This way, optimal operating regimes have been defined for both turning and milling operations resulting in the use of large depths of cut and large spindle speeds simultaneously. Some researchers went even further and developed computer programs for chatter prediction [5, 6].

Drilling has not been researched to the same extent as turning and milling processes. This is due to the fact that drilling tools have a more complex geometry compared to milling and turning tools – chisels create a ploughing action, cutting lips have varying rake angles, chip thickness changes along the edge and so on. All this creates a number of problems when modelling the process. Chatter in drilling often involves large amplitude torsional vibrations, while they are not present in the cases of milling or turning. When analyzing problems of chatter in drilling, several models have been investigated throughout the literature: torsional – axial model, bending model and the combination of axial and bending models. Apart from the different models, several different influential factors have to be taken into consideration like: friction, chisel edge, margin engagement, grinding errors, misalignment of the tool, drill flank, pilot hole size, length of the drills, and so on.

As suggested in [8], there are two different approaches for detecting chatter used throughout the literature. The first approach is the Frequency Domain Method, which is based on the appearance of increase in frequency or equivalent signals used like acceleration or sound, with the onset of chatter. The second approach covered in the literature is the Time Domain Method based on on-line determination of a time series. Both of these techniques still encounter certain problems in determination of the proper chatter onset threshold especially if the cutting conditions show variability. The analytical part of research is covered in a number of papers that deal with analysis of stability and bifurcations of non-linear differential equations [9]. Stability curves obtained by these analyses give a significant increase in efficiency of cutting operations, and the boundary conditions of chatter onset still remain extremely important. Determination of whether the bifurcation is subcritical or supercritical is considered to be crucial.

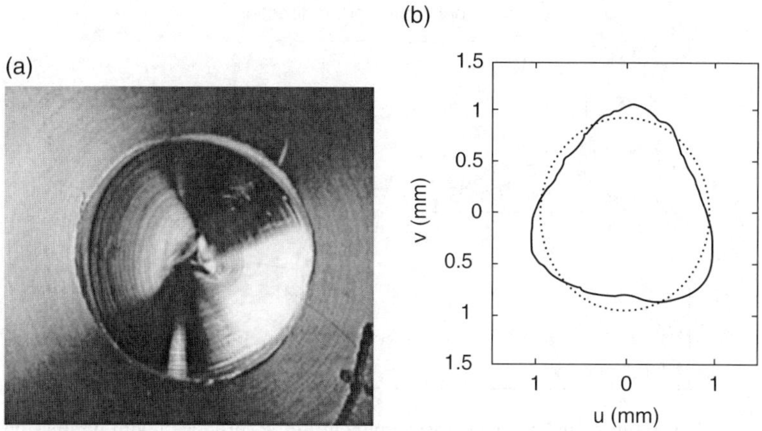

Figure 9.2 Photograph and measured profile of a three-lobed hole. Reproduced with permission from [14]. Copyright 2002 ASME

Ema and his colleagues started some of the first work on the issues of chatter problems in drilling in 1986 and 1988, [10–12], analyzing the issues of chatter by keeping the tool stationary and feeding it into the rotating workpiece. These papers investigated the effect of the drill geometry, chisel edge, drill flank and pilot hole. They have showed that the whirling vibrations, which are a result of the regeneration effect at the major cutting edges, result in distorted polygonal holes with an odd number of sides, which was later investigated by Bayly and his team (Figure 9.2). They showed that the chisel edge and flank surfaces cause damping in the system. However, due to the fact that these investigations have been conducted in an unconventional way, the results are not completely trustworthy since there is rarely a situation where one would have to rotate the workpiece instead of the drill. By using the stationary tool, the effect of the spindle and bearings was neglected, neglecting the conventional problems in drilling [13].

The state of the art in turning, milling, drilling and grinding has been presented by [15], covering present mathematical models and chatter suppression methods. The first part of the paper deals with the dynamics of the orthogonal metal cutting and chatter stability lobes, defining the linear dependency of the cutting forces on the width of cut and feed per revolution (or static chip thickness):

$$\begin{aligned} F_t &= K_t a h_0 \\ F_r &= K_r a h_0 \end{aligned} \tag{9.1}$$

Where: a – width of cut, h_0– feed per revolution or static chip thickness, K_t, K_r – tangential and radial cutting coefficients.

The dynamic chip thickness and transfer function of the closed loop system have been defined as:

$$h(t) = h_0 - \delta[y(t) - y(t-T)] \tag{9.2}$$

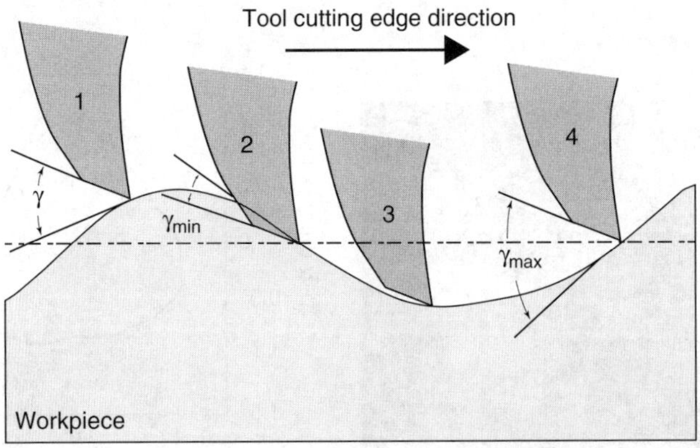

Figure 9.3 Process damping mechanism in dynamic cutting. Reproduced with permission from [16]. Copyright 1986 ASME

$$\frac{h(s)}{h_o(s)} = \frac{1}{1+(1-e^{-sT})K_r a\Phi(s)} \tag{9.3}$$

Where: T – spindle rotation period, δ – overlap factor, y(t) and y(t – T) – present and past vibration amplitudes in the radial direction, $\Phi(s)$ – frequency response function.

Stability lobes have been presented, showing that if the depth of cut (DOC) and spindle speed are selected below the stability lobe, the process is stable and if they are above the lobe, the process becomes unstable with growing vibrations.

The issue of stability at high and low spindle speeds has been discussed stressing the difficulty of defining the stability region for the low speeds due to the complexity of modeling process damping (clearance angle becomes zero or negative if there is a lot of waviness in the surface) (Figure 9.3).

Authors of this paper also deal with the complexity of the drilling process suggesting that in drilling, the chisel has a ploughing action and the cutting lips have a varying rake angle and chip thickness along the edge as well as explaining that drilling forces and torque are a function of chip thickness, width of cut, material properties, drill geometry and drill tip vibrations at the current time and one tooth earlier. It has been stated that in order to predict the cutting load when there are vibrations in the drill, the chip thickness distribution along the cutting edge has to be identified considering vibrations in every direction. These vibrations influence the cutting forces in drilling therefore making it very challenging to solve the equation of motion:

$$[M]\begin{Bmatrix} \ddot{x}(t) \\ \ddot{y}(t) \\ \ddot{z}(t) \\ \ddot{\theta}(t) \end{Bmatrix} + [C]\begin{Bmatrix} \dot{x}(t) \\ \dot{y}(t) \\ \dot{z}(t) \\ \dot{\theta}(t) \end{Bmatrix} + [K]\begin{Bmatrix} x(t) \\ y(t) \\ z(t) \\ \theta(t) \end{Bmatrix} = \begin{Bmatrix} F_x \\ F_y \\ F_z \\ T_c \end{Bmatrix} \tag{9.4}$$

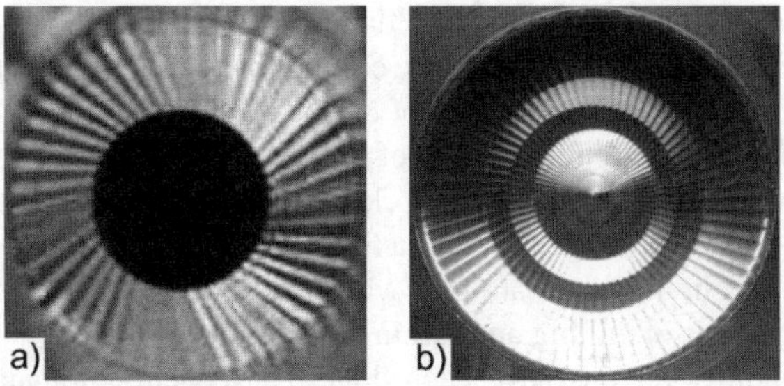

Figure 9.4 (a) chatter surface on the bottom surface of a hole for a twist drill (with pilot hole); (b) for an indexable drill. (Source: UBC – M.A.L.) [15] Reproduced from [15]. Copyright 2004 Elsevier

Where: x, y – lateral, z – axial, θ – torsional deflection of the drill, matrices M, C, K – mass, damping and stiffness characteristics at the drill tip, Fx, Fy lateral forces, thrust Fz force and Tc – torque.

It has been suggested that one of the ways to start the analysis of a process as complex as drilling is to consider individual vibration mechanisms. Therefore, torsional-axial, lateral and whirling vibrations have been covered individually in this paper [15].

9.2.1 Torsional-axial Model

The coupling of torsional and axial vibrations happens when the drill starts to lengthen and shorten respectively due to the torsional vibration creating a wavy surface at the bottom of the hole (Figure 9.4). As the previous material surface profile influences the next cut, a regenerative effect of thrust and torque acts on the tool. This can lead to an increase in the vibrations of the system, leading to unstable vibrations during machining. This is why it is necessary to determine stability boundaries for the torsional-axial model.

Bayly and his colleagues started research on the torsional axial model in 2001, [3] by developing a mathematical model of the torsional-axial coupling that causes chatter in drilling. In this work, researchers have taken into consideration three translations and one rotation, leaving the bending moments out of the equation.

The equation of motion can be written as:

$$M\ddot{x} + C\dot{x} + Kx = F$$

where: (9.5)

$$x = [x_1, x_2, .., x_N, y_1, y_2, .., y_N, z_1, z_2, .., z_N, \theta_1, \theta_2, .., \theta_N]^T$$
$$F = [F_{x1}, F_{x2}, .., F_{xN}, F_{y1}, F_{y2}, .., F_{yN}, F_{z1}, F_{z2}, .., F_{zN}, M_1, M_2, .., M_N]^T$$

x_i, y_i, z_i – deflections in x, y, z (i = 1–N) , θ– twist about the axis.

The time varying component of the modal equation can be written as:

$$M_p\ddot{\eta}_p + C_p\dot{\eta}_p + K_p\eta_p = -\alpha C_1 b[\eta_p - \eta_p(t-\tau)] \tag{9.6}$$

where: M, K, C – modal mass, stiffness and damping, $\alpha = C_2/C_1 + \theta_{Np} R_{av}$, θ_{Np}-torsional axial coupling parameter, R_{av} – average radius of the cutting force, C_1, C_2-tangential and thrust forces per unit area of uncut chip, b – radial depth of cut, η_p– modal coefficient of the pth mode, τ – time delay between passage of cutting edges of the drill.

The difference between this equation and classical chatter equations is in the constant α which includes the effects of tangential and axial forces as well as torsional axial coupling. Another difference lies in the fact that $-\alpha C_1$ which presents the effective cutting pressure is not always positive as it is in milling and turning operations.

Predictions of stability regions have been done by frequency domain analysis and confirmed by both experimental work and computer simulations. The drill dynamics were described by the Frequency Response Function (FRF), which is defined as displacement at the given location in response to a unit force, as a function of excitation frequency.

Radial depth of cut is expressed as:

$$b = \frac{-1}{\alpha C_1 2\,\mathrm{Re}[\Phi p(\omega)]} \tag{9.7}$$

Where: $\mathrm{Re}[\Phi(\omega)]$ is the real part of the frequency response function for the pth mode. The absolute depth of cut can be obtained when considering the minimum value of the real part of the frequency response function.

Equation (9.7) has been used to determine stability boundaries shown in Figure 9.5 enabling the comparison of simulation results and analytical results. The consequence of the obtained derivations which can be understood from Figure 9.5 is that unlike the onset of chatter in turning and milling, it is necessary for the chatter frequency to be slightly below the natural frequency of the corresponding mode, not slightly above, for the chatter episodes to occur. This has also been experimentally confirmed by cutting tests. In their experiment, Bayly *et al.* have used a symmetric 'winged' steel collar, attached 25 mm from the tip of the drill. The collar was designed symmetrically to avoid inertial coupling between bending and torsion. However, the added inertia of the collar lowered the natural frequencies of the drill so that the phenomena usually apparent at high speeds would be observed at lower speeds. This causes the natural frequencies in the test to be much lower than the frequencies that actual drills have. Apart from the collar effect, the authors pointed out that they have neglected to include some effects which influence the model: burnishing, ploughing, rubbing and process damping due to the clearance face of the tool interfering with the cut surface. The process damping should be especially significant at lower speeds. Owing to the fact that this paper deals with the existing piloted hole, the phenomenon of damping was neglected. Further, there is the inevitable effect of the friction between the rake face and the chip as well as the margin effect when the margin of the drill interacts with the sides of the hole. The effect of the axial loading of the drill during drilling has been neglected even though it would cause some change in the torsional natural frequency. The fact that the workpiece deforms during the drilling process has been neglected, even though it will influence the surface and therefore the chip load. Other possible causes of torsional vibration in drilling that have not been covered are considered to be: stick slip interaction between tool and wall of the hole, built-up edge, and so on.

Stone and Askari continued on Bayly's work in 2002 [17] developing a model which incorporated the friction effect into the axial-torsional vibrational model by studying the stick – slip interaction mode between the tool and workpiece material. The stick-slip mode

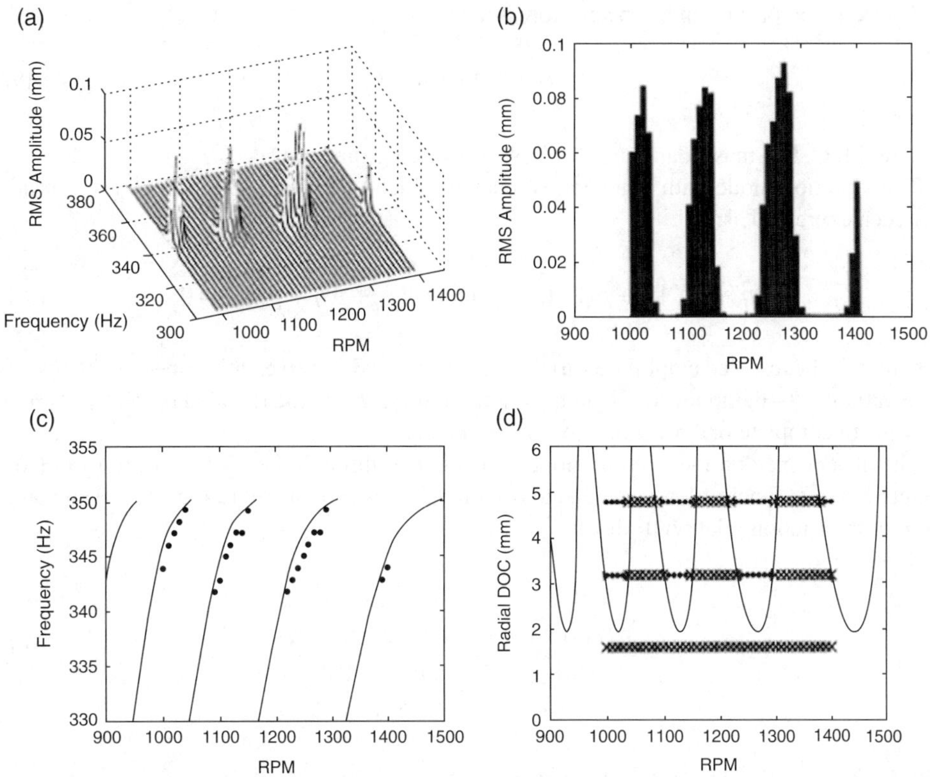

Figure 9.5 Simulation results and comparison to analytical predictions of chatter stability and frequency. (a) Vibration amplitude as a function of spindle speed and chatter frequency; (b) Amplitude vs. spindle speed; (c) Chatter frequency vs. spindle speed; (d) Comparison of results of numerical integration (• unstable, x stable) with stability boundaries predicted from Equation (9.7). Reproduced with permission from [3]. Copyright 2001 ASME

provides non-linear process damping to the system. By interacting with the vibrational mode it produces instability in the non-regenerative chatter case.

The idea that the drag of the chip on the rake phase of the cutter causes the drill to experience a stick-slip friction effect has been incorporated into the model as well, and the law has been mathematically defined. The friction law used to incorporate the effect of friction in the derivations is defined as:

$$\mu(u) = \mu_1 + (\mu_2 - \mu_1)\exp\left[\left(u/U_o\right)^2\right] \tag{9.8}$$

Where: u – chip speed, U_o – friction cutting speed, μ_1, μ_2 – threshold friction coefficients.

For obtaining the dynamic model of the drill reverse engineering has been used, where the drill has been laser scanned and the obtained 3D model was imported into ABAQUS program for analysis. The motion of the tip during vibration has been obtained from ABAQUS.

The equation of motion has been expressed as:

$$M\ddot{\eta}+C\dot{\eta}+K\eta=F_{\eta} \tag{9.9}$$

Where: M, C, K – mass, damping, stiffness, F_{η}–forcing function.

The vibration mode with regenerative chatter which includes the chip thickness variation has been expressed as:

$$\eta''+\gamma\eta'+\eta+\beta(1-(\eta-\eta_T)\cos\theta)(p_o+\tilde{p}_1\eta'+\tilde{p}_2\eta'^2)=0 \tag{9.10}$$

Where: η is the rescaled amplitude variable, and η' is the derivative with respect to the rescaled time variable, γ – damping coefficient, $\beta=w\tau/k$ and P_1, P_2 are the rescaled p_1, p_2, $\eta_T=\eta\,(t-T)$, T - time to complete one revolution of the workpiece.

The dependence on the friction model is explained through the coefficient p_1 of G (force penetration rate function), which depends on the rake angle α and vibration angle θ and on the form of the friction relation $\mu=h(u)$:

$$p_1=\frac{-2\cos(\theta+\phi_0)}{\sin^3\phi_0}\frac{1+h_0^2\dfrac{-h_0'}{\cos(\phi_0-\alpha)}\dfrac{-\cos(\phi_0-\theta)}{}}{2+\dfrac{h_0'}{1+h_0^2}\dfrac{V\cos\alpha}{\cos^2(\alpha-\phi_0)}} \tag{9.11}$$

Where: θ – vibration angle, ϕ – shear plane angle, α – rake angle and

$$\frac{du}{d\dot{\eta}}=\frac{-\cos(\phi_0-\theta)}{\cos(\phi_0-\alpha)} \tag{9.12}$$

$$\frac{du}{d\phi}=\frac{V\cos\alpha}{\cos^2(\alpha-\phi_0)} \tag{9.13}$$

The dependence of the coefficient p_1 on vibration effects and rake angles can be graphically seen in Figure 9.6.

The critical depth of cut is expressed as:

$$\omega_{crit}=\frac{k}{2\tau p_0\cos\theta}(\Gamma\pm 2)\Gamma \tag{9.14}$$

Where: + is used for $p_o\cos\theta>0$ and – for $p_o\cos\theta<0$, p_o – constant coefficient of G, G – force penetration function, Γ – dimensionless damping coefficient, τ – shear stress of the material. As the rake angle is increased from $\alpha<\alpha_{crit}$ to $\alpha>\alpha_{crit}$ the response frequency becomes less than the natural frequency and a shift in the stability threshold occurs. This mathematical phenomenon is seen on the shop floor as the tool starts digging into the material as opposed to being forced away from it.

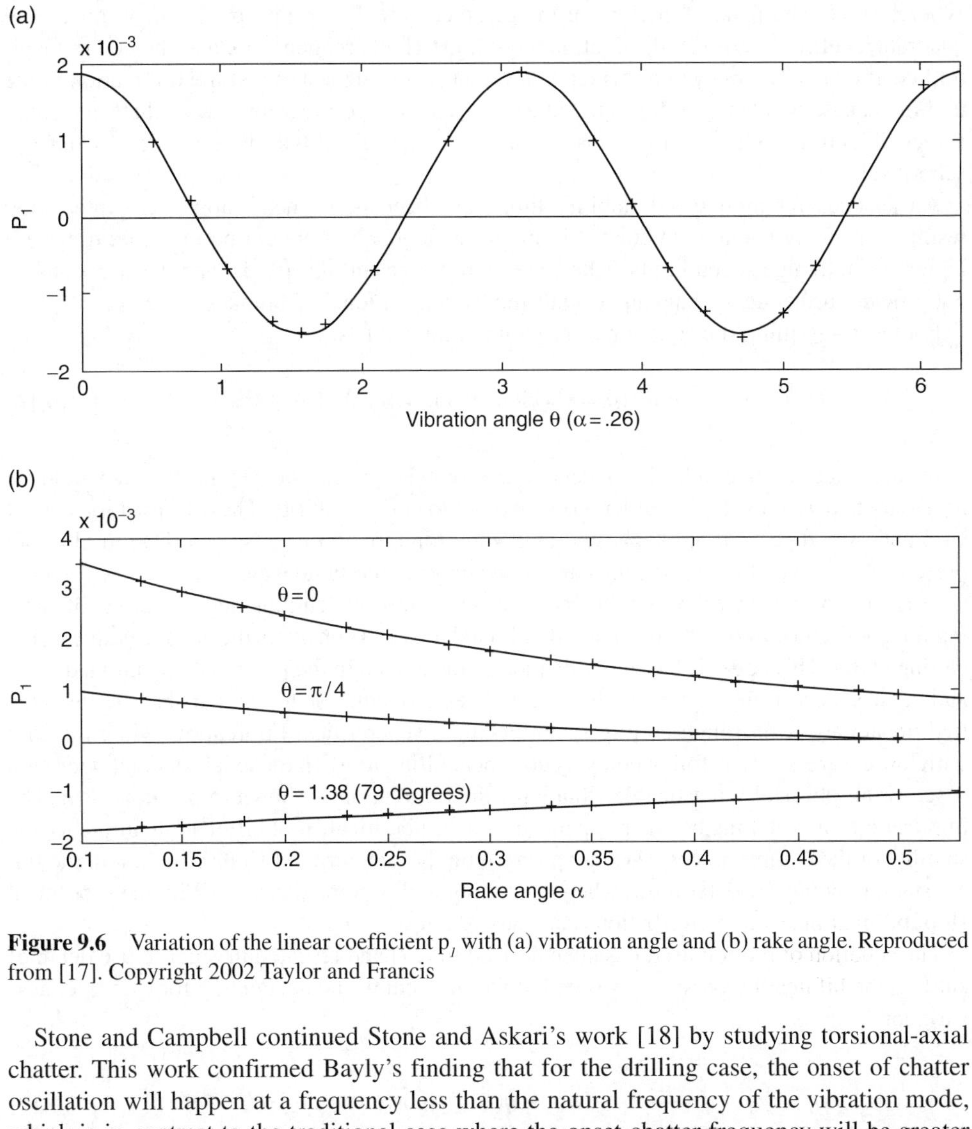

Figure 9.6 Variation of the linear coefficient p_l with (a) vibration angle and (b) rake angle. Reproduced from [17]. Copyright 2002 Taylor and Francis

Stone and Campbell continued Stone and Askari's work [18] by studying torsional-axial chatter. This work confirmed Bayly's finding that for the drilling case, the onset of chatter oscillation will happen at a frequency less than the natural frequency of the vibration mode, which is in contrast to the traditional case where the onset chatter frequency will be greater than the natural frequency. Authors pointed out that the previous research had an approximation that $\beta p_1 < \gamma$ (β – bifurcation parameter, p_1 – linear coefficient of G, γ – dimensionless damping coefficient) which causes the difference between the cases, resulting in the presence of closed loops in the traditional case.

The criticality of the Hopf bifurcation has been analyzed considering the sign of the quantity:

$$a = \frac{1}{8}[\Psi_{12}(0)(3f_{111} + f_{122}) + \Psi_{22}(0)(f_{112} + 3f_{222}]$$
$$- \frac{1}{8\omega}[(\Psi_{12}(0)^2 - \Psi_{22}(0)^2)f_{12}(f_{11} + f_{22}) + 2\Psi_{12}(0)\Psi_{22}(0)(f_{22}^2 - f_{11}^2)] \quad (9.15)$$

Where: The f_{jk} and f_{ijk} are functions of the parameters β, T, γ, p_0, p_1, p_2 (constant, linear and quadratic coefficients of G), θ(vibration angle), the Hopf frequency ω, and the centre manifold coefficients. By analyzing this equation it has been suggested that the bifurcation in the traditional case is subcritical for high values of T and supercritical for intermediate and small values of T. In the axial-torsional case it has been determined that the bifurcation is always subcritical.

An analysis for high-speed stability limits has been performed stating very interesting results, especially if one is interested in micro drilling, which would be in the field of very high speeds. In high speed limits, T becomes zero (large cutting speeds) and the equations do not provide further delay, but can be analyzed by conventional techniques.

For the T→0 limit, the equation of motion is expressed as:

$$\eta'' + \gamma\eta' + \eta - \beta(p_o + \tilde{p}_1\eta' + \tilde{p}_2\eta'^2) = 0 \tag{9.16}$$

If calculated, a=0 pointing out degeneracy of the system, suggesting that it would be necessary to determine higher order terms in order to define stability. These bifurcations re not well understood yet and are suggested as a good topic for future research. Due to all afore presented, macro and micro drilling cannot be analyzed the same way.

Stone and Askari's work was followed by research [19] which illustrates the effect of varying thickness and speed with linear stability boundaries and considers the change with varying cutting width. This paper is based on the models developed in the previously mentioned work and defines the stability regions with respect to depth of cut. The researchers have defined the traditional case with a high cutting speed having a supercritical bifurcation, while the case with low cutting speed is followed by a subcritical bifurcation. In the axial-torsional vibration case, for all physically reasonable situations, the bifurcation is shown to be subcritical. The implications of switching between supercritical and subcritical Hopf bifurcation, influence the stability of the cutting process. When approaching the supercritical Hopf bifurcation, the linear stability boundary determines when the cutting will become unstable. With the subcritical Hopf bifurcation, small perturbations can cause the instability.

The equation of motion used has been derived in [17] and rescaled in order to use depth of cut t_1 as the bifurcation parameter instead of width of cut w and accounting for chip thickness variation:

$$\eta'' + \gamma\eta' + \eta = \frac{w\tau}{k}\{t_1 - [\eta(t) - \eta(t-T)]\cos(\theta)\}(p_0 + \tilde{p}_1\eta' + \tilde{p}_2\eta'^2) \tag{9.17}$$

Where: $\tilde{p}_1 = \omega_0 p_1$ and $\tilde{p}_2 = \omega_0^2 p_2$.

Further, Dilley *et al.* studied the effect of the chisel edge on the chatter frequency in drilling in [20]. Owing to the problem of complex boundary conditions of the margin and chisel edge, most researchers have been dissuaded from studying such problems. A bending motion has been seen due to the movement of the chisel and therefore it is proven to be wrong to use fixed end conditions that would make the chisel stationary. This paper considers a simplified model where a spring end condition, which represents the chisel edge effect is used (Figure 9.7).

Experimental and analytical results of the natural frequency of the fixed-free drill have been compared and presented in Table 9.1.

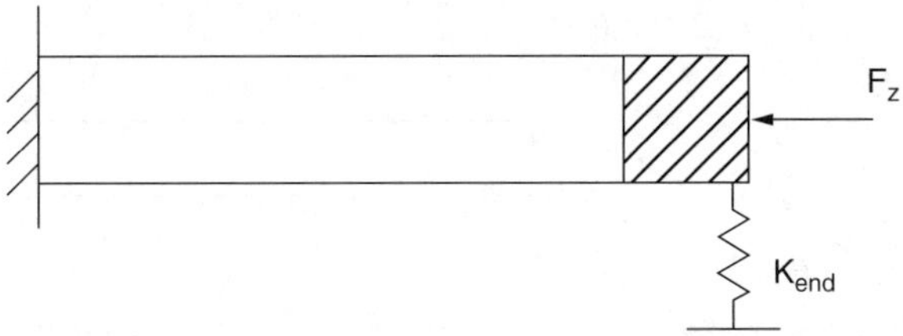

Figure 9.7 Single 1 degree of freedom model of the tool. Reproduced from [20]. Copyright Elsevier 2005

Table 9.1 Frequencies for the first bending mode

	Drill 1	Drill 2	Drill 3	Drill 4	Drill 5
Experimental modal analysis fixed-free	192	466/506	554/595	442	825
Analytical fixed-free	206	540	577	481	851
	(7.3%)	(15.9%, 6.7%)	(4.2% – 3.0%)	(8.8%)	(3.2%)
Analytical fixed-free (adj.)	192	466/506	554/595	442	825
Experimental modal analysis fixed-embedded	591/549	1945/2150	2288/2400	1382	3150
Analytical fixed-pinned (adj)	542	1857/2016	2206/2369	1178	3284
	(–8.3%, –1.3%)	(–4.5%, –6.2%)	(–3.6%, –1.3%)	(–14.8%)	(4.3%)
Analytical fixed-fixed (adj)	954	2980/3236	3540/3802	2123	5270
	(61.4%, 73.8%)	(53.2%, 50.5%)	(54.7%, 58.4%)	(53.6%)	(67.3%)

It can be seen that the analytical results of the fixed-free frequencies differ by 16% from the experimentally obtained frequencies, however, if the diameter and fluted section adjustments are conducted, the results are almost the same. Values from the fixed-pinned solution yields a 4.3–14.8% error, while the fixed-fixed model shows a significant error of 50.5–73.8%. The experimental natural frequency can be matched by attaching the spring to the tip of the fixed-free model.

As discussed in the previous paragraph, in order to have proper frequency matching with the experimental data, it is considered necessary to use a spring end condition to represent the contact between the chisel edge and workpiece. To accommodate this fact, a transverse spring and axial thrust force were added to the simple fixed-free beam, approximating the equation of motion using the Rayleigh-Ritz procedure to:

$$[M]\ddot{q}(t)+[K]q(t)=0 \tag{9.18}$$

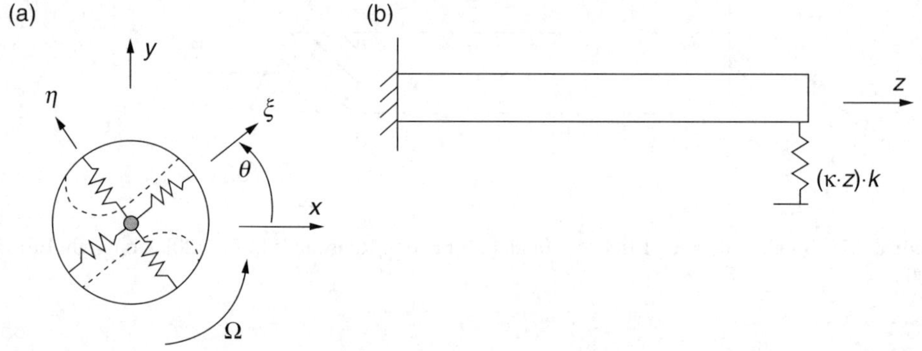

Figure 9.8 Drill model diagrams: a. 2-DOF rotating single-mode system and .b. simplified drill model with spring-end condition representing effects of margin engagement. Tool fixed frame: ξ - η. Space fixed frame: x-y, where z is the depth into the hole. Reproduced with permission from [21]. Copyright 2005 ASME

$$M_{ij} = \int_0^L \mu \psi_i(z)\psi_j(z)dz \tag{9.19}$$

$$K_{ij} = \int_0^L EI_y \psi_i''(z)\psi_j''(z)dz + K_{end}\psi_i(L)\psi_j(L) - \int_0^L F_z \psi_i'(z)\psi_j'(z)dz \tag{9.20}$$

Where: M – mass, K– stiffness, ψ – trial function, q – generalized coordinates, $(F_zL^2)/(EI_y)$ – dimensionless quantity.

After analyzing the effect of the chisel, Dilley *at al.* studied the problem of margin engagement and its effect on the frequency shift in drilling [21]. This paper showed that there is a significant frequency shift in the case of margin engagement.

The drill model has been presented in Figure 9.8:

The effect of margin engagement is incorporated into the equation by adding a spring constant, making the model a function of the depth of penetration into the hole:

$$M_r \begin{bmatrix} \ddot{\xi} \\ \ddot{\eta} \end{bmatrix} + C_r \begin{bmatrix} \dot{\xi} \\ \dot{\eta} \end{bmatrix} + K_{rM} \begin{bmatrix} \xi \\ \eta \end{bmatrix} = \vec{F}_{ext}$$

$$M_r = \begin{pmatrix} m_{\xi\xi} & 0 \\ 0 & m_{\eta\eta} \end{pmatrix}, \quad C_r = \begin{pmatrix} c_{\xi\xi} & -2m_{\xi\xi}\Omega \\ 2m_{\eta\eta}\Omega & c_{\eta\eta} \end{pmatrix} \tag{9.21}$$

$$K_{rm} = \begin{pmatrix} k_{\xi\xi}\left(1-k_{\xi\xi}z\right) - m_{\xi\xi}\Omega^2 & -c_{\xi\xi}\Omega \\ c_{\eta\eta}\Omega & k_{\eta\eta}\left(1-k_{\eta\eta}z\right) - m_{\eta\eta}\Omega^2 \end{pmatrix}$$

Where: M, C, K – mass, damping, stiffness, ξ – η – tool fixed frame.

The frequency shift due to the margin engagement can be seen from the waterfall plots shown in Figure 9.9. The plots show power spectral density functions as a function of drill depth.

It has been experimentally confirmed that the drill natural frequency increases with the depth of the hole. By showing that this does not happen when machining a tube, it has been proven that the margin effect is the sole reason of this frequency increase. Both of Dilley's papers have been aimed at obtaining and constructing predictive models for engineers with the aim of drilling process optimization.

9.2.2 *Bending Model*

Apart from the axial and torsional vibrations, bending vibrations have been found to be common in the drilling process as well. It has been suggested that bending vibrations have a more detrimental effect on hole quality due to the fact that the hole form changes drastically with the effect of bending (Figure 9.10).

Arvajeh and Ismail focused on the bending models and investigation of regenerative chatter in bending of both short and long drills (Figure 9.11). The first of two papers [22] is based on the model delivered by Ulsoy [23]. However, unlike the cited paper, the cutting forces with the regenerative effect have been added.

The drilling equation of motion in this case is:

$$\begin{pmatrix} M & 0 \\ 0 & M \end{pmatrix}\begin{pmatrix} \ddot{x} \\ \ddot{y} \end{pmatrix}+\begin{pmatrix} C & -2M\Omega \\ 2M\Omega & C \end{pmatrix}\begin{pmatrix} \dot{x} \\ \dot{y} \end{pmatrix} + \begin{pmatrix} K-M\Omega^2 & -C\Omega \\ C\Omega & K-M\Omega^2 \end{pmatrix}\begin{pmatrix} x \\ y \end{pmatrix} = -b\begin{pmatrix} P & -Q \\ -P & Q \end{pmatrix}\begin{pmatrix} x(t)+x(t-\tau) \\ y(t)+y(t-\tau) \end{pmatrix} \tag{9.22}$$

Where: M, C, K – bending dynamic properties of the drill, Ω – rotational speed of the drill, τ-time between each cutting pass, the matrix with P and Q is the matrix of bending cutting pressures.

Owing to the results presented in [24] which show that the chatter frequency is equal to the bending natural frequency when the tip of the drill is simply supported, the adjustment of the drill being presented as a clamped-pinned beam with lumped mass at the middle has been used. The analytical model developed by Rincon [25] has been used for obtaining cutting pressure coefficients for bending stability analysis. The authors have developed the relations between the cutting angles and cutting forces as well as rotational angles at the tip and deflections at the middle of the drill:

$$\begin{pmatrix} F'_x \\ F'_y \end{pmatrix} = b\begin{pmatrix} C'_{11} & C'_{12} \\ C'_{21} & C'_{22} \end{pmatrix}\begin{pmatrix} \Gamma_x \\ \Gamma_y \end{pmatrix} \qquad \Gamma_x = kx, \ \ \Gamma_y = ky \tag{9.23}$$

Where $C'_{11,\,12,\,21,\,22}$ are cutting coefficients, $\Gamma_{x,y}$ are the rotational angles at the tip, and b is the radial width of cut, x, y– deflections at the middle of the drill, k– constant.

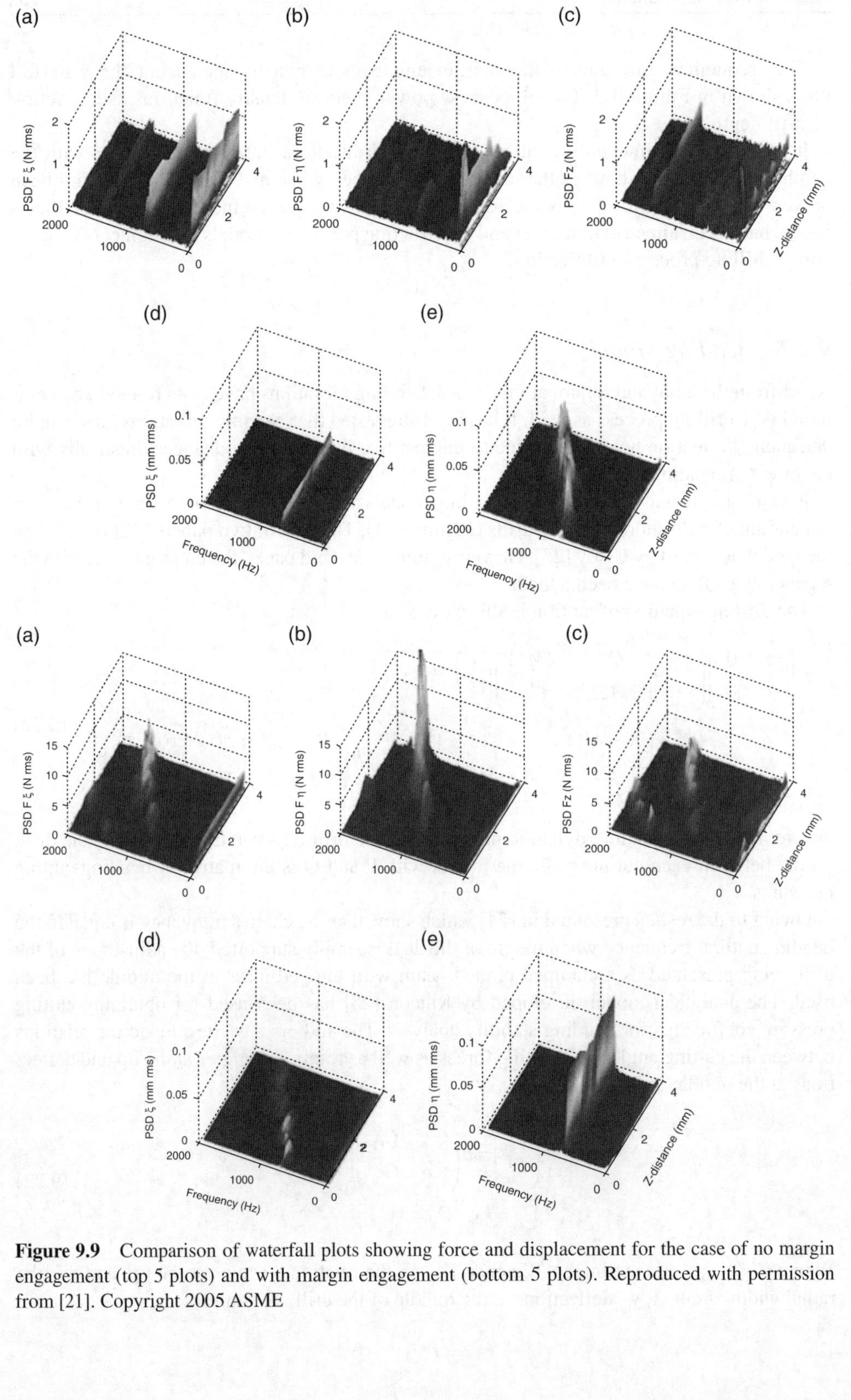

Figure 9.9 Comparison of waterfall plots showing force and displacement for the case of no margin engagement (top 5 plots) and with margin engagement (bottom 5 plots). Reproduced with permission from [21]. Copyright 2005 ASME

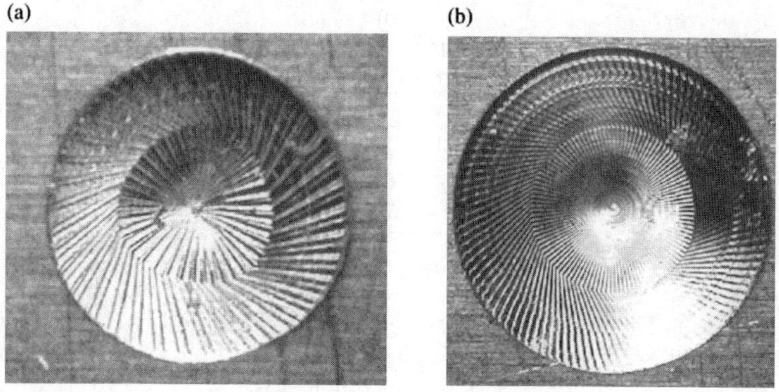

Figure 9.10 Striations in the bottom of the hole caused by vibration of the drill: (a) bending chatter; (b) torsional chatter. Reproduced from [20]. Copyright Elsevier 2005

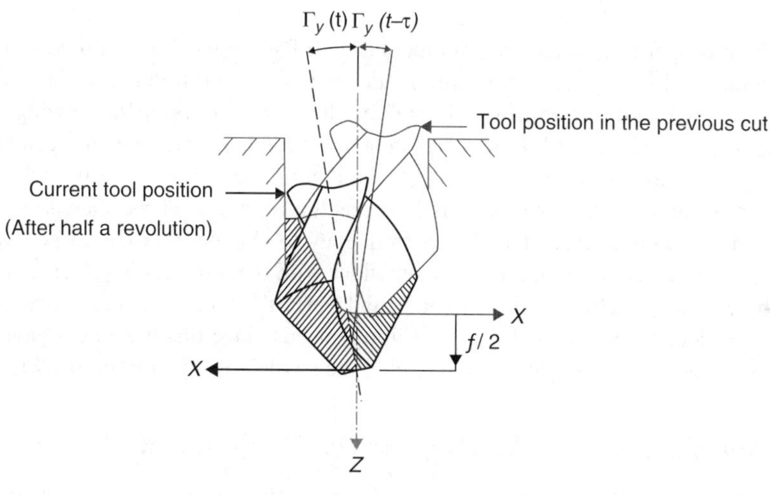

Figure 9.11 Regeneration effect in bending vibration. Reproduced from [22]. Copyright 2006 Elsevier

It has been shown that in this case the gyroscopic effect taken into consideration causes the natural frequencies of the system to vary according to the drill rotational speed. However, the drill rotational speed needs to be determined for the stability boundary. In this case, due to the dependency between the natural frequency and the speed at which the drill rotates, it is necessary to use the stationary drill natural frequency. Therefore, the relationship between the chatter frequency and natural frequency of the stationary drill has been defined:

$$f_c = rf_n \pm \frac{2\pi r^2 f_n}{4\pi N + 2\varphi \mp 2\pi r} \tag{9.24}$$

Where: r – ratio, N – integer number of full waves, φ – phase difference.

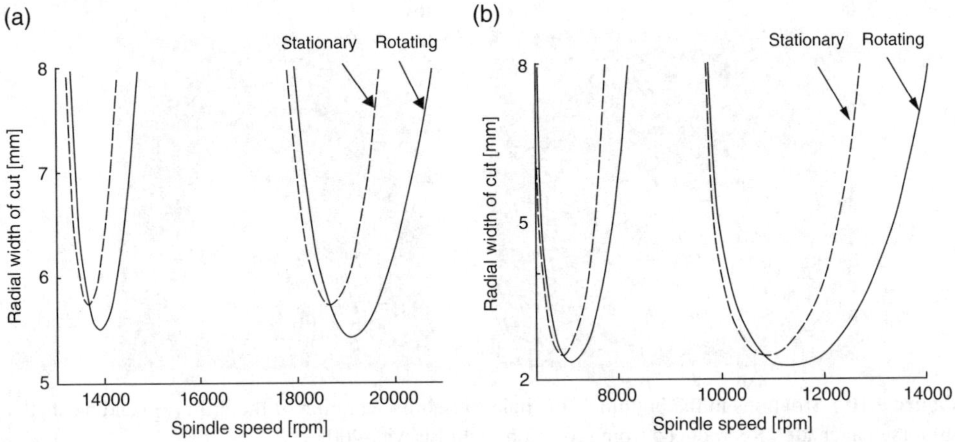

Figure 9.12 Stationary and bending stability lobes for (a) short, (b) long drill, showing widening and lowering of the lobes with increase in speed. Reproduced from [22]. Copyright 2006 Elsevier

Three-dimensional laser scans in combination with Patran and Nastran models have been used to obtain the drill surface and acquire mode shapes, natural frequencies and stiffness of drills. After all the modelling and analysis, stability lobes were developed, showing results for both the short and the long drill. It has been shown that the higher frequencies can be ignored since they are less critical. It has also been shown that the bottom of the lobes which defines the minimum stability of the process gets lower as the rotational speed increases due to the gyroscopic force, which causes the chatter frequency to change with the speed. Apart from lowering the stability minimum, the lobes get wider with the gyroscopic effect. Owing to the fact that the gyroscopic effect will both widen and lower the lobes, it is necessary to take this effect into consideration when using high drilling speeds since this would extremely change the stability boundaries for those cases, reducing the stable area to a minimum (Figure 9.12).

9.2.3 Combination of the Bending and Torsional-axial Models

As the continuation of the previously presented work, reference [26] presents a combination of the effects of torsion and bending on chatter in drilling using 'short' and 'long' drills. The analytical work regarding the torsional limits of stability in drilling, have been developed using Bayly's work [3].

Bending equations with two degrees of freedom have been combined with the equation of motion in the axial direction (z):

$$\begin{aligned} &M\ddot{x}(t)+C\dot{x}(t)+(K-M\Omega^2)x(t)-2M\Omega\dot{y}(t)-C\Omega y(t)=-F_x'(t)\\ &M\ddot{y}(t)+C\dot{y}(t)+(K-M\Omega^2)y(t)-2M\Omega\dot{x}(t)-C\Omega x(t)=-F_y'(t)\\ &M_a\ddot{z}(t)+C_a\dot{z}(t)+K_a z(t)=-F_z(t) \end{aligned} \tag{9.25}$$

Where: $F_x'(t)$, $F_z(t)$ $F_y'(t)$ -variations of cutting forces, x(t), y(t), z(t) - drill deflections in the X, Y and Z directions, x(t), y(t) - deflections of the middle of the drills, z(t) - axial elongation of the tip, M, C, K – dynamic properties of the bending modes, M_a, C_a, K_a – dynamic properties of the axial mode, Ω – drill rotational speed.

A time domain simulation of chatter drilling has shown strong agreement between the experiments and analytical derivations for the 'long' drills, but not for the 'short' drills due to inaccurate mathematical modelling. It has been experimentally determined that there were two sets of chatter marks for the 'long' drill, corresponding to the bending mode and the torsional mode. For the 'short' drill all the chatter marks correspond to the torsional mode. This was explained by the fact that the support in the case of the 'short' drill acts more like a spring rather than a rigid pin and therefore makes the bending mode less important. The comparison between the experimental and theoretical stability lobes has been presented in Figure 9.13 where squares show stable cuts, circles show unstable cuts due to torsional chatter and crossed out circles show unstable cuts due to both bending and torsion.

There are several parameters that have not been included in the study: process damping due the interaction between the flank faces of the drill and the just cut surface (the 'rubbing' mechanism tends to dampen the vibration and it plays an important role in the stability of the tool), and the chisel edge in cases where there is no previous pilot hole as well as edge forces.

Dynamics, kinematics and stability in drilling operations have been covered in a two-part paper by Roukema and Altintas. The first part of the paper [7] deals with the problem of combined lateral drill deflection, lateral chatter vibrations and combined lateral and torsional-axial vibrations. The presented model was used for prediction of lateral chatter and prediction of the combination of lateral and torsional-axial vibrations.

The general equation of motion for the dynamic drilling system in the stationary frame are:

$$[M]\begin{Bmatrix} \ddot{x}(t) \\ \ddot{y}(t) \\ \ddot{z}(t) \\ \ddot{\theta}(t) \end{Bmatrix} + [C]\begin{Bmatrix} \dot{x}(t) \\ \dot{y}(t) \\ \dot{z}(t) \\ \dot{\theta}(t) \end{Bmatrix} + [K]\begin{Bmatrix} x(t) \\ y(t) \\ z(t) \\ \theta(t) \end{Bmatrix} = \begin{Bmatrix} F_x \\ F_y \\ F_z \\ T_c \end{Bmatrix} \tag{9.26}$$

Where: x, y - lateral, z - axial, θ - torsional deflection of the drill, matrices M, C, K - mass, damping and stiffness characteristics at the drill tip, Fx, Fy lateral forces, thrust Fz force, and Tc - torque.

The total displacement in any direction can be calculated by summing vibrations in all natural modes in a chosen direction, giving the overall effect:

$$q(t) = \sum_{i=1}^{k_m} \Delta q(t) \tag{9.27}$$

These deflections will influence the chip thickness, changing the force distribution and effecting the vibrations in all directions, therefore the cutting force system has to be coupled through the dynamic chip thickness distribution along the cutting edges.

The influence of pilot hole size, spindle speed, torsional-axial chatter on lateral forces was experimentally observed and compared with simulated results. As the authors point out, this paper has a fundamental contribution to the investigation of drilling by creating a complex numerical model that incorporates cutting force models, rigid body motion as well as vibrations that occur in the process of drilling. Apart from combining these effects, the authors include both grinding errors and tool misalignment effects, (Figure 9.14).

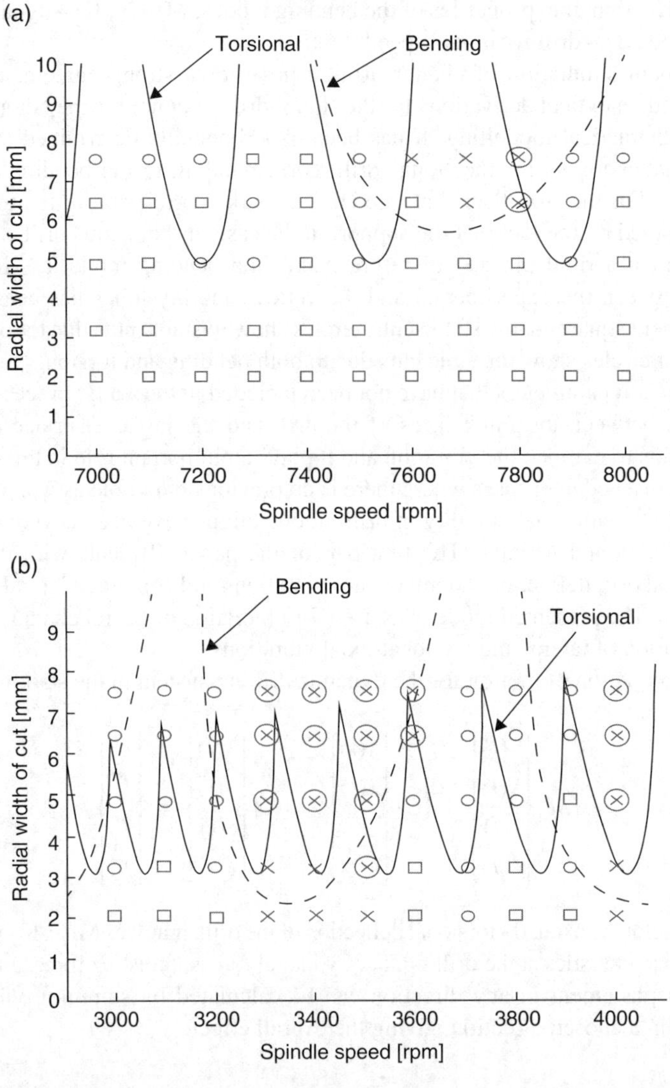

Figure 9.13 Comparison between the experimental and theoretical results for the short (left) and long (right) drills. Reproduced from [26]. Copyright 2006 Elsevier

From the data collected from the experiments, it has been shown that the coupling of lateral and torsional vibrations does not happen; torsional-axial chatter is seen as dominant, while lateral chatter is absent. The lateral chatter could become more important in longer, slender drills. On the contrary to the experimental work, the model actually predicts a small value for the coupling of vibrations and actually predicts both torsional-axial and lateral chatter vibrations (Figure 9.15).

Therefore, it can be seen that the model does not actually agree with the experimental results. The predicted values for stiffness and damping are not in agreement with the obtained

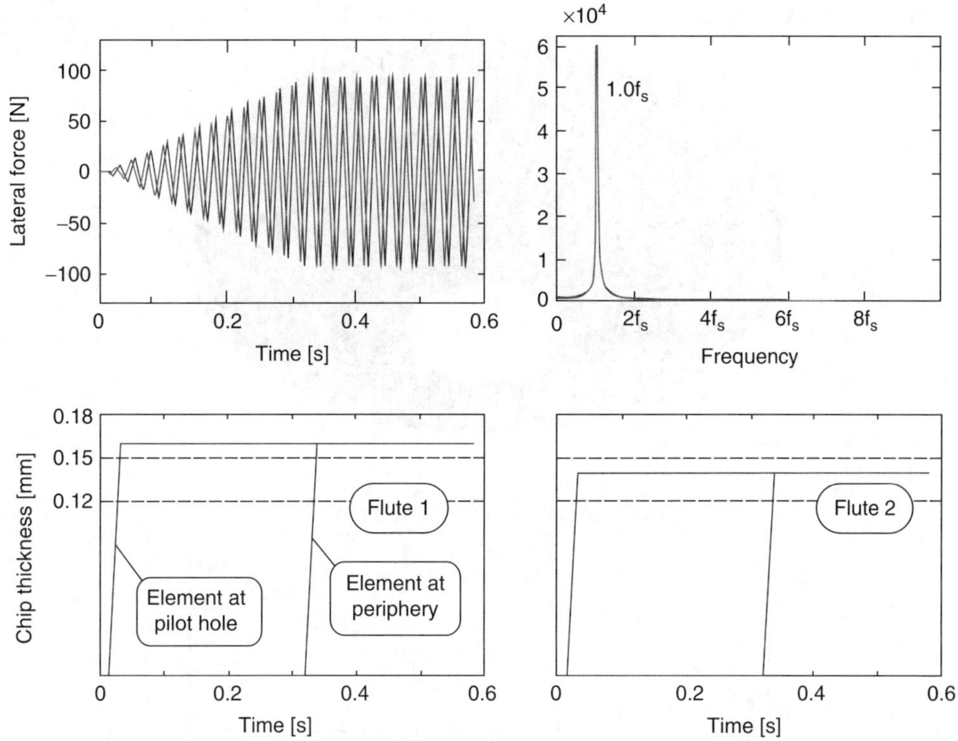

Figure 9.14 Simulation of the chip thickness, thrust forces and Fourier spectra for a drill with a grinding error presented through the lip hight difference of 10 μm. Parameters used: 24000 rpm, 0.30 mm/rev, pilot hole diameter 4 mm, hole depth 15 mm. Reproduced from [7]. Copyright 2007 Elsevier

experimental data, and issues like the rubbing of the cutting lips against the bottom of the hole which is wavy in the first place, hole wall contact with drill flutes, the change in the contact boundaries as the drill penetrates deeper in the hole, the friction effect all add to the complexity of accurate prediction of chatter vibrations.

The second part of the two-part paper [6] presents a general solution of drilling chatter stability in the frequency domain. In order to determine the regenerative chip thickness, the drill's flexibility in torsional, axial and lateral directions has been analyzed.

All the mentioned vibrations have an influence on the regenerative chip thickness. Mathematically, the change in chip thickness due to regenerative displacements can be defined as:

$$
\begin{aligned}
dh_1 &= \frac{dx\cos\Omega t - dy\sin\Omega t}{\tan\kappa t} \\
dh_2 &= -\frac{dx\cos\Omega t - dy\sin\Omega t}{\tan\kappa t}
\end{aligned}
\tag{9.28}
$$

Where: κt is the tip angle of the drill and Ωt is the tool rotation angle.

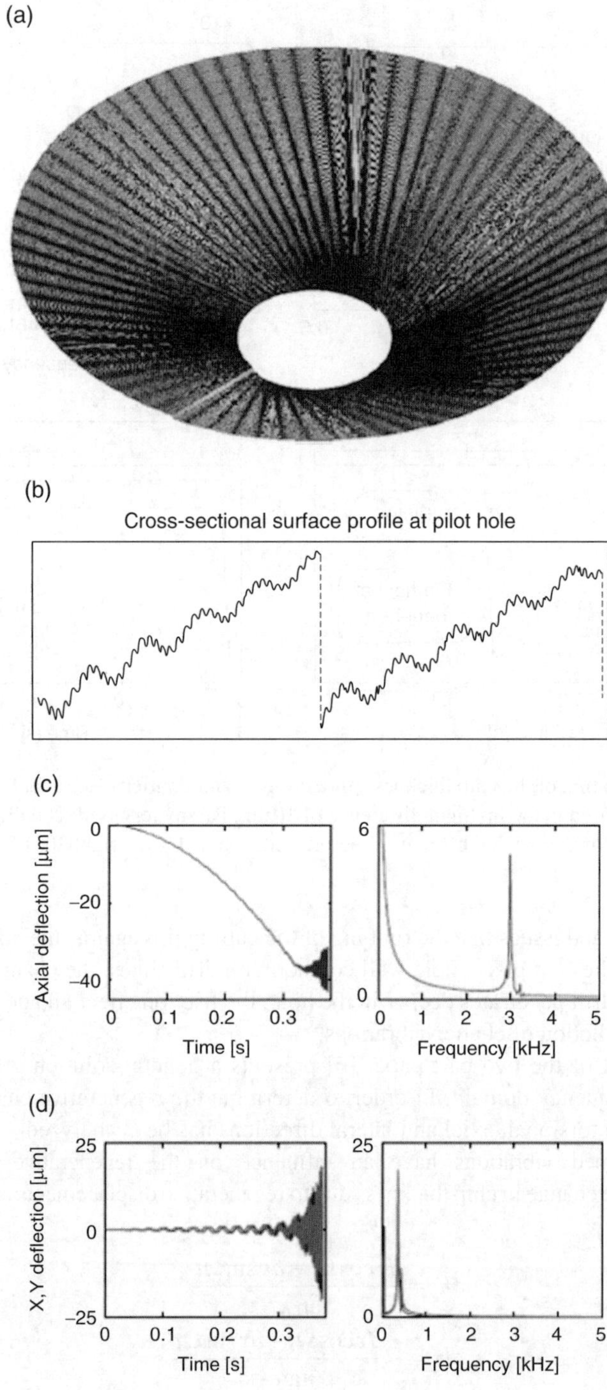

Figure 9.15 Simulation for a flexible drill in torsional and axial directions: (a) bottom of workpiece, (b) cross sectional surface profile at pilot hole, (c) axial tool deflection and Fourier spectrum, (d) lateral tool deflection and Fourier spectrum. Reproduced from [7]. Copyright 2007 Elsevier

The effect of the torsional vibrations on the chip thickness can be defined as:

$$dh_1 = dh_2 = -d\theta \times \frac{1}{2\pi} \times f_r \tag{9.29}$$

Where f_r is feed per revolution, θ – torsional tool tip deflection.

The effect of the axial vibrations on the chip thickness can be defined as:

$$dh_1 = dh_2 = -dz \tag{9.30}$$

When all the vibrations are taken into consideration, the effect of the whole system on the change in chip thickness is:

$$\begin{Bmatrix} dh_1 \\ dh_2 \end{Bmatrix} = \begin{Bmatrix} \frac{1}{\tan \kappa_t}(dx\cos\Omega t - dy\sin\Omega t) - dz - \frac{f_r}{2\pi}d\theta \\ -\frac{1}{\tan \kappa_t}(dx\cos\Omega t - dy\sin\Omega t) - dz - \frac{f_r}{2\pi}d\theta \end{Bmatrix} \tag{9.31}$$

The dynamic forces can therefore be expressed as:

$$\begin{Bmatrix} Fx \\ Fy \\ Fz \\ Tc \end{Bmatrix} = \begin{Bmatrix} k_{tc}b(dh_1 - dh_2)\sin\Omega t - k_{tc}k_{rc}(dh_1 - dh_2)\cos\Omega t \\ k_{tc}b(dh_1 - dh_2)\cos\Omega t - k_{tc}k_{rc}(dh_1 - dh_2)\sin\Omega t \\ k_{tc}k_{ac}b(dh_1 - dh_2) \\ \{k_{tc}b(dh_1 + dh_2)\}(1 - k_{rc})R_t \end{Bmatrix} \tag{9.32}$$

Where: Fx, Fy, Fz – cutting forces acting on the tool tip, Tc – cutting torque acting on the tool tip, R_t – torque arm – from tangential and radial forces, k_{tc} – tangential cutting stiffness, k_{rc}, k_{ac} – radial and axial force cutting stiffness factors, b- radial depth of cut.

The dynamic cutting forces can be summarized as:

$$\{F(t)\} = bk_{tc}[B(t)]\{\Delta r\} \tag{9.33}$$

Where, [B(t)] is the dynamic drilling coefficient matrix dependent on time and contains the negative regenerative displacements.

In this paper, the frequency response method has been used and compared with results from other models. The frequency response matrix at the tool tip has been defined as:

$$\Phi(i\omega) = \begin{bmatrix} \Phi_{xx}(i\omega) & 0 & 0 & 0 \\ 0 & \Phi_{yy}(i\omega) & 0 & 0 \\ 0 & 0 & \Phi_{zz}(i\omega) & \Phi_{z\theta}(i\omega) \\ 0 & 0 & \Phi_{\theta z}(i\omega) & \Phi_{\theta\theta}(i\omega) \end{bmatrix} \tag{9.34}$$

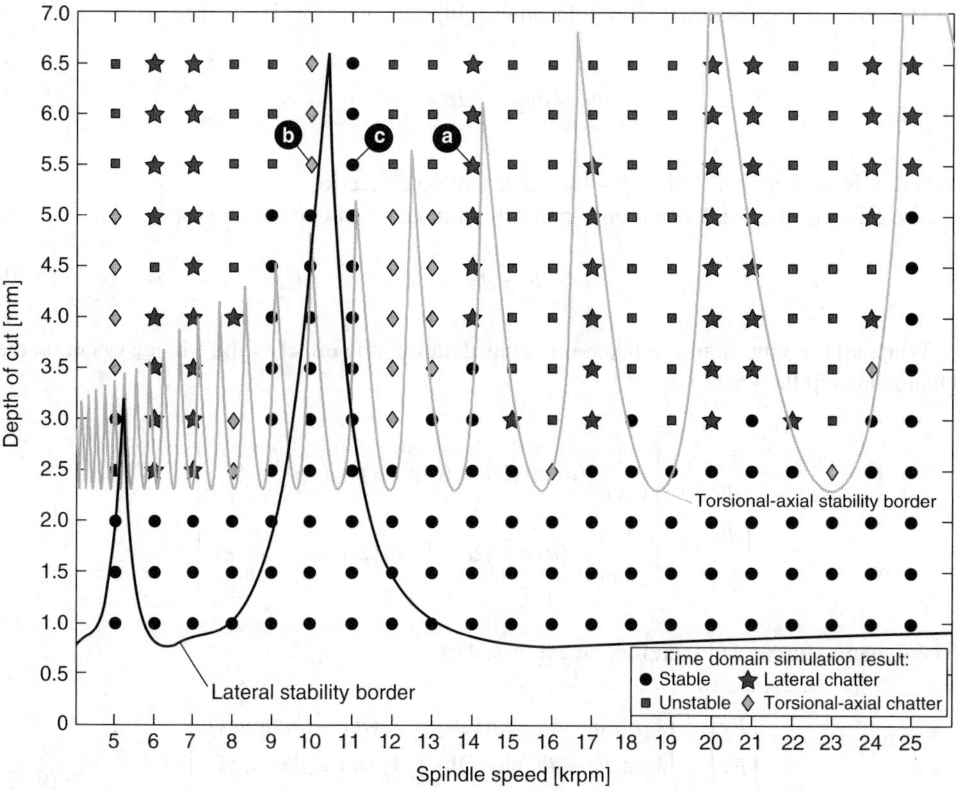

Figure 9.16 Comparison of the proposed model with the time domain model. Reproduced from [6]. Copyright 2007 Elsevier

Where: Φaa (iω) is the direct frequency response function and Φab (iω) is the cross frequency response function.

After extensive computation the authors have expressed the eigenvalues as:
Lateral:

$$\Lambda_1 = \frac{-\gamma_1 + \sqrt{\gamma_1^2 - 4\gamma_0}}{2\gamma_0}, \quad \Lambda_2 = \frac{-\gamma_1 + \sqrt{\gamma_1^2 - 4\gamma_0}}{2\gamma_0} \tag{9.35}$$

Torsional - axial:

$$\Lambda_1 = \frac{-\gamma_3 + \sqrt{\gamma_3^2 - 4\gamma_2}}{2\gamma_2}, \quad \Lambda_2 = \frac{-\gamma_3 + \sqrt{\gamma_3^2 - 4\gamma_2}}{2\gamma_2} \tag{9.36}$$

Where $\gamma_0, \gamma_1, \gamma_2, \gamma_3$ – coefficients of the characteristic equation.

The results show that frequency domain solutions actually give a more conservative stability estimate than the time domain model (Figure 9.16).

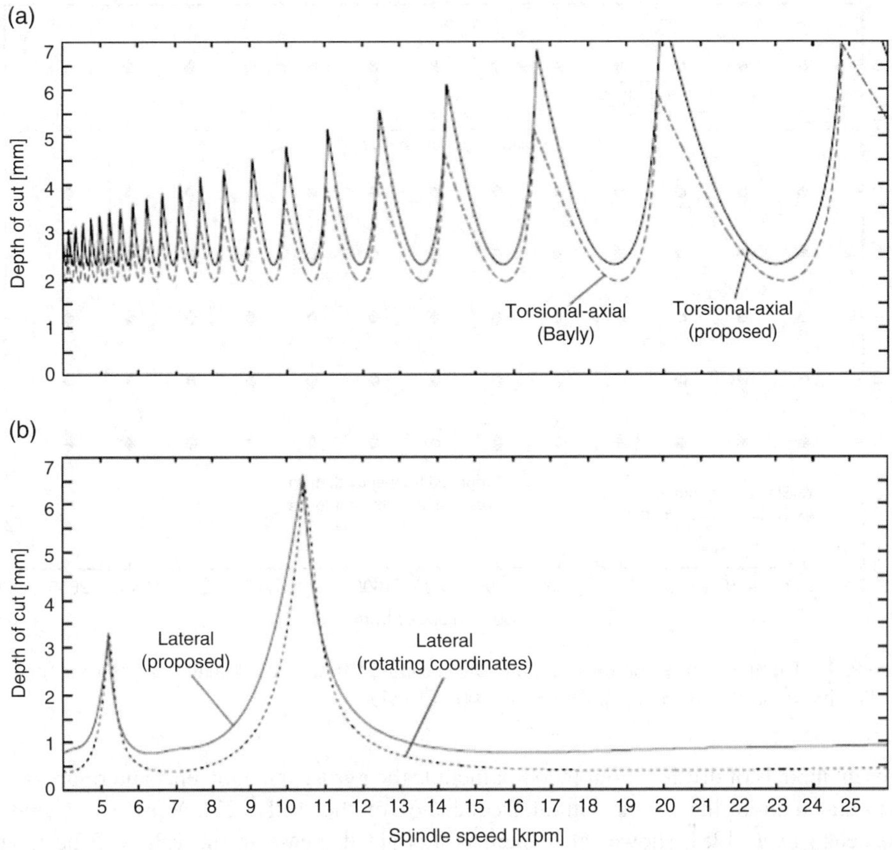

Figure 9.17 (a) Comparison of the proposed solution to the torsional-axial stability model (b) Comparison of the proposed lateral stability model with the rotating coordinates model. Reproduced from [6]. Copyright 2007 Elsevier

By comparing the model with the results from other researchers [3] it has been shown that the values for critical depth of cut are less conservative in this particular model, while the shape of stability lobes is different. The lateral chatter stability was compared to a rotating coordinate approach, where the difference is in using an average of time varying coefficients instead of time invariant cutting coefficients (Figure 9.17).

This comparison shows good agreement. The frequency domain solutions have been compared with the experimental data to establish the agreement and validity of the model used. Unfortunately, the stability lobes obtained are both too small and too narrow to be used for selection of actual cutting conditions (Figure 9.18). The experiments are not in agreement with the linear stability prediction laws and show that if the spindle speed is below 2000 rpm drilling is always stable and stability rises with decrease in spindle speed. It has even been stated that stability increases with chisel edge engagement. Although this paper has the most comprehensive model and covers the largest number of influential parameters, it is not in agreement with the experimental results and needs further work.

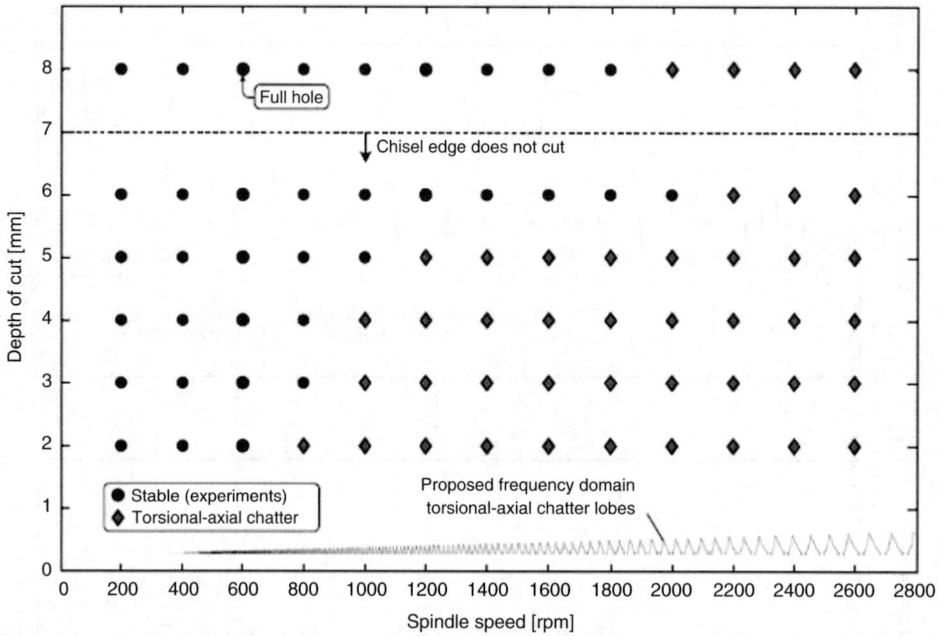

Figure 9.18 Comparison of the experimental and proposed frequency domain solution for torsional axial stability. Reproduced from [6]. Copyright 2007 Elsevier

Present models of drilling motion do not include the gyroscopic moments and rotary inertia effects and their influence on natural frequencies of the drills [27]. Work conducted by Timoshenko *et al.* [28], showed that there is a slight decrease in the value of the natural frequency calculated with the effect of rotary inertia. The non-linear equations of motion have been linearized, including gyroscopic moments and rotary inertia:

$$
\begin{aligned}
& m\ddot{v}_1 - 2m\omega_0\dot{v}_2 - m\omega_0^2 v_1 - J_1\ddot{v}_{1zz} + J_p\beta_0\ddot{v}_{2zz} + J_2\beta_0^2\ddot{v}_1 + J_2\omega_0^2\beta_0^2\dot{v}_1 \\
& \quad + J_p\omega_0^2\beta_0 v_{2z} - J_1\omega_0^2 v_{1zz} - EI_1 v_{1zzzz} - 2E(I_1 + I_2)\beta_0 v_{2zzz} \\
& \quad - E(2I_1 + 4I_2)\beta_0^2 v_{1zz} + 2E(I_1 + I_2)\beta_0^3 v_{2z} + EI_1\beta_0^4 v_1 \\
& \quad + Fz[v_{1zz} - 2\beta_0 v_{2z} - \beta_0^2 v_1] = F_\Xi \\
& m\ddot{v}_2 - 2m\omega_0\dot{v}_1 - m\omega_0^2 v_2 - J_2\ddot{v}_{2zz} + J_p\beta_0\ddot{v}_{1zz} \\
& \quad + J_1\beta_0^2\ddot{v}_2 + J_1\omega_0^2\beta_0^2\dot{v}_2 + J_p\omega_0^2\beta_0 v_{1z} - J_2\omega_0^2 v_{2zz} \\
& \quad + EI_2 v_{2zzzz} + 2E(I_1 + I_2)\beta_0 v_{1zzz} - E(4I_1 + 2I_2)\beta_0^2 v_{2zz} \\
& \quad - 2E(I_1 + I_2)\beta_0^3 v_{1z} + EI_2\beta_0^4 v_2 + Fz[v_{2zz} - 2\beta_0 v1 - \beta_0^2 v_2] \\
& \quad = F_\Psi
\end{aligned}
\tag{9.37}
$$

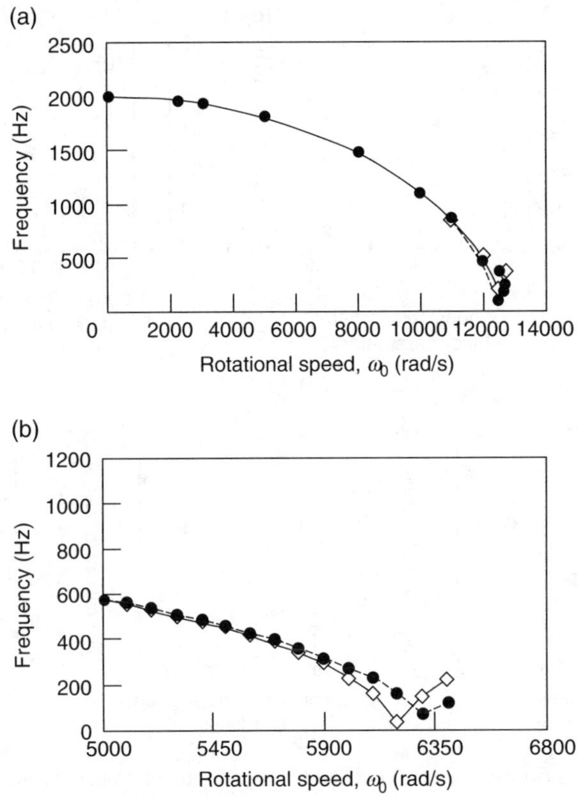

Figure 9.19 Effect of the rotary inertia and gyroscopic moment on the fundamental frequency for the (a) short drill and (b) long drill. ◊ - no rotary inertia and gyroscopic moment, • - with rotary inertia and gyroscopic moment. Reproduced from [27]. Copyright Elsevier 1994

Rincon *et al.* [27] confirm Timoshenko's statement showing that the effects of the gyroscopic moment and rotary inertia are not significant if the rotational speed is lower than 60 000 rpm for longer drills and even 120 000 rpm for short drills (Figure 9.19).

Authors have also explored the effect of drill geometry – coolant holes, web taper and number of flutes – showing that increasing the web taper causes a slight decrease in bending frequency, the increase in the number of flutes increases the stiffness of the drill bit and the increase in the coolant hole size slightly increases the bending frequency of the drill, Figure 9.20 (a, b, d). The authors also investigated the effect of material properties along the drill bit Figure 9.20 (c), showing that the natural frequency of a steel tool increases by 26% if steel is replaced by carbide and 20% if the drill was made as a combination of steel and carbide.

9.2.4 Chatter Suppression

Chatter suppression has been one of the main directions of research in the area of vibrations in machining. Chatter control in turning and milling has been extensively covered in [29–38]. Spindle speed variation methods have been successfully applied for chatter suppression in

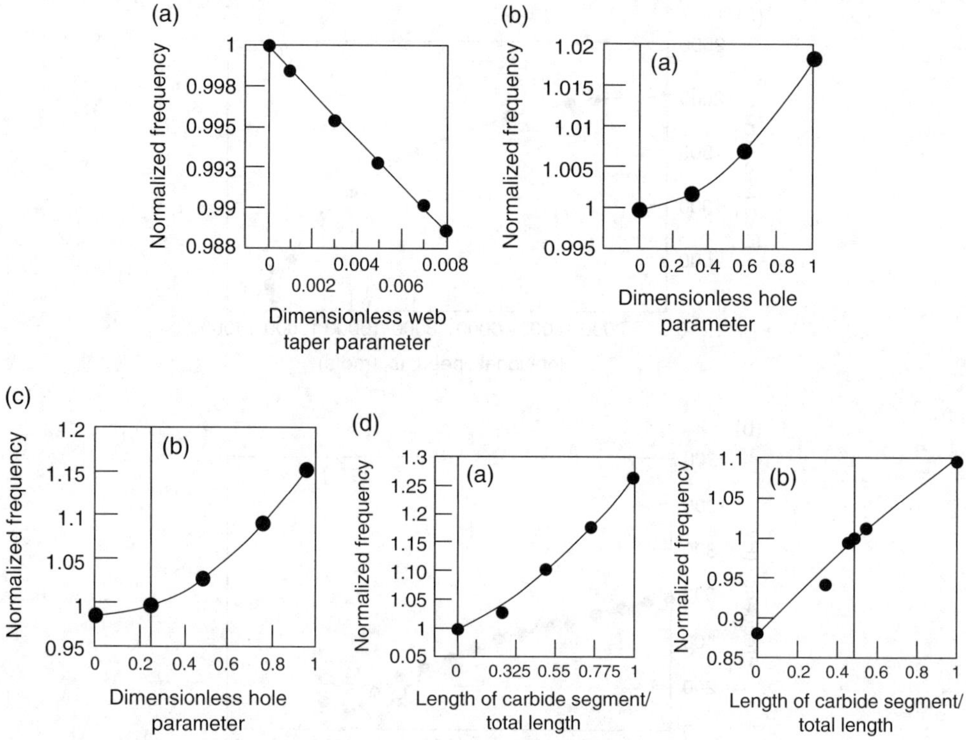

Figure 9.20 Effect of (a) web taper (b) coolant holes (c) material properties and (d) number of flutes on the normalized frequency. Reproduced from [27]. Copyright Elsevier 1994

turning and milling operations. Reference [39] presents a spindle speed variation method for chatter suppression in drilling. It has been shown that based on the stability charts chatter can be suppressed by changing the spindle speed to a certain range. Using thrust force as the basis for chatter detection, authors have presented both theoretical and experimental results. The predicted stable spindle speeds are considered to be:

$$n_s = \frac{30fc}{N} \tag{9.38}$$

Where N = 1, 2, 3, …, and f_c is the chatter frequency.

Figures 9.21 and 9.22 show the effect of the spindle speed variation method on chatter suppression.

Further, in reference [40], the relationship between the spindle speed and the phase shift has been studied. It has been shown that phase shift can be modified by changing spindle speed. Therefore, the cut width limit can be increased or decreased by the change of the spindle speed as well. Chatter occurs when the width of cut is greater than the limiting value and is suppressed if the spindle speed is changed so that the width of cut is smaller than the limiting value. Two strategies are investigated: the first one that states that the regulation of the spindle speed to

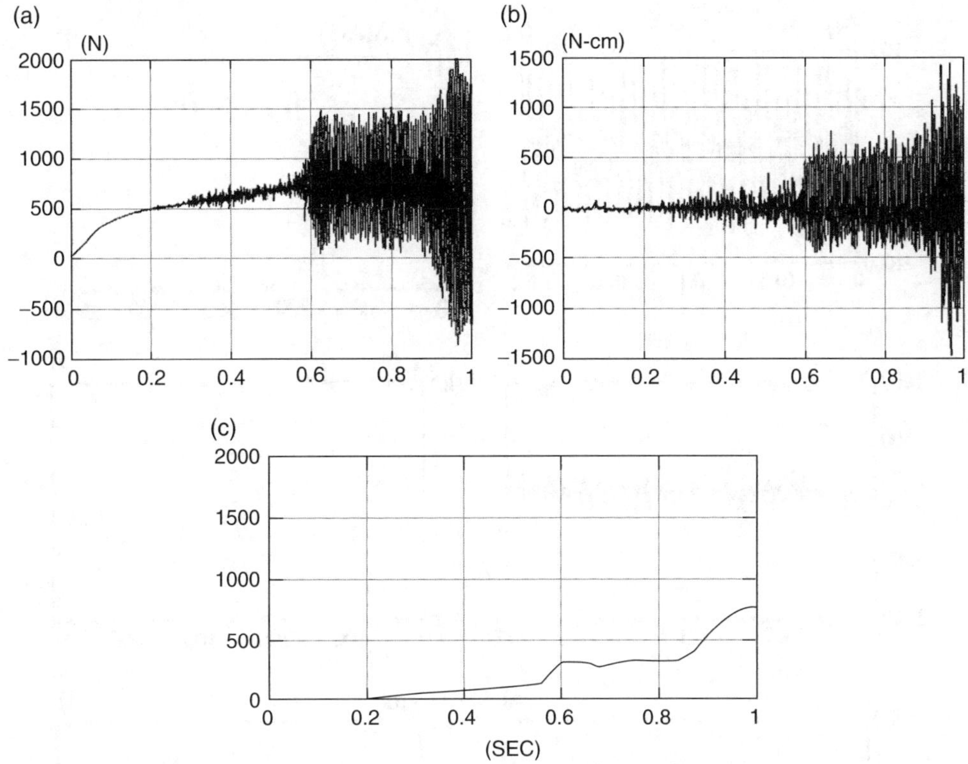

Figure 9.21 Thrust foce signal, torque signal and the standard deviation of the torque signal indicating strong chatter for the initial spindle speed 1500 rpm, feed per revolution 0.266 mm. Reproduced with permission from [39]. Copyright 1994 ASME

phase shift will be optimal at 2π, and the second one at π/2. It has been experimentally determined that the idea of the optimal shift being π/2 is impossible while the 2π phase shift has been confirmed.

The value of the width of cut at the stability limit has been defined as:

$$b_{\lim} = \frac{-1}{k_s G_m (jf_c)(1-e^{-j\varepsilon})} \tag{9.39}$$

Where, G_m (jω) is the frequency response function of the machine tool structure, f_c (rad/s) is the chatter frequency, and $\varepsilon = \omega T$ is the phase shift between the inner and outer modulation.

The relationship between the tooth passing frequency and chatter frequency has been defined as:

$$\frac{f_c}{f_t} = N + \frac{\varepsilon}{2\pi} \tag{9.40}$$

Where, $f_t = zn_s$ (z – number of cutting edges, ns – spindle speed), N is the largest integer to satisfy the equation and $0 < \varepsilon/2\pi \leq 1$. $0 < \varepsilon/_{2\pi} \leq 1$

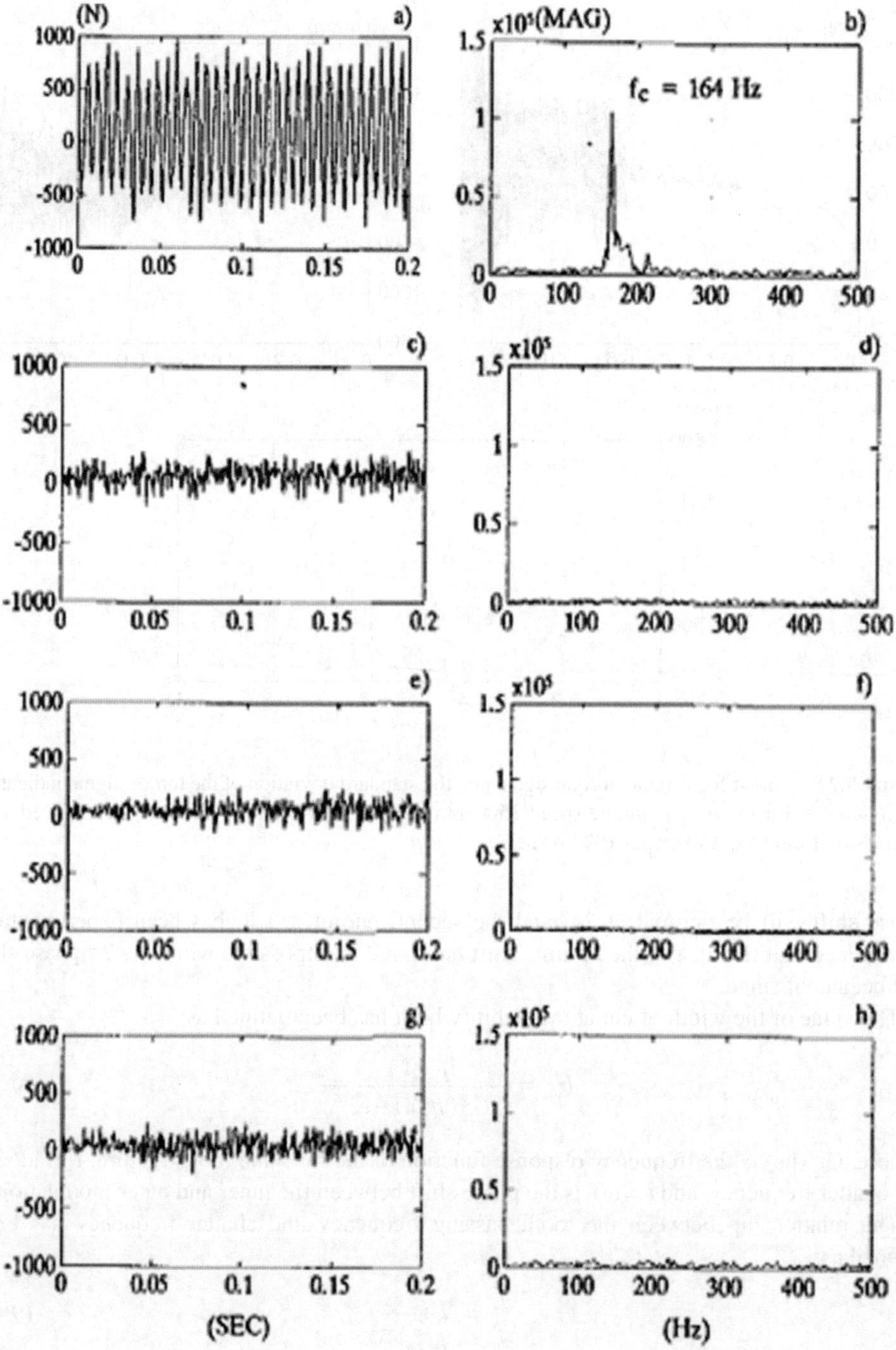

Figure 9.22 Results of the chatter suppression showing elimination of chatter: Thrust force signal and coresponding spectrum. Reproduced with permission from [39]. Copyright 1994 ASME

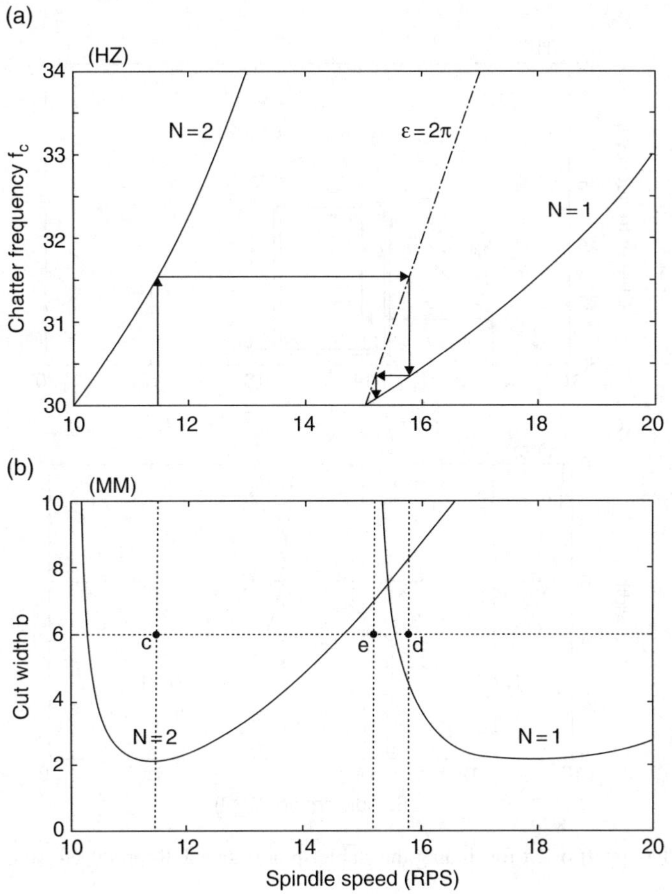

Figure 9.23 Phase shift of 2π for finding the stable spindle speed. Reproduced from [41]. Copyright 1997 Elsevier

When the phase shift was chosen to be 2π, the stable spindle speed was calculated to be:

$$n_s = \frac{f_c}{(N+1)z} \tag{9.41}$$

In this case it was shown that chatter frequency curve and the stable spindle speed curve intersect close at the place close to the largest stability limit of the width of cut b_{lim} (Figure 9.23).

When the phase shift is chosen as π/2, the stable spindle speed is:

$$n_s = \frac{f_c}{(N+1/4)z} \tag{9.42}$$

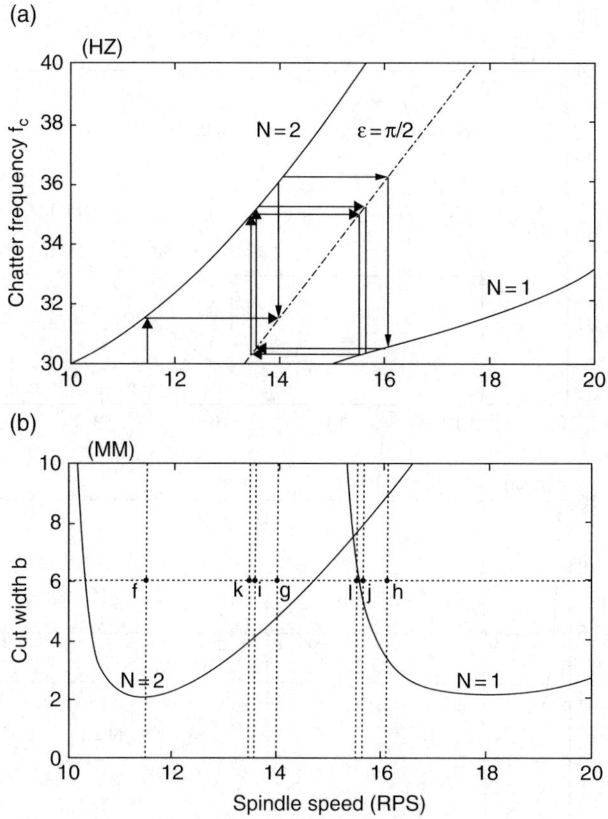

Figure 9.24 Phase shift of 2π for finding the stable spindle speed. Reproduced from [41]. Copyright 1997 Elsevier

It has been shown that in this case, the chatter frequency curve and the stable spindle speed curve do not intersect, therefore, proving that it is impossible to use this combination (Figure 9.24).

After defining the chatter frequencies and the optimal shifts, the authors continued their research in the area of chatter detection working on neural networks for reduction of chatter in drilling [8].

9.2.5 Research Challenges

There are many research challenges to be considered in chatter recognition and suppression in drilling operations. Modelling the process becomes harder with the complexity and modifications of the chisel edge of the drill. Further, prediction of cutting forces in this region is difficult due to the variety of cutting speeds and rake and oblique angles along the cutting edge. As presented in this review paper, in order to solve the problems many researchers have simplified them by analyzing one model or one problem at a time. The fact is that the axial, torsional and lateral vibrations can be acting at the same time in drilling, the effect of the chisel edge, margin effect, pilot hole size, and so on, are going to be present as well and the optimal model would need to have the option to incorporate all these effects in order for it to be successfully used in industrial applications.

9.3 Investigation of Chatter in Micro Drilling

In machining operations it is well known that chatter presents one of the problems affecting the surface of the part as well as the tool used. Chatter issues have been investigated since the 1950s in various ways, but what is common for all the investigations is that it is necessary to have accurate measurements of the Frequency Response Function (FRF) at the tip of the tool when it is attached to the tool holder and spindle. For macro testing, the impact test is sufficient for dynamic measurements, but it is impossible to apply the same system to micro tools due to their size and fragility. Impact hammer forces would be detrimental to the tools used in micro machining. The procedure is also impossible to be applied due to the mass loading necessary if an accelerometer is to be used. Many times, tools have been modelled as cantilever beams, incorrectly representing the actual dynamics of the tool tip.

Receptance coupling technique developed by Park *et al.* [42] suggests coupling of analytical and experimental FRFs of the components with the aim of obtaining the response of the assembly as a whole (Figure 9.25).

The accuracy of the receptance coupling technique depends on accurate identification of the joint dynamics of the substructures at the assembly joint, and the FRFs of each substructure. Any noise in the FRF and inaccuracies in the joint dynamic parameters amplify the errors in obtaining the FRF for the assembled tool-spindle system. This paper proposes an algorithm that allows analytical extraction of rotational dynamics at the joints from linear displacements and impact force tests. The authors have analytically identified the rotational dynamic response of the spindle-holder by substituting the direct and cross FRF measurements taken at the free end of the assembly and joint of the blank cylinder in the receptance coupling expressions. Combining FE analysis and experimental work led to acquisition of the overall dynamics of the system.

The FRF of the substructure A can be defined as:

$$\begin{Bmatrix} X_1 \\ X_{A,2} \end{Bmatrix} = \begin{bmatrix} H_{A,11} & H_{A,12} \\ H_{A,21} & H_{A,22} \end{bmatrix} \begin{Bmatrix} F_1 \\ F_2 \end{Bmatrix} \tag{9.43}$$

The FRF of the substructure B can be defined as:

$$\{X_{B,2}\} = [H_{B,22}]\{F_{B,2}\} \tag{9.44}$$

Where: X, F – displacement and force vectors applied on the structure. $H_{A,ij}$ vectors - FRFs between points i and j.

Boundary conditions used for coupling substructures A (free – free model of the end mill) and B (spindle) can be defined as:

$$\begin{aligned} F_2 &= F_{A,2} + F_{B,2} \\ X_2 &= X_{A,2} = X_{B,2} \end{aligned} \tag{9.45}$$

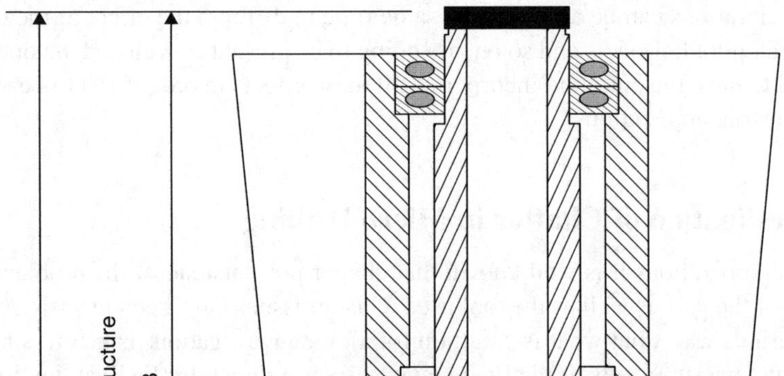

Figure 9.25 Spindle/tool holder and tool substructures. Reproduced from [42]. Copyright 2003 Elsevier

The displacements X_1 and X_2 can be expressed in a matrix form as:

$$\begin{Bmatrix} X_1 \\ X_2 \end{Bmatrix} = \begin{bmatrix} (H_{A,11} - H_{A,12} H_2^{-1} H_{A,21}) & (H_{A,12} - H_{A,12} H_2^{-1} H_{A,22}) \\ (H_{A,21} - H_{A,22} H_2^{-1} H_{A,21}) & (H_{A,22} - H_{A,22} H_2^{-1} H_{A,22}) \end{bmatrix} \tag{9.46}$$

The direct and cross receptances at the tool tip can be expressed as:

$$\begin{aligned} \frac{X_1}{F_1} &= H_{A,11} - H_{A,12}(H_{A,22} + H_{B,22})^{-1} H_{A,21} = H_{11} \\ \frac{X_1}{F_2} &= H_{A,12} - H_{A,12}(H_{A,22} + H_{B,22})^{-1} H_{A,22} = H_{12} \end{aligned} \tag{9.47}$$

Since each FRF contains translational and rotational displacements, the FRF expressed earlier needs to be expanded as:

$$\begin{Bmatrix} x_1 \\ \theta_1 \end{Bmatrix} = \begin{bmatrix} h_{11,ff} & h_{11,fM} \\ h_{11,Mf} & h_{11,MM} \end{bmatrix} \begin{Bmatrix} f_1 \\ M_1 \end{Bmatrix} \rightarrow \{X_1\} = [H_{11}]\{F_1\}$$
$$\begin{Bmatrix} x_1 \\ \theta_1 \end{Bmatrix} = \begin{bmatrix} h_{12,ff} & h_{12,fM} \\ h_{12,Mf} & h_{12,MM} \end{bmatrix} \begin{Bmatrix} f_2 \\ M_2 \end{Bmatrix} \rightarrow \{X_1\} = [H_{12}]\{F_2\} \tag{9.48}$$

Where: each h_{ij} is evaluated from the receptance coupling expression by including both translational (x) and rotational (θ) displacements due to lateral force (f) and moment (M).

The direct and cross receptances at the tool tip need to be expanded as well as:

$$[H_{11}] = \begin{bmatrix} h_{A11,ff} & h_{A11,fM} \\ h_{A11,Mf} & h_{A11,MM} \end{bmatrix} - \frac{1}{h_{2,Mf}^2 - h_{2,MM}h_{2,ff}} \begin{bmatrix} h_{A12,ff} & h_{A12,fM} \\ h_{A12,Mf} & h_{A12,MM} \end{bmatrix} \begin{bmatrix} -h_{2,MM} & h_{2,fM} \\ h_{2Mf} & -h_{2,ff} \end{bmatrix} \begin{bmatrix} h_{A21,ff} & h_{A21,fM} \\ h_{A21,Mf} & h_{A21,MM} \end{bmatrix}$$
$$[H_{12}] = \begin{bmatrix} h_{A12,ff} & h_{A12,fM} \\ h_{A12,Mf} & h_{A12,MM} \end{bmatrix} - \frac{1}{h_{2,Mf}^2 - h_{2,MM}h_{2,ff}} \begin{bmatrix} h_{A12,ff} & h_{A12,fM} \\ h_{A12,Mf} & h_{A12,MM} \end{bmatrix} \begin{bmatrix} -h_{2,MM} & h_{2,fM} \\ h_{2Mf} & -h_{2,ff} \end{bmatrix} \begin{bmatrix} h_{A22,ff} & h_{A22,fM} \\ h_{A22,Mf} & h_{A22,MM} \end{bmatrix} \tag{9.49}$$

The elements that represent the rotational degrees of freedom are complex to determine and they can be expressed through rewritten equations with common FRF terms. After extensive rewriting, rotational degrees of freedom can be expressed as:

$$h_{B22,Mf} = \beta - h_{A22,Mf}$$
$$h_{B22,MM} = \delta - h_{A22,MM} \tag{9.50}$$

where: β,δ are explained in detail in [42] through a symbolic non-linear analytical toolbox and will not be further analyzed here.

The comparison of the predicted and measured results can be seen in Figure 9.26 showing good agreement between the two. When comparing results with previously suggested approaches, it has been shown that the design has improved accuracy in experiments carried out with various sizes of end mills, successfully indentifying chatter free conditions in milling operations.

Developing capability for tool point frequency response prediction for micro end mills has been covered by reference [43]. In order to predict the response of the whole system as an assembly, receptance coupling substructure analysis was used. Response functions of the tool, tool holder and spindle were coupled showing an integrated system. The model geometry used is presented in Figure 9.27.

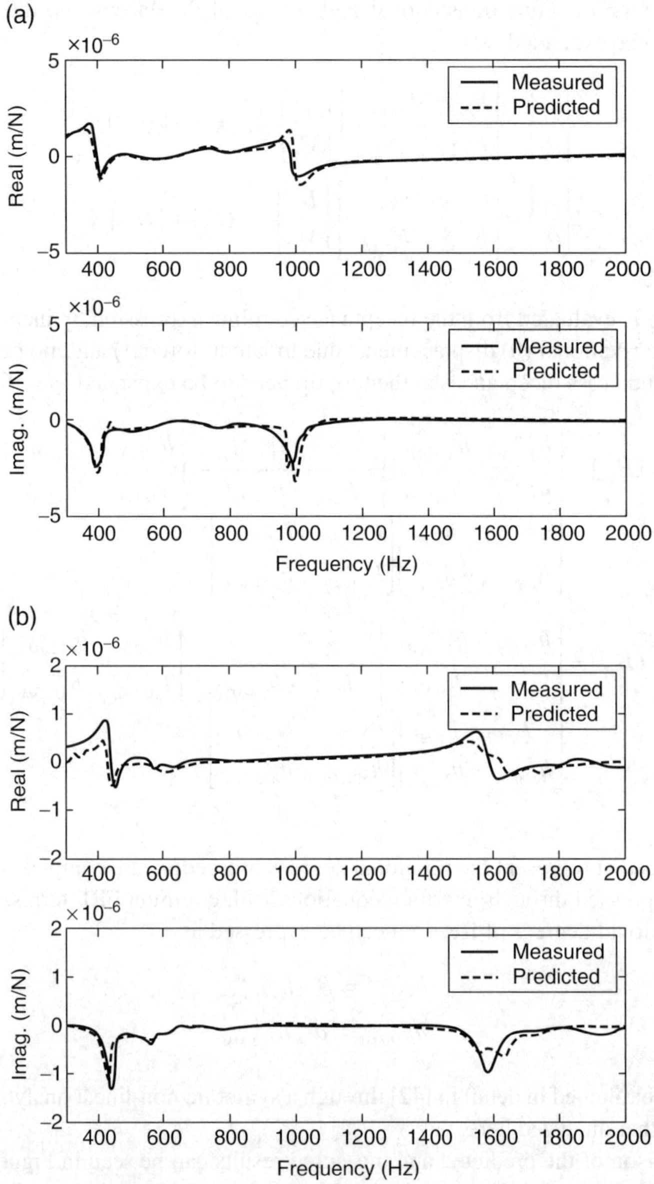

Figure 9.26 Comparison of the predicted and measured FRF of the (a) blank, (b) end mill attached to the spindle, when considering rotational dynamics. Reproduced from [42]. Copyright 2003 Elsevier

Receptances of the tool and collet holder were obtained from using Timoshenko beam models due to their size and fragility, while the spindle translational and rotational receptances were obtained experimentally.

The spindle holder base receptances have been determined by defining the receptance for the standard tool holder:

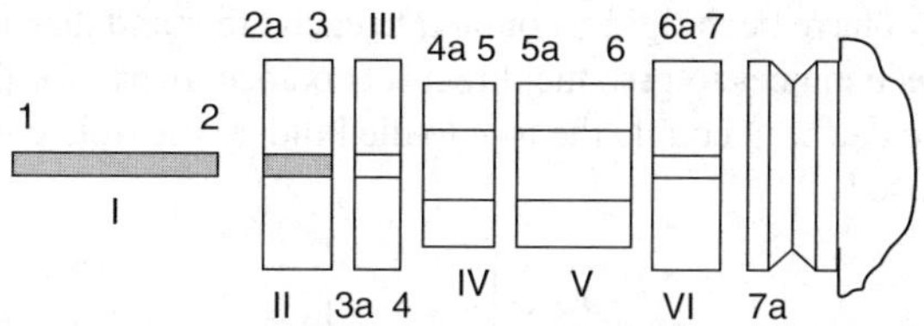

Figure 9.27 Model of the tool (I), collet holder (II–VI) and spindle (flange shown at coordinate 7a). Reproduced with permission from [43]. Copyright 2005 ASME

$$L_{55}(\omega) = \frac{H_{55} - H_{55a}}{S} \tag{9.51}$$

Where, H_{55} is the direct FRF at coordinate 5, H_{55a} is the cross FRF at coordinate 5 and S was considered to be 50.5 mm.

The rotation to moment receptance was determined as:

$$P_{55} = \frac{L_{55}N_{55}}{H_{55}} = \frac{L_{55}^2}{H_{55}} \tag{9.52}$$

Where H_{55} was measured and N_{55} was assumed equal to L_{55} by reciprocity.

An assembly receptance matrix has been defined as:

$$G_{55}(\omega) = \begin{bmatrix} H_{55} & L_{55} \\ N_{55} & P_{55} \end{bmatrix} \tag{9.53}$$

Further, Timoshenko beam models were used for substructures I and II, and the tool and tool holder have been rigidly coupled using:

$$\begin{aligned} RS_{55} &= R_{55} - R_{56}(R_{66} + R_{6a6a})^{-1} R_{65} \\ RS_{77} &= R_{77} - R_{76a}(R_{66} + R_{6a6a})^{-1} R_{6a7} \\ RS_{57} &= R_{56}(R_{66} + R_{6a6a})^{-1} R_{6a7} \\ RS_{75} &= R_{76a}(R_{66} + R_{6a6a})^{-1} R_{65} \end{aligned} \tag{9.54}$$

Where RS – coupled substructures receptance matrices, R – substructure matrices.

The tool holder –spindle assembly receptances have been defined through beam models:

$$\begin{aligned} G_{41} &= \begin{bmatrix} H_{41} & L_{41} \\ N_{41} & P_{41} \end{bmatrix} \\ G_{44} &= \begin{bmatrix} H_{44} & L_{44} \\ N_{44} & P_{44} \end{bmatrix} \end{aligned} \tag{9.55}$$

Where, H, L, P and N are different assembly responses. Subassembly receptance matrices have been developed through Timoshenko beam models for substructures I–VI and spindle holder dynamics.

Substructures IV–VI have been rigidly coupled to each other and then rigidly coupled to the spindle holder base receptances to give the direct receptance matrix for the subassembly.

In order to combine data and couple the tool to the holder a flexible connection and a scalar stiffness matrix was used:

$$Kc = \begin{bmatrix} k_{xf} & k_{xm} \\ k_{\theta f} & k_{\theta m} \end{bmatrix} \tag{9.56}$$

Where: k's represent stiffness values from displacement (x) and rotation (θ) due to force (f) and moment (m).

Mascardelli *et al.* [44] applied receptance coupling to predict the FRF at the tip of the micro tool and consequently use the FRF for chatter analysis. The authors have divided the machining system into two subsections in their research. The dynamics of the first subsystem, the cutter and the taper have been performed using FE analysis due to the fragile and slender nature of the tool, while the dynamics of the rest of the structure including the tool and the tool holder have been defined experimentally. By identifying the rotational degree of freedom (RDOF) at the joint, these two subsystems were linked. This way, determination of stability lobes that provides information on the stability of the system for a given axial depth of cut and spindle speed was accomplished. In FE analysis, two different models have been used, a perfectly constrained cantilever beam as well as an imperfectly constrained beam. Micro milling tests were performed in order to confirm the chatter-stability analysis based on the receptance coupling (RC) method. Information on high-frequency forced response tool dynamics was obtained by using acoustic emission sensors.

The response of the substructures has been defined as:

$$\begin{bmatrix} X_1 \\ X_2 \end{bmatrix} = \begin{bmatrix} H_{11} & H_{12} \\ H_{21} & H_{22} \end{bmatrix} \begin{bmatrix} F_1 \\ F_2 \end{bmatrix}$$

$$[X_3] = [H_{33}]F_{33} \tag{9.57}$$

$$X = \{x, \theta\}$$

$$F = \{f, M\}$$

Where: H_{ij} - transfer function of the substructures, subscripts indicate the location of the displacement and force, X - represented by translational displacement, x, as well as rotation θ and F -represented by force f and moment M.

The matrices of the assembled structures are defined as:

$$\begin{bmatrix} X_1 \\ X_2 \\ X_3 \end{bmatrix} = \begin{bmatrix} G_{11} & G_{12} & G_{13} \\ G_{21} & G_{22} & G_{23} \\ G_{31} & G_{32} & G_{33} \end{bmatrix} \begin{bmatrix} F_1 \\ F_2 \\ F_3 \end{bmatrix} \tag{9.58}$$

Where: Gij are the transfer functions of the assembly. The TF matrices are obtained either using EMA (experimental modal analysis) or FEA. The responses at the beginning and end of the substructure 1 can be predicted mathematically taking both translational and rotational degrees of freedom into account. The rotational dynamics at the end of substructure 2 are identified using the technique covered in [42] giving the result:

$$\begin{aligned}
b_{mf} &= \frac{(-b_{ff}G_{11,ff}h_{22,mf} + b_{ff}h_{21,mf}G_{12,ff} + b_{ff}h_{11,ff}h_{22,mf} - b_{ff}h_{h21,mf}h_{12,ff} + h_{21,mf}h_{22,ff}h_{12,ff} - h_{21,ff}h_{12,ff}h_{22,mf}}{h_{11,ff}h_{22,ff} - G_{11,ff}h_{22,ff} - h_{21,ff}h_{12,ff} + h_{21,ff}G_{12,ff}} \\
b_{mm} &= \frac{1}{h_{11,ff}h_{22,ff} - G_{11,ff}h_{22,ff} - h_{21,ff}h_{12,ff} + h_{21,ff}G_{12,ff}} b_{ff}h_{21,mf}^{2}G_{12,ff}^{2} + (2b_{ff}h_{11,ff}h_{22,mf}h_{21,mf} + \\
&\quad h_{12,ff}h_{21,mf}^{2}h_{22,ff} - 2b_{ff}G_{11,ff}h_{22,mf}h_{21,mf} - h_{12,mf}h_{21,ff}^{2}h_{22,mf} + h_{22,ff}h_{21,ff}h_{21,mf}h_{21,ff} - 2h_{12,ff}b_{ff}h_{21,mf}^{2} - \\
&\quad h_{12,ff}h_{21,mf}h_{21,ff}h_{22,mf})G_{12,ff} - h_{22,ff}^{2}h_{12,mf}h_{21,mf}G_{11,ff} + h_{22,ff}^{2}h_{12,mf}h_{21,mf}h_{11,ff} + \\
&\quad h_{22,mf}^{2}b_{ff}h_{11,ff}^{2}h_{22,ff}h_{12,mf}h_{21,ff}h_{22,mf}h_{11,ff} + h_{22,ff}h_{12,mf}h_{21,ff}h_{22,mf}G_{11,ff} + h_{22mf}^{2}\, b_{ff}G_{11,ff}^{2} + \\
&\quad h_{12,ff}h_{22,ff}h_{22,mf}h_{21,mf}h_{11,ff} - 2h_{22,mf}^{2}b_{ff}G_{11,ff}h_{11,ff} + h_{12,ff}h_{12.mf}h_{21,ff}^{2}h_{22,mf} - h_{12,ff}h_{22,ff}h_{22,mf}h_{21,mf}G_{11,ff} \\
&\quad - h_{12,ff}h_{22,ff}h_{12,mf}h_{21,mf}h_{21,ff} + 2h_{12,ff}b_{ff}G_{11,ff}h_{22,mf}h_{21,mf} + h_{12,ff}^{2}b_{ff}h_{21,mf}^{2} - 2h_{12,ff}b_{ff}h_{11,ff}h_{22,mf}h_{21,mf} + \\
&\quad h_{12,ff}h_{22,mf}^{2}h_{21,ff}G_{11,ff} + h_{12,ff}^{2}h_{21,mf}h_{21,ff}h_{22,mf} - h_{12,ff}^{2}h_{21,mf}^{2}h_{22,ff} - h_{12,ff}h_{22,mf}^{2}h_{21,ff}h_{11,ff}
\end{aligned} \tag{9.59}$$

Where: $h_{ff} = x/f$, $h_{fm} = x/M$, $h_{Mf} = \theta/f$, $h_{MM} = \theta/M$, Gij – transfer functions of the assembly, x – translational displacement of beams, θ - rotational displacement of beams.

It has been suggested that there is a 3% difference between the experimental and receptance coupling results, which is considered to be acceptable for chatter prediction purposes. It has been confirmed that adding the effect of the shank, taper and tip bending modes can be ignored for chatter prediction purposes because the additional lobes did not have any effect until the speed reached 120 000 rpm (Figure 9.28). At that depth the allowed axial depth of cut exceeded 100 microns, which would break the tool. The lobes did reduce the depth of cut to 38 microns at speeds in excess of 180 000 rpm, which are beyond the speed ranges of available spindles.

Owing to this analysis, it has been concluded that only the primary bending mode of the tool must be evaluated. This paper only covers the case where the feed rate is greater than the critical chip thickness, which means that it is only applicable to soft materials. Therefore, the RC technique for chatter prediction is applicable only to shearing-dominant cutting mechanisms, which occur only in machining soft materials. The ploughing would bring non-linearity effects into the system, making RC less suitable to use.

Paper [45] presents a review of the state of the art in micro machining with micro milling being mostly covered. A comparison between micro and macro scales has been done, pointing out how the knowledge from one scale can be applied to the other. There are several problematic issues when dealing with micro machining. First, the accurate measurement of forces is extremely challenging due to the size of tools and small values of the forces themselves. Noise if present, greatly disturbs the signals and gives distorted results. The issue of applying the Merchant's cutting model that is used for macro cutting processes is problematic to say the least. This is due to the effects of edge radius that will result in high negative rake angle and elastoplastic effects. Presence of high negative rake angles will also affect the magnitude of shearing and ploughing resulting in a substantial increase in the axial force and increased friction on the rake face of the tool. The shearing

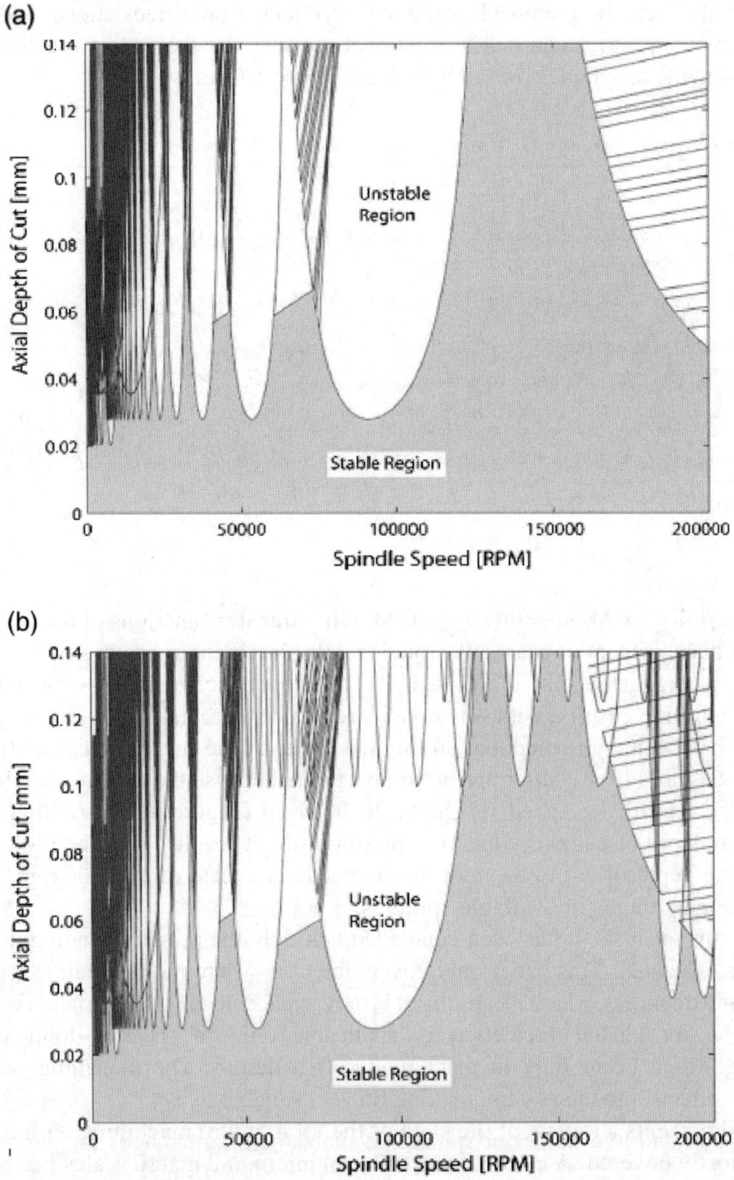

Figure 9.28 Chatter stability lobes for (a) low and (b) high frequencies. Reproduced with permission from [44]. Copyright 2008 ASME

and ploughing effects would result in a relatively large volume of material having to become fully plastic for a relatively small amount of material to be removed, which would cause significant increase in the specific energy.

Apart from the aforementioned problems, there is also a challenge regarding the interaction between tool and the workpiece. In micro cutting the depth of cut needs to be larger than the

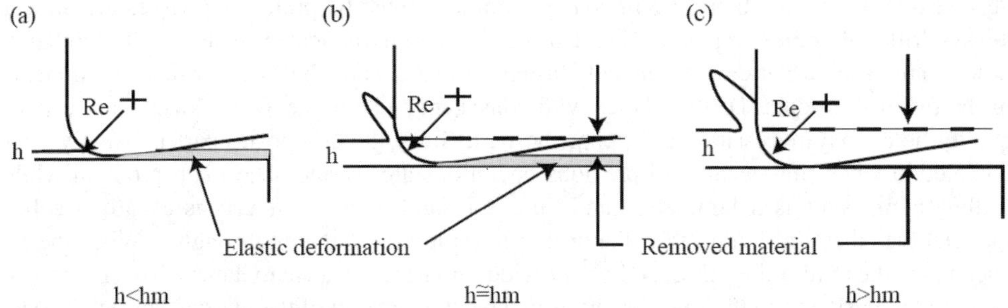

Figure 9.29 Schematic of the effect of the minimum chip thickness (Re, radius of cutting tool; h, undeformed chip thickness; hm, minimum chip thickness). Reproduced from [45]. Copyright 2006 Elsevier

critical chip thickness in order for chips to be formed due to the elastic effect present when machining such small thicknesses. The relationship between the tool radius and minimum chip thickness depends on the cutting edge radius and the material of the workpiece. This minimal chip thickness, which has been studied extensively [46–48] can be obtained either by experiments or FE analysis (Figure 9.29).

Regenerative chatter presents an important phenomenon that needs to be taken in consideration, however, accurate dynamic measurement between the tool and the workpiece is extremely challenging [43]. It has been suggested by various authors that chatter, tool wear, monitoring and work handling are one of the issues that have not been sufficiently investigated on a micro scale.

Chatter presents a very significant problem due to the fact that vibrations can have a catastrophic impact in micro systems leading to failure. Micro machining has to incorporate two aspects of chatter, the fact that contrary to macro machining cases, the feed rate will have an influence due to the elasticity of the workpiece and the regenerative aspect resulting in dynamic change of the chip thickness.

Another important issue in chatter in micro drilling is the fact that homogeneity of the material machined cannot be assumed. In macro drilling, the effect of grains, grain boundaries and anisotropy of the material can be neglected, while in micro drilling due to the size of the tool, they will have a significant effect on the process.

Apart from the aforementioned problems when investigating chatter in micro machining, there are also the issues of tool run-out caused by misalignment of the axis of symmetry between the tool and tool holder or spindle, available sensing methods for monitoring and controlling high speed machining processes, thermal expansion and the application of adaptive control techniques.

9.4 Case Study: Micro Drilling Medical Polymer Materials and Composites

Recent developments have led to an increase in composite use. Many different composites are currently in use today. Composites may be formed from fibre glass, carbon, metal, and polymer fibres combined a wide variety of binders. Composites cannot usually be formed in the desired final shape and therefore finishing operations must be preformed. Holes are almost always drilled after the composite is bonded because it is impractical to create them beforehand. Since composites are often mechanically fastened to other parts, holes are critical components of the finished product. Drilling holes with consistency is a major issue. Misplaced or poor quality holes may necessitate scrapping the part. Since the composite part is usually near completion when this occurs, all previous operations are wasted. The main problem while drilling composites is delamination and fibre pull out. Since the matrix is usually a softer material than the reinforcing fibre, the drill or mill bit will easily cut through it. When the bit encounters the reinforcing fibre the fibre provides greater resistance to the cut. This resistance may tend to push or pull the fibres away from the cutter and out of the matrix. This leads to an uneven surface once the fibres relax. In mild situation this can led be seen as discolouration on the surface. In more extreme cases the fibres may bind to the tool and pull out or separate from the matrix and crack the part. While this resistance may easily overcome with a sharp tool, the abrasive nature of the fibres works against the cutting tool. Glass and carbon fibres are very abrasive, leading to tool wear. This quickly results in a dull cutting edge that tears the fibres instead of cutting cleanly. If heat builds up on the tool surfaces due to the increased friction the polymeric matrix may melt or burn also resulting in a failed part.

9.4.1 Tooling Selection

Tools used to machine composites must be wear resistant and be capable of dissipating heat. The abrasive nature of the material makes tool wear a major concern. A sharp cutting edge is necessary in order to make a clean cut. High-speed steels (HSS) are commonly used when machining other materials. While they may initially produce satisfactory results, the material quickly abrades and the cutting surfaces dull. Even when coated these tools are not resistant enough to abrasion to have useful tool life when milling composites. High-speed steel bits are often used, however, in drilling operations. Standard twist drills may be used, but specially designed drills are recommended. Ceramic tools are also not suitable for machining composites. Ceramic bits are too brittle and crack when used on composites. Carbide and diamond tools are recommended for use because of their resistance to abrasive wear as well as thermal and mechanical shock. Carbide tools are formed out of various carbide compounds sintered into the desired shape. Carbide tools are created using powder metallurgy. They are usually formed out of a combination of cobalt, titanium carbide (TiC), tantalum carbide (TaC) and tungsten carbide (WC) powders. Once formed into the desired bit or insert; various coatings may be applied. Titanium compounds such as TiN, TiC, and TiAlN are used to increase surface hardness, reduce friction and improve edge retention. Coating layers are built up by chemical vapour deposition (CVD) or by physical vapour deposition (PVD). The coating may be formed out of a single compound or by alternating layers of two or more compounds. Thicker coatings are desired when machining composites. A thicker coating lasts longer before wearing through and once the coating is penetrated the surrounding layer supports the exposed area.

Diamond tools are formed by depositing, or sintering, diamond crystals to a HSS substrate or a tungsten carbide base. Polycrystalline diamond (PCD) tools are formed by sintering diamond and cobalt particle together at high temperatures. The resulting material is then brazed onto an insert or bit. PCD tools are limited to simple geometries in order to facilitate easy attachment. PCD tools have been shown to wear 120 times slower that carbide tools. Use of PCD is limited however by its high cost, low toughness and the difficulty to form it into appropriate cutting geometries. PCD tools may be 6 times more expensive than carbide tools but may only last 3 times as long. Another option is diamond-coated tools. Diamond abrasive particles are deposited on the HSS or carbide body and then bonded to the surface by an electrodeposited nickel layer. The characteristics of this coating depend on diamond fraction, diamond particle size and embedment. The best coatings have a high volume fraction of large (>125 micron) diamond particles with an embedment of ½ the particle diameter.

9.4.2 Cutting Mechanisms and Considerations

The mechanism of chip formation in composites is highly dependent on fibre orientation. Various fibre alignments have been analyzed. At 0° alignment the fibres are compressed and have a tendency to buckle. When using a sharp tool, large chips tend to form. When a dull tool is applied, large fragments may break off from the part. At 90° alignment the fibres are subject to shearing. An acute edge will cleanly shear the fibres, while a more rounded edge will cause the fibres to bend before shearing. The bending motion pulls the fibres out of the supporting matrix and forms cracks in the composite. At 45° alignment the fibres are aligned in the direction of the cut (+45°) or opposing the cut (–45°). When the fibres are aligned in the direction of the cut (+45°) the fibres are depressed before they are cut. This results in fibre spring-back after the cut is made. The cut fibres then rub against the clearance face resulting in greater abrasive wear. When the fibres oppose the direction of the cut, the fibres are pushed away from the cutting edge forming cracks in the workpiece and a stepped pattern on the workpiece. In both cases lower edge acuity exacerbates the deformation. The cutting edge radius should be approximately the same as the fibre diameter. When the cutting edge radius wears to 5 to 10 times the fibre diameter, the cut is poor. Abrasive wear is highest at +45° fibre orientation and lower at 90° and 0° alignments. This abrasion combined with a low acuity cutting edge increases friction and generates heat.

In all cases the greatest damage usually occurs when beginning or ending the cut. When drilling, as the tool enters the workpiece the flutes tend to pull the fibres up leading to ply separation. The same issue occurs when exiting the workpiece as the fibres are no longer supported by the material. Higher feed rates correspond with mechanical damage. Varying feed rates when entering and exiting a cut can prevent delamination and linting. When cutting glass or carbon fibre composites, the brittle nature of the material allows for easy shearing. The material hardness does however cause abrasion and resistance to the cut. Aramid (Kevlar®) fibres are ductile which possess a challenge when making a cut. The fibre must first be pulled taught and then severed. If the fibre is stretched too much it may pull out of the matrix causing linting. Tools designed to cut aramid composites often incorporate both positive- and negative-angle cutting edges so that the fibres are pulled out.

Heat build-up should be prevented during machining. High temperatures can melt or char the supporting matrix and fibres. If adjusting cutting speeds cannot prevent heat build-up, cutting fluids should be used to cool the tool and workpiece. A good cutting fluid will not be absorbed by or react with the composite material. Improper fluid selection can lead to chemical damage. Microlubrication, or gas cooling may prevent this damage.

9.4.3 Drilling

Drilling operations are often performed on composites. This is because composite components are usually connected using mechanical fasteners. Care must be taken when developing specifications for drilling processes. A misplaced or malformed hole can ruin a nearly finished part. Poorly cut holes cause about 60% of part rejection, during quality checks. The two primary concerns are cutting speed and feed rate. High cutting speeds lead to higher friction that can result in thermal damage. High feed rates lead to delamination and other mechanical damage. Pre-drilling holes may reduce damage by lowering the load on the drill. Adding a reinforcing backing also prevents damage. The backing provides support to the composite preventing fractures during drill exit. Variable feed rates may be used while drilling. At tool entrance and exit, lower feed rates may be used to prevent damage. Higher feeds may be used while the drill passes through the center portion of the material.

Standard twist drills are often used on composites. High-speed steel and carbide tools are currently favoured. High-speed steel wears quickly, carbide is more resistant to wear. Diamond tools are currently being developed, but it is difficult to bond the material onto the complex geometry of a drill bit. Polycrystalline diamond coatings are currently in use. Other drill types are also used. Diamond core (saw) drills produce very good holes. Due to the abrasive cutting action, core drills require greater power input. Negative point angle drills are useful for cutting aramid composites. These bits create lower quality holes than twist drills, but they wear at a much slower rate. Step drills are also in use. Negative point drills and step drills produce the least amount of delamination. This is due to the distribution of the thrust force. Since only a small portion of the drill tip is in contact with a given layer of the composite, the composite layers are less likely to separate. Twist drills designed for composites use the same principle. The point angles of these drills are more acute than typical drill bits.

In experiments conducted on the drilling of composites, the delamination factor (F_d) is used to quantify the level of damage. The factor is the ratio between the maximum diameter of the delaminated area and the drill diameter. The delaminated area can be determined by visual, ultrasonic, or x-ray analysis. As tool wear increases the required torque and thrust force increase; the damaged area around the hole also increases. Delamination is always higher on the exit side of the hole, because the fibres are unsupported. According to one researcher, delamination does not occur at low feed rates. In a drilling experiment recently conducted, various drills types were examined. The twist drill showed an increase in delamination factor only when the feed rate was increased. The negative point angle drill's delamination factor was increased only by drill diameter. The saw drill's factor was affected by cutting speed, feed rate and drill diameter. Both the negative point angle and saw drills produced lower delamination factors than the twist drill.

When compared to twist drills with 85° and 115° point angles, a negative point angle drill shows superior characteristics at high speeds. The delamination factor for the two twist drills is shown to increase with feed speed. At high speeds the negative point angle drills factor remains unaltered. The twist drill with the larger point angle (115°) showed higher delamination at all feeds. At high spindle speeds the negative point angle drill performs the best, but at low spindle speeds the 85° twist drill produces less delamination. At high spindle speeds (40 000 RPM), increasing the feed rate does not cause an increase in delamination. Feed rates should be lowered as the drill exits the cut to prevent delamination.

High speeds may result in thermal damage. Because the composite material is usually non-conductive, the drill and chips carry all the heat from the cut. Polycrystalline diamond coatings are especially useful because of their high hardness and conductivity. Cutting fluids should be used if the desired material removal rate creates enough friction to damage the matrix. Cryogenic cooling is particularly useful since it also lowers the ductility of the composite. Aramid based composites in particular benefit greatly from cryogenic cooling, since the cold temperature reduces the elasticity of fibres allowing for a cleaner cut.

9.4.4 Burr Elimination when Drilling Polymers

A recent industrial application where drilling is critical to the functionality of a product is concerned with the processing of a medical tool holding rack. This unit is likely used as packaging and as a sanitation rack for medical implements. The part is composed of Propylux® a polypropylene resin designed for toughness, chemical resistance and resistance to abrasion. The part is machined from a solid block of the material. Pockets are cut out of the block; most likely to reduce weight, allow for drainage and allow air to flow while drying. Over 100 holes are drilled in the part to hold the implements or to allow for air to flow. Many of the holes are countersunk, providing a smooth surface around the holes. The finished product has many burrs. The milled surfaces appear to be mostly burr free, but the holes often have burrs.

Burrs are problematic for two reasons. First of all, burrs affect the overall appearance of the product, making it less attractive. Second, it is likely that the burrs may provide an area that may harbour bacteria or flake off and become a contaminant. Currently the parts are deburred manually; this requires both time and personnel. The majority of the burrs appear to form at points where the drill bit exits the material. Some burrs also form mid-cut as the material adheres to the face of the cutting tool.

The majority of burr formation occurs when the tool exits the workpiece. As the tool exits the part; the amount of material supporting the cutting zone decreases. The thrust force from the tool pushes the material away from the cutting edge. This uncut material forms the burr. This effect can usually be prevented by adding a back support to support the workpiece during drilling. On this particular part, however, this is not possible due to the intersection of many of the holes inside the product. Another way to prevent burr formation is to lower the feed rate while the tool exits the workpiece. This can be accomplished by programming the CNC controls to lower the feed rate while the tool edge is entering and exiting the material. The thrust force can also be reduced by spreading it out over a larger area. Specially designed drill bits with sharply ground (60°) points accomplish this. The sharp point reduces the force on the tip and spreads the cut over a larger area. Plastic cutting drill bits are available from many suppliers. To ensure a clean cut, the drills should be kept as sharp as possible so that the cutting edges shear the material instead of ploughing through it.

When machining plastics, high temperatures can cause the polymer to soften and deform. Therefore the cutting temperatures should be regulated so that the polymer remains in a rigid condition. Liquid coolants should be carefully selected since the fluid may react with or be absorbed by the polymer. Cryogenic gas coolants eliminate this problem. Cryogenic cooling may be used to pre-chill the part, cool the cutting tool, or as a spray while cutting. Pre-chilling the workpiece requires large volumes of fluid and therefore may not be economical. Since the machining is performed at low temperatures, part dimensions may be altered when the material is brought back to room temperature. The lower temperature makes the material more brittle allowing for easier chip removal. Cooling the cutting tool is usually accomplished by means of coolant flow through the tool or a reservoir in the tool holder. The tool material must be conductive for this technique to work. Cryogenic spraying directs a jet of coolant towards the cutting zone. This small spray should not significantly change the temperature of the product and therefore does not affect the final dimensions. Since the cutting face is cooled, the possibility of chip adhesion is reduced. Cutting chips are embrittled by the low temperatures, and are easier to break and remove. Cryogenic temperatures increase tool hardness during the cutting process. Harder tool surfaces are more resistant to wear, leading to longer tool life.

Finished products are currently deburred by hand. While the goal of any machining process is to eliminate burr formation; this is not always possible or economical. There are other deburring options available that may be less expensive than the manual process. Water-jet deburring removes excess material using a high-pressure stream. The pressure stream easily removes small burrs and flushes contaminants off the part. Since there is no abrasive involved, the product's surface finish is not affected and there is no risk of abrasive embedding in the surface of the workpiece. Since the jet size is adjustable, very small holes can be reliably cleaned. Cryogenic deburring uses non-abrasive media and cold temperatures to remove material. The low temperature reduces the material's ductility and increases its hardness protecting the surface finish. Since the burrs have a larger surface area but a small attachment zone, they are affected to a greater extent. Non-abrasive polycarbonate media is then blasted onto the part, removing the imperfections. Processing time is very fast, some parts may only need a few seconds exposure. This method does not affect surface finish or dull part edges. The process may be used on metals and plastics.

9.5 Conclusions

The chapter summarizes the effect and suppression of chatter in drilling and micro drilling operations. Research in the area of chatter recognition and suppression has been conducted since the 1950s. There have been substantial advances in the field of modelling and many of the theoretical conclusions have been applied in practice especially in milling and turning operations. Drilling is a process that has not been researched to the same extent and still requires theoretical and experimental development. It has been seen that it is necessary to incorporate axial-torsional, lateral and bending vibrations into one system as well as incorporating other influential parameters such as the chisel edge, margin effect, pilot hole size, tool grinding errors, misalignment, and so on.

Micro drilling operations are yet to be addressed in the literature and require a significant amount of attention in order to model the process accurately and obtain explanations regarding the influential parameters. The idea of applying theory that has been developed for mesoscale drilling operations has been shown to be inappropriate but the analysis, made on the similarities and differences, does give important information. It is necessary in future work to develop the mathematical models for micro drilling operations and further work on the receptance coupling techniques for obtaining the overall dynamic models of the tool-workpiece assembly.

Owing to the nature of composite materials, extra care must be taken when designing machining operations. The composite materials behave differently depending on their construction and the angle at which they are machined. Tooling and cutting parameters must be adjusted to handle this. The primary concern while cutting composites is that the cutting edge is sharp enough to cleanly shear the fibres.

For milling operations high-speed steel, carbide, and diamond tools may be used. High-speed steel tools are not recommended due to the abrasive nature of composite materials. Carbide tools and diamond tools are preferred. Carbide tools wear out more quickly than diamond tools, but are currently more economical. Coated carbide tools used to machine composites should have thicker than normal coatings since they are subject to high amounts of abrasion. Polycrystalline tools provide superior tool life but are currently limited to simple geometries and by production cost. Diamond coated tools are less expensive than polycrystalline diamond tools. While they do not provide the same tool life, they are less expensive and therefore a good alternative.

When milling composites a sharp edge is necessary for a clean cut. The edge of the cutting tool must have a smaller radius than that of the fibres being cut. Once the tool edge has blunted; the tendency for the edge to push or pull the composite fibres out of the matrix increases and delamination occurs at a greater rate. High cutting speeds lead to thermal damage of the composite. High cutting feeds cause mechanical damage. Both parameters must be monitored to produce a clean cut.

When drilling composites high-speed steel, carbide, and diamond tools may be used. The cleanest cuts are made with core (saw) drills but this requires higher power input. Standard twist drills may be used with good results, but their performance suffers at higher feed rates. Sharper point angles produce better results. Negative point angle drills are useful when high feed rates are desired since the quality of the surface finish is not affected. With all drilling operations, the feed rate when entering and exiting the cut effects the delamination of the composite. Lower feed rates during these portions of the cut prevent damage to the composite. Backing materials may also be used to protect the composite during tool exit, but this adds to production time and cost.

Cutting fluid may be used in cases where temperature control is needed. High temperatures may melt or char the composite. Cutting fluids should be selected with care. Chemicals in the fluid may react with the composite. Some composites may also absorb water, if water based fluids are used. Cryogenic cooling is an effective solution since there is little chance for reactivity and because of the effects on material ductility. Cryogenic temperatures reduce the flexibility of the fibres and the matrix; allowing for cleaner cuts. This is especially useful when machining aramid (Kevlar®) composites as the elasticity of the fibres is reduced.

Burr formation can be controlled by lowering feed rates during tool entrance and exit during drilling operations. Drill bits with more acute points lower thrust force. Sharp tool edges help ensure clean burr-free cuts. Dulled tools should be replaced as soon as hole quality deteriorates. Other deburring options are available either from off-site service or in-house processing. These options may be less expensive than manual cleaning. A selection of possible drills are commercially available that may significantly reduce, or eliminate, burr formation in medical grade plastic materials. Another option is to use coated tools that remove heat away from the cutting zone.

Acknowledgements

The authors wish to thank Springer Verlag publishers for allowing the authors to re-publish material that was originally published as a paper in the *International Journal of Advanced Manufacturing Technology* entitled, 'Chatter Problems in Micro and Macro-Cutting Operations, Existing Models and Influential Parameters – A Review', 2010, 47:597–620, Re-printed with kind permission of Springer Science+Business Media B.V.

References

[1] Tobias, S.A. (1965) *Machine Tool Vibration*. New York: Wiley.

[2] Koenigsberger, F., and Tlusty, J. (1970) *Machine Tool Structures*. Vol. 1. New York: Pergamon Press.

[3] Bayly, P.V., Metzler, S.A., Schaut, A.J., and Keith, A.Y. (2001) Theory of torsional chatter in twist drills model stability and composition to test. *Journal of Manufacturing Science and Engineering*, 123: 552–561.

[4] Minis, I., and Yanusheshevsky, R. (1993) A new theoretical approach for the prediction of chatter in milling. *Journal of Engineering for Industry*, 115: 1–8.

[5] Altintas, Y., and Budak, E. (1995) Analytical prediction of stability lobes in milling. *Annals of the CIRP*, 44(1):357–362.

[6] Roukema, J.C., and Altintas, Y. (2007) Generalized modeling of drilling vibrations. Part II: Chatter stability in frequency domain. *International Journal of Machine Tools & Manufacture* 47: 1474–1485.

[7] Roukema, J.C., and Altintas, Y. (2007) Generalized modeling of drilling vibrations. Part I: Time domain model of drilling kinematics, dynamics and hole formation. *International Journal of Machine Tools & Manufacture*, 47: 1455–1473.

[8] Tarng, Y.S., and Li T.C. (1995) Adaptive pattern recognition of drilling chatter. *Journal of Materials Processing Technology*, 48: 247–253.

[9] Kalmar, T., Stepan, G., and Moon, T. (2001) Subcritical Hopf bifurcation in the delay equation model for machine tool vibrations. *Nonlinear Dynamics*, 26:121–142.

[10] Ema, S., Fujii, H., and Marui, E. (1988) Whirling vibration in drilling. Part 3: Vibration analysis in drilling workpiece with a pilot hole. *Journal of Engineering for Industry*, 110: 315–321.

[11] Fujii, H., Marui, E., and Ema, S. (1986) Whirling vibrations in drilling. Part 1: Cause of vibration and role of chisel edge. *ASME Journal of Engineering Industry*, 108(3): 157.

[12] Fujii, H., Marui, E., and Ema, S. (1986) Whirling vibration in drilling. Part 2: Influence of drill geometries, particularly of the drill flank on the initiation of vibration. *ASME Journal of Engineering Industry*, 108(3):163.

[13] Bayly, P.V., Lamar, M.T., and Calvert, S.G. (2002) Low frequency regenerative vibration and the formation of lobed holes in drilling. *Journal of Manufacturing Science and Engineering*, 124(2): 275–285.

[14] Bayly, P.V., Lamar, M.T., and Calvert, S.G. (2002) Low-frequency regenerative vibration and the formation of lobed holes in drilling. *Journal of Manufacturing Science and Engineering*, 124:275–285.

[15] Altintas, Y., and Weck, M. (2004) Chatter stability of metal cutting and grinding. *CIRP Annals – Manufacturing Technology*, 53(2):619–642.

[16] Tlusty, J. (1986) The dynamics of high-speed milling. *Journal of Engineering for Industry*, 108:50–67.

[17] Stone, E., and Askari, A. (2002) Nonlinear models of chatter in drilling processes. *Dynamical Systems*,17(1): 65–85.

[18] Stone, E., and Campbell, S.A.(2004) Stability and bifurcation analysis of a nonlinear DDE model for drilling. *Journal of Nonlinear Science*, 14:27–57.

[19] Campbell, S.A., and Stone, E. (2006) Analysis of chatter instability in a nonlinear model for drilling. *Journal of Computational and Nonlinear Dynamics*, 1:294–306.

[20] Dilley, D.N., Bayly, P.V., and Schaut, A.J. (2005) Effects of the chisel edge on the chatter frequency in drilling. *Journal of Sound and Vibration*, 281:423–438.

[21] Dilley, D.N., Stephenson, D. ., Bayly, P.V., and Schaut, A.J. (2005) Frequency shift in drilling due to margin engagement. *Journal of Manufacturing Science and Engineering*, 127:271–276.

[22] Arvajeh, T., and Ismail, F. (2006) Machining stability in high-speed drilling—Part 1: Modeling vibration stability in bending. *International Journal of Machine Tools & Manufacture*,46:1563–1572.

[23] Ulsoy, A.G. (1983) A lumped parameter model for the transverse vibration of the drill bits. *Control of Manufacturing Processes and Robotic Systems*, 15–25.

[24] Ema, S., Fujii, H., and Marui, E. (1998) Chatter vibration in drilling. *Journal of Engineering for Industry*, 110: 309–314.

[25] Rincon, D.M. (1993) *Coupled force and vibration modeling of drills with complex cross sectional geometries, in Department of Mechanical Engineering and Applied Mechanics*. University of Michigan: Michigan.

[26] Arvajeh, T., and Ismail, F. (2006) Machining stability in high speed drilling—Part 2: Time domain simulation of a bending-torsional model and experimental validations. *International Journal of Machine Tools & Manufacture*, 46: 1573–1581.

[27] Rincon, D.M., and Usloy, A.J. (1994) Complex geometry, rotary inertia and gyroscopic moment effects on drill vibration. *Journal of Sound and Vibration*, 188: 701–715.

[28] Timoshenko, S., Young, D.H., and Weaver, W. (1974) *Vibration Problems in Engineering*. New York: John Willey and Sons, Inc.

[29] Altintas, Y., and Chan, P.K. (1990) In-process detection and suppression of chatter n milling. *International Journal of Machine Tools & Manufacture*, 32(3): 329–347.

[30] Inamura T. (1974) Stability analysis of cutting under varying spindle speed. *Annals of the CIRP*, 23(1): 119–120.

[31] Li, C.J., Ulsoy, A.G., and Endres, W.J. (2006) The effect of spindle speed variation of chatter suppression in rotating tool machining. *Materials Science Forum*, 505–507: 859–864.

[32] Radulescu, R., Kapoor, S.G., and DeVor, R.E. (1997) An investigation of variable spindle speed face milling for tool-work structures with complex dynamics, Part 2: Physical explanations. *Journal of Manufacturing Science and Engineering*, 119:273–279.

[33] Sastry, S., Kapoor, S.G., and DeVor, R.E. (2000) Compensation of progressive radial run-out in face-milling by spindle speed variation, *International Journal of Machine Tools & Manufacture*,40:1121–1139.

[34] Smith, S., and Tlusty, J. (1990) Update on high-speed milling dynamics. *ASME Journal of Engineering for Industry*, 112:142–149.

[35] Smith, S., and Delio, T. (1992) Sensor-based chatter detection and avoidance by spindle speed selection. *ASME Journal of Dynamic Systems, Measurement, and Control*, 114:486–492.

[36] Takemura, T., Kitamura, T., and Hoshi, T. (1974) Active suppression of chatter by programmed variation of spindle speed. *Journal of the Japan Society of Precision Engineering*, 42(11): 1049–1055.

[37] Tsao, T., McCarthy, M.W., and Kapoor, S.G. (1992) A new approach to stability analysis of variable speed machining systems. *International Journal of Machine Tools & Manufacture*, 33(6):791–808.

[38] Smith, S., and Tlusty, J. (1993) Efficient simulation programs for chatter in milling. *Annals of the CIRP*, 42(1):463–466.

[39] Tarng, Y.S., and Li, T.C. (1994) Detection and suppression of drilling chatter. *Journal of Dynamical System Measurement and Control*, 116: 729–734.

[40] Tarng, Y.S., and Lee, E.C. (1997) A critical investigation of the phase shift between the inner and outer modulation for the control of machine tool chatter. *International Journal of Machine Tools and Manufacture*, 37(12): 1661–1672.

[41] Tarng, Y.S., and Lee, E.C. (1997) A critical investigation of the phase shift between the inner and outer modulation for the control of machine tool chatter. *International Journal of Machine Tools and Manufacture*, 37(12):1661–1672.

[42] Park, S.S., Altintas, Y., and Movahhedy, M. (2003) Receptance coupling for end mills. *International Journal of Machine Tools & Manufacture*, 43:889–896.

[43] Cheng, C-H., Arakere, N., Schmitz, T.L., andDuncan, S. (2005) An approach for micro end mill frequency response predictions. *American Society of Mechanical Engineers, Manufacturing Engineering Division, MED*, 16(2):1139–1145.

[44] Mascardelli, B.A., Park, S.S., and Freiheit, T. (2008) Substructure coupling of microend mills to aid in the suppression of chatter. *Journal of Manufacturing Science and Engineering*,130(0110101–01101012).

[45] Chae, J., Park, S.S., and Freiheit, T. (2006) Investigation of micro-cutting operations. *International Journal of Machine Tools & Manufacture*, 46:313–332.

[46] Kim, B., Schmittdiel, M.C., Degertekin, F.L., and Kurfess, T.R. (2004) Scanning grating interferometer for MEMS metrology. *Journal of Manufacturing Science and Engineering*, 126:807–812.

[47] Vogler, M.P., Devor, R.E., and Kapoor, S.G. (2003) Microstructure-level force prediction model for micro-milling of multi-phase materials. *Journal of Manufacturing Science and Engineering*, 125:202–209.

[48] Weule, H., Huntrup, V., and Tritschler, H. (2001) Micro-cutting of steel to meet new requirements in miniturization. *Annals of CIRP*,50:61–64.

10

Micro Grinding Applications

Han Huang
School of Mechanical and Mining Engineering, The University of Queensland, Australia

10.1 Introduction

Miniaturization is one of the frontier technologies in the twenty-first century, which is making an impact on the lives of human beings. The miniaturization of devices associated with a number of fields demands the production of mechanical components with manufactured features ranging from several to several hundred micrometers, while still holding tight tolerances, in a broad range of engineering materials. Particularly in recent years, various micro scale components are heavily needed for applications in optics, electronics, avionics, communications, bio-technology and medicine. The rapidly growing need to make products from macroscopic to microscopic scales from those industries echoes the international research direction of micro/nano-manufacturing in the new millennium.

The fabrication of micro devices started from the adoption of fabrication technologies for microelectronic and micro-electro-mechanical systems. Those technologies are limited to 2 and 2.5 dimensional features and a narrow class of materials, mainly of silicon-based components, with very limited relative accuracy (ratio of feature tolerance to size) [17, 43]. However, the escalating demand for micro devices [11, 16, 18, 25, 26, 45, 51, 55] normally requires the use of three-dimensional micro features in a wide range of structural materials, and therefore cannot be created by Si-based fabrication methods alone. To gain the full potential of multi-scale technologies, which rely on integrated nano, micro and meso scale features and systems, the limitations of current technologies need to be overcome and new complementary capabilities need to be developed [16, 19, 35, 45, 57].

Micro grinding is a machining process that has a material removal unit at nanometric scale and is capable of machining 3D components with feature size ranging from several to several hundreds of microns. The machined products must be of high dimensional accuracy and high surface and subsurface integrity, simply because in most of the cases, such a process is used for the final fabrication procedure of micro components. Different from the other micro/nano

Micro-Cutting: Fundamentals and Applications, First Edition. Edited by Kai Cheng and Dehong Huo.

mechanical machining methods, such as cutting that is usually used for ductile or less hard materials, micro grinding can be used to machine a large variety of structural materials for electronic, optic/photonic and biomedical applications, typically including glass and germanium for optical and photonic components [21, 25, 29, 35], ferrite/quartz for recording heads and resonators [18], hard and tough metals for moulds [33], bearings, gears, cranks, camshafts and engine blades [19, 42, 44, 45], carbides for dies and moulds [12, 16, 48] and ceramics for motion-transmission components [55].

Early in the 1990s, driven by the need of the optics/photonics industries, Koehler *et al.* [33] used grinding as a finishing process for spherical mould inserts of 100 to 500 μm in diameter. The ground silicon mould insert was used to fabricate micro lenses employed in a miniaturized endoscope. Great efforts were made in the past decade to develop grinding technologies for machining spherical/aspherical micro lenses and mould inserts [13, 35, 36, 48, 51, 52, 61]. The smallest aspherical tungsten carbide mould with a diameter of about 200 μm was fabricated [16, 26].

Previous works were largely concerned with the study of the effect of micro grinding conditions, including truing and dressing methods and grinding types, on grindability and ground surface quality [19, 45, 53, 58]. The research outcomes demonstrated that grinding could produce a much better surface finish than other micro machining methods. Micro grinding was also capable of machining 3D micro features. The application of micro/nano grinding technologies has become increasingly more widespread in recent years. For example, micro grinding was shown to be capable of fabricating micro gear shafts with minimum feature size of 50 μm [58]. Tönshoff *et al.* [55] developed micro grinding technologies to machine an alumina aerostatic guide in a micro actuator. The alumina guide has a feature size of approximately 300 μm with a tight dimensional accuracy and straightness. Ohmori *et al.* [44] successfully ground micro grooves on germanium substrate for advanced mid-infrared image analyzers, which led to better performance than other micro machining means. Huang *et al.* [15, 25, 63, 64] developed nanometric grinding technologies to produce optic fibre master patch cords, with mirror surface finish and sub-micron form accuracy. The ground master patch cords produced better optical performance than those polished. In the above-mentioned applications, micro grinding could be the only solution for such components.

In the grinding of brittle materials, there are basically three types of removal modes: brittle, ductile and the combination of both [30, 40, 42, 60]. In brittle mode grinding, brittle fracture owing to micro indentation or scratching produced in the grinding process is responsible for material removal, which results in fragmented surface and subsurface damage with micro cracks. The surface generated by ductile mode grinding consists of a compacted top layer sitting above the bulk material. Extremely small surface and subsurface damage occurs in the top layer. Ductile grinding was often reported when individual grit depth of cut was at nanometric scales. However, no convincing evidence was demonstrated on the occurrence of completely ductile deformation for brittle materials. Transition from brittle to ductile removal mode under certain grinding conditions has also been the subject of debate [3, 6, 22–24, 60]. Apparently, the mechanisms of material removal under ductile grinding are not well understood because significant impediments to a common understanding of those mechanisms exist. The reasons could be multi-fold. First, the removal mechanism is dependent on the micro structure and mechanical properties of the material being machined [27, 65, 70, 71]. Contrary to ductile materials, there is no universal removal mechanism in brittle materials. Second, there might be multiple removal mechanisms working simultaneously [24, 31, 69].

Third, grinding involves many factors and uncertainties, so the process is too complex to accurately characterize its variables and to control consistent interactions between abrasives and the workpiece material. Fourth, systematic and comprehensive investigations of removal mechanism using robust material analytical techniques are still lacking.

The volume of material removal associated with micro grinding is considerably small compared with that of conventional grinding. Therefore, it is of importance to adopt the 'one step' grinding strategy [66, 67]. In other words, grinding and subsequent polishing procedures usually employed for conventional grinding need to be combined into one procedure or performed using one grinding wheel in a micro grinding process. Grinding is the final process for the product. To realize 'one step' grinding, workpiece material must be removed in the ductile regime, which results in a mirror surface finish and a damage-free subsurface. The realization of ductile removal requires a comprehensive understanding of the deformation and removal mechanism under micro grinding conditions.

In a micro grinding process, the penetration depth of an abrasive grit into workpiece depends on its geometry and kinetic motion, as well as the workpiece material properties [32]. The depth is thus affected by the wear of the grit. Though abrasive grits for grinding brittle materials are made of diamond, their wear is significant. This is because wheels for micro grinding are rather smaller and active cutting grits in such wheels are fewer than in conventional grinding wheels. The worn surface of abrasive grits on a grinding wheel and the subsequent grit pulling-off from the wheel have a significant effect on the wheel topography and grit geometry, and hence on the grinding quality and efficiency [13, 15, 26]. This can introduce subsurface defects detrimental to the product. To control the tool wear and its effect on grinding requires a deep understanding of wear mechanism of abrasive grits on the workpiece material at nanometric scales, and detailed knowledge of grinding wheel, such as wheel shape, grit height distribution and wheel wear rate. Measurements of grinding wheel topography and abrasive wear are thus an important issue in micro grinding processes [12, 27]. Such a measurement need poses new challenges as component dimensions get smaller and the machining accuracy goes beyond conventional wisdoms.

The understanding of micro wheel topography that is characterized by abrasive grit exposure and height distribution and the accurate determination of wheel wear is especially important to micro grinding process, in particular for the profile grinding process. In recent years, the demand for micro moulds that are required for mass production of optical lenses via glass injection moulding is substantially increased due to the rapid development of biomedical, optics/photonics and telecommunication industries. For the fabrication of micro moulds made of hard and brittle materials, micro grinding must be employed in order to achieve high dimensional accuracy and excellent surface integrity. In such cases, extra attention must be paid to compensate for the profile error caused by the uncertainties generated by the grinding wheel and the machine tool, as well as the measurement of the ground profile.

This chapter presents an overview of the micro grinding technologies developed for hard and brittle materials. The principle and methodologies associated with the grinding processes will be summarized. In particular, attention will be paid to the processes that are used to fabricate micro moulds made of cemented tungsten carbide. In this case, the effect of micro structure of work material on a removal mechanism involved in the micro grinding process and the pragmatic technologies for making aspherical micro moulds will be discussed in detail. This chapter will also focus on the development of truing and dressing technologies for micro grinding wheels and the error compensation method for the fabrication of aspherical

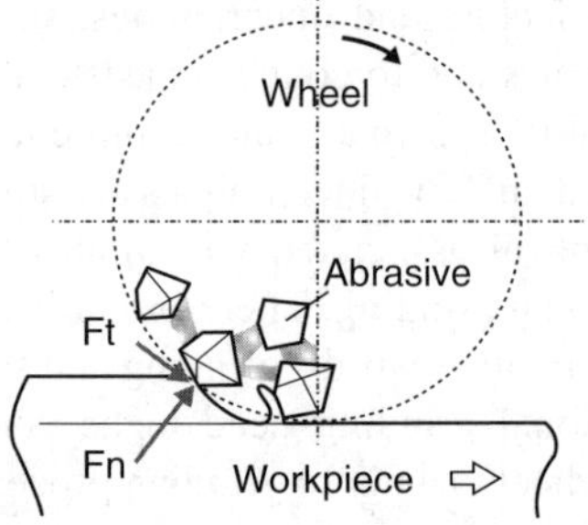

Figure 10.1 Illustration of interaction between abrasive grit and work material

profiles as the wheel preparation and tool wear compensation play critical roles in the micro grinding processes. The fabrication of micro/meso aspherical profiles will be demonstrated.

10.2 Principles and Methodologies

10.2.1 Removal Mechanism in the Grinding of Brittle Materials

In a micro grinding process, as illustrated in Figure 10.1, the removal of a work material is conducted through repeated penetrating, rubbing and scratching of abrasive particles on the work material at a relatively high speed. If the work material is brittle and hard, diamond abrasives are normally used as the cutting tool. The severity of interactions between diamond particles and the work material determines the mode of material removal involved in the grinding process. The removal modes of brittle materials can be categorized as brittle removal, ductile removal and the combination of both.

Grinding force can be used to characterize the material removal characteristics associated with a grinding process [6, 41]. It is known that for the machining of brittle materials, particularly ceramics, normal grinding force plays a critical role in the determination of removal modes. The material removal modes involved in the grinding of ceramics could be categorized as lateral crack chipping, grain dislodgement and micro cutting. The specific normal grinding forces for the three removal modes are relative to the wheel geometry parameters, work material properties and machining conditions, which can be estimated using the following equations [6],

$$F_{n1} = \beta_1 H \left(v_w / v_s \right) a_e \quad \text{for micro cutting,} \tag{10.1}$$

$$F_{n2} \approx \beta_2 \left(T^{1/2} H / E^{2/5} \right) \left(v_w / v_s \right)^{3/4} a_e \quad \text{for grain dislodgement,} \tag{10.2}$$

$$F_{n3} \approx \beta_3 \left(T^{4/9} H^{4/3} / E^{2/3} \right) \left(v_w / v_s \right) a_e \quad \text{for lateral crack chipping,} \tag{10.3}$$

where β_1, β_2 and β_3 are constants related to the wheel topography, H, T and E the hardness, the toughness and the elastic modulus of the work material, respectively, and v_w, v_s and a_e are grinding conditions, namely work speed, wheel speed and depth of cut, respectively.

For stock removal of brittle materials, the achievement of a high removal rate relies on brittle chipping through lateral crack propagation [24, 30, 42]. So a rough grinding process will lead to damage on the surface and subsurface, such as generating cracks in the ground workpiece. Surfaces created by brittle fracture are usually rough. Radial cracks may be generated on the subsurface too, which deteriorates the overall surface integrity and may cause catastrophic failure when the final products undertake cyclic loading or vibration. Therefore, for micro grinding processes that normally have small material volume being removed, brittle removal mode should be avoided despite its large removal rate. As a consequence, researchers have been making great efforts to seek the appropriate grinding conditions that enable the material being removed in a 'ductile' manner through dealing with brittle materials [3–6].

Ductile removal was reported when a brittle material was removed at an extremely small scale [5, 6, 40, 65, 68] or under high pressures [7, 8]. It is well documented that there exists a critical depth of cut, below which the brittle material being removed is in the ductile regime. Bifano *et al.* [4] found that the critical depth of cut for ductile removal in a brittle material was related to the intrinsic mechanical properties of the material. The critical indentation depth (d_c) for transition from brittle and ductile removal is expressed as,

$$d_c \sim (E/H)(T/H)^2 \tag{10.4}$$

where E, H and T denote the elastic modulus, the hardness and the fracture toughness of the material being machined, respectively.

Practically, the Bifano model suggested that the 'ductile' removal occurs in the grinding of a brittle material if the depth of cut in a removal event is smaller than d_c. The depth of cut of individual abrasives can be estimated using the maximum undeformed chip thickness that is determined by the grinding parameters and wheel conditions. For a surface grinding of a work material with translational movement, h_{max} is expressed as [41]

$$h_{\max} = (3/C\tan\alpha)^{1/2}\left(v_w/v_s\right)^{1/2}\left(a_e/d_s\right)^{1/4} \tag{10.5}$$

where C is the density of the active cutting points, α the semi-included angle for the undeformed chip cross section, d_s the wheel diameter, v_w the workpiece feed rate, v_s the wheel speed and a_e the wheel depth of cut. For face grinding of with rotational movement, $h_{\max}$ is written as [41]

$$h_{\max} = \left[(2/Cr)\left(v_w/v_s\right)\right]^{1/2} \tag{10.6}$$

where r the ratio between the width and depth for an undeformed chip thickness.

The Bifano model was valid for certain materials. Nevertheless, the generality of this model has been debated for many years because it was found invalid in some grinding applications, where the value of d_c was one or two orders smaller than the predicted [23, 40, 62, 66]. As stated by Blaedel *et al.* [6], the isotropy assumption of the material in the Bifano model might cause the error in prediction, as many of the brittle materials, such as polycrystalline ceramics, are locally highly anisotropic. Under those circumstances, the interactions between the work material and diamond abrasives must be understood and the effect of material micro structure on the removal must be included to determine the 'ductile' removal mode. In those cases, no simple model can be used to determine the critical depth of cut for damage-free grinding. Therefore, simulated grinding processes, such as indentation, scratching and single grit grinding have to be used to explore the fundamental mechanisms of removal in the ductile regime and to determine the transition of removal from brittle to ductile for a specific material. With them, detailed information on the effect of micro structure of workpiece materials on deformation and removal mechanisms could be obtained.

As indicated in the previous research on the grinding of hard or brittle materials [27, 31, 32, 50, 60], the penetration of an abrasive grit into the workpiece surface in a grinding process can be envisaged as an indentation process, more closely, as a scratching process. The penetration depth represents the interaction between the abrasive grit and the workpiece surface. The indentation and scratching have thus been used for studies on removal mechanisms involved in micro grinding [20, 37, 50]. The benefit of such studies is that indentation, scratching and single grit grinding can simulate an individual removal event in grinding, with controllable parameters, such as load and penetration depth. Nevertheless, the key issue for the simulated studies is concerned with how the removal mechanisms involved in indentation, scratching and single grit grinding apply to the transition between ductile and brittle modes in micro grinding of brittle materials. The microscopic view of an abrasive grit penetration or sliding over a surface and causing micro cracks to form in the subsurface need to be combined with more macroscopic geometry to determine what grinding conditions could be used to produce a surface without residual damages. In other words, how the pragmatic grinding technologies could be improved with the understanding of materials removal mechanism. This requires systematic investigations of micro grinding technology.

10.2.2 Interaction Between a Work Material and Diamond Abrasives

Nano indentation can be used to study deformation characteristics of brittle materials under the static normal loading condition. In this method, an indentation made with a diamond pyramid is characterized by an extremely small load, with a value of below several milli-Newtons, by impression that remains in the material. The impression process of the diamond tip on the material is analogous to the loading condition in a micro grinding process, where a small load of similar magnitude is applied to the workpiece material through a diamond abrasive grit. After indenting, surface characteristics of the impression can be examined using an *in situ* atomic force microscope (AFM) [31] and an *ex situ* high resolution scanning electron microscope (HRSEM) [65]. The penetration depth and pile-up area can also be measured from the AFM topographies. Figure 10.2a is an atomic force microscopic image of the indentation impression made on MgO substrate, where slipping lines can be clearly seen. Figure 10.2b shows the symptom of plastic deformation when indenting on SiN substrate at a very small load of 1 mN. If needed, plasma etching can be used to remove the uppermost material in the pile-up area layer by layer for further AFM and HRSEM examinations to study the subsurface damage. The threshold load for initiating cracks can be observed by gradually increasing the load. Micro-hardness tester can be used for the cases where the load in nano indentation may not be sufficiently great to initiate micro cracks in some brittle materials.

The effect of micro structure of a work material on the mechanical properties, such as elastic modulus and hardness, can be measured using the nano indentation method. Table 10.1 shows the modulus and hardness of cemented tungsten carbide measured under different indentation loads. It can be seen that both elastic modulus and hardness were influenced by the indentation load (or penetration depth). A greater load or penetration depth led to the decrease in modulus and hardness. As the cemented tungsten carbide is a composite material that consists of tungsten

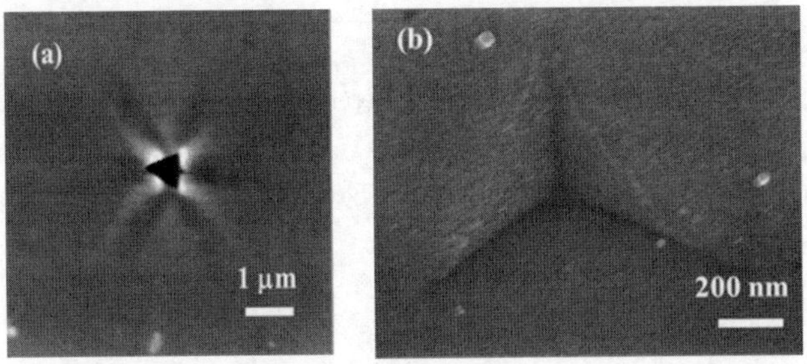

Figure 10.2 (a) A 2D AFM image of an indentation impression on MgO substrate, showing slipping lines induced by indentation. Reproduced with permission from [20]. Copyright 2010 Sage Publications; (b) A SEM image of a small impression on SiN material, showing characteristics of ductile deformation

Table 10.1 Elastic modulus and hardness of cemented tungsten carbide

Indentation load (mN)	Penetration depth (nm)	Elastic modulus (GPa)	Hardness (GPa)
3	112.34	604.83	26.85
6	222.93	606.31	23.48
9	336.28	522.43	19.74
12	422.28	495.28	19.60
20	568.39	441.27	18.88

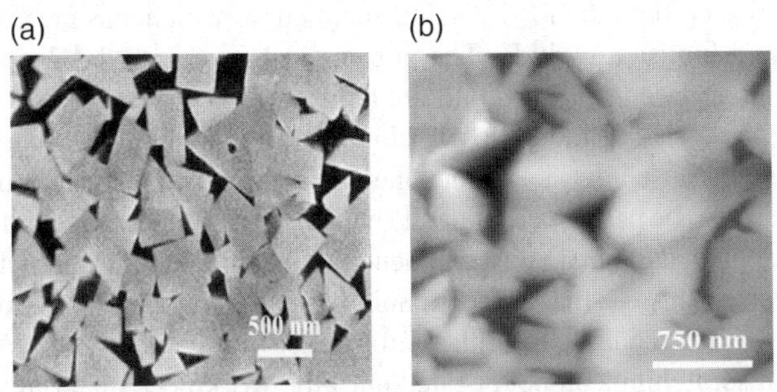

Figure 10.3 (a) Micro structure of cemented tungsten carbide. (b) An AFM image of indents made on the WC grain and in the cobalt-binder rich area. The indentation load is 4 mN. Reproduced with permission from [31]

carbide particles and cobalt bonds [56], and it can also be seen in Figure 10.3a. A larger indentation load would include more effect of cobalt boundaries. As shown in Figure 10.3b, the size of the indent on a tungsten carbide particle is much smaller than that on the cobalt-binder area, indicating that the micro structure of this material significantly influence the elastic and plastic properties of the materials. Provided that the Bifano model was employed in this material, the critical depths of cut would be significantly different for grinding tungsten carbide grains and

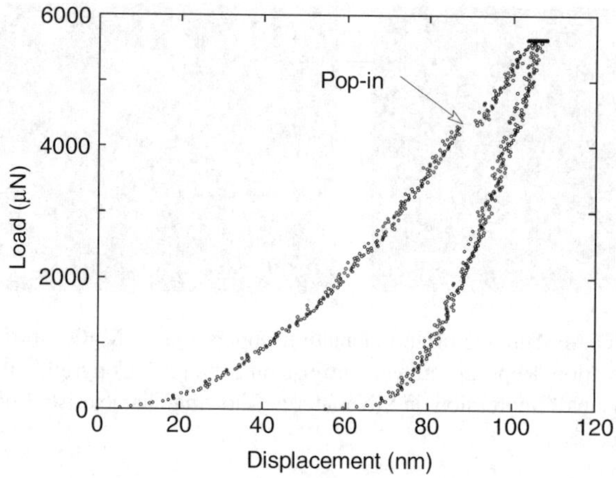

Figure 10.4 A pop-in event observed on the indentation load-displacement curve, indicating the occurrence of brittle fracture induced

cobalt binder regions. The insights gained from the investigations of deformation mechanics thus enable the accurate determination of critical depth of cut for damage-free grinding.

Detailed analysis of indenting processes can provide further insights into the deformation characteristics of a work material being investigated. With the state-of-the-art nano mechanical testing instrument, besides indenting on the intentionally selected area by use of *in situ* AFM imaging technique, the load-displacement curve recorded during indenting also gives extra information on the fracture behaviour of the work material. As shown in Figure 10.4, a pop-in event was observed on the load-displacement curve which was measured when indenting on the cemented tungsten carbide. The pop-in load was around 4.2 mN, suggesting that above the load damage-free grinding could not be achieved.

A scratching process should more closely simulate an individual grit-workpiece interaction event than indenting because it has taken the kinetics of a removal process into account [37, 50]. Nano scratching can also be performed using the nano indentation instrument. During scratching, various loads are employed to obtain different penetration depth of the diamond tip into the brittle material. The scratch can also be made by continuingly increasing the penetration depth or normal load, so the transition from ductile to brittle deformation can be observed. While in the ductile regime of removal, scratching can be repeated until the workpiece material is being removed. The critical number of passes required for initiation of material removal should provide very useful information for optimizing grinding depth of cut. Figure 10.5 shows the effect of normal load on the scratched surface characteristics of cemented tungsten carbide. It is seen that the scratch generated at relatively low load is smooth, while micro cracks were formed on the surface scratched at relative large load. The effect of grain size of various brittle materials on the deformation process and removal mechanism was investigated in the previous research [50]. In particular, parallel or multiple scratching tests can be carried out to investigate the interactions of scratches.

To examine the subsurface micro structure, scratching can also be performed on samples that are prepared using the bonded-interface sectioning technique [24]. The sectioned surfaces are thus examined using AFM, HRSEM and transmission electron microscopy (TEM) to confirm if micro cracks are not generated and the scratching is undertaken in the ductile removal regime.

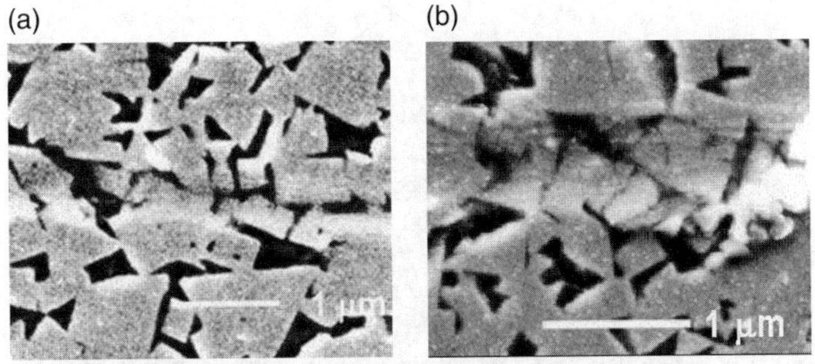

Figure 10.5 (a) A scratch made using a normal load of 2 mN showed no fracture evidence. (b) A scratch made at a normal load of 4 mN generated many micro fractures on the scratched surface

Figure 10.6 Several cracks (as indicated by the arrows) were induced on a tungsten carbide grain underneath the scratch made a relative large normal load of 9 mN

In the case of the scratching of cemented tungsten carbide, micro fracture or cracking can be formed underneath the scratch if a relatively large load is used. Figure 10.6 shows the example in which several cracks, as indicated by the arrows, were formed on the tungsten carbide grain, which is some distance away from the scratch groove. Surface characteristics and morphology of debris or chips produced by scratching can also be collected and then examined using optical microscope and HRSEM. Such examination can provide insights to the mechanisms for plastic deformation and chip formation, as well as the evidence on ductile removal. Scratches are then examined using *in situ* AFM, as shown in Figure 10.7. The previous study also found that the interaction of multiple scratches was substantially affected by wear particles and volume removal rate [1]. The study employed AFM and an optical microscope to investigate scratch topography including its depth, width and pattern of protuberances.

As nano scratching can only use limited scratching speeds which are far below the speed that diamond particles scratch work material, single grit grinding tests can be performed. Using this approach, a diamond particle can be electroplated on the tip of a conically shaped steel pen which is mounted on the periphery of a steel wheel in an ultraprecision machine. The workpiece is then mounted on the work table with a small slope. The combination of the slope and the positioning accuracy of the machine tool enable the grinding depth of cut of a grit

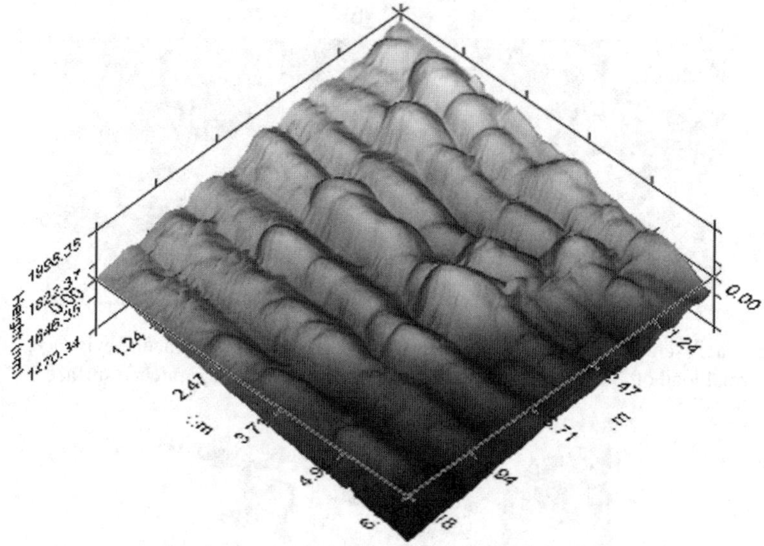

Figure 10.7 A 3D image of multi-scratched surfaces of cemented tungsten carbide. Reproduced from [50]. Copyright 2010 Elsevier

to reach several to several hundreds of nanometres. When the diamond grit slides over the workpiece surface, its depth of penetration is gradually and continuously increased on the work material. Single grit grinding is similar to scratching, but the diamond grit scratches the workpiece surface at the same sliding speed as abrasive grits have in grinding. Such experiments provide direct visual observation of transition of material removal from ductile to brittle modes. The experiment also characterizes aspects in a grinding process, such as the thermal influence caused by high sliding velocity. To study the interaction of abrasive grits, two or multiple diamond grits could be mounted on the wheel periphery. Similar specimen preparation and analytical methods used in the scratching test can be applied to examine surface and subsurface characteristics of workpiece materials and chip formation.

In indentation, scratching or single grit grinding of brittle materials at nanometric scales, it is important to identify the conditions that involve the full development of plastic deformation underneath the diamond tip or abrasive grit. Though critical load or depth of cut can be measured experimentally, it is the state of stress which determines whether facture occurs or not. As reported by Bridgman [9], brittle materials, like glass, are capable of ductile behaviour under high hydrostatic pressures. It is thus necessary to understand the effect of grit geometry, depth of cut and workpiece material properties on the resulting force system and stress status which acts at the surface and determines the deformation mechanics. The best means to explore this effect is modelling and simulation, combined with experiments. With the combination of nanoscratch experiments and modelling, the relation between lateral force and applied load on the deformation of materials being scratched could be established [37]. Finite element studies also indicated that the stress distributions during scratching and indenting were somewhat different though the same normal loads and the same indenter were employed.

Molecular dynamics (MD) adopts a classical point of view, typically representing atoms or molecules as point masses interaction through forces that depend on the separation of these

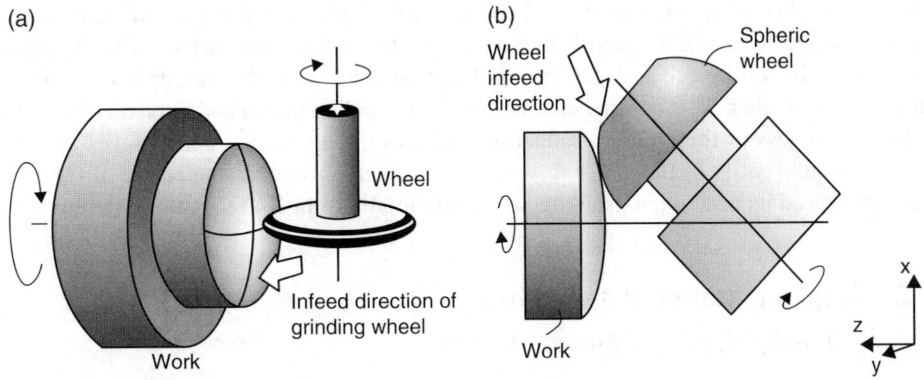

Figure 10.8 Schematic illustration of (a) the cross and (b) parallel grinding methods. Reproduced from [35]. Copyright 1996 Elsevier

objects [47]. Simulation using MD principles thus appears to be best suited to illustrate the mechanics where material deformation and removal occurs at nanometric scales. MD simulations of nano indentation, nano scratching and single grit grinding were performed [34]. In the MD simulation, the diamond indenter or abrasive grit was usually modelled as an infinitely hard object. The workpiece material is divided into three different zones, namely moving, peripheral and boundary zones. MD simulations could include the effect of workpiece materials of realistic characteristics, that is, with a defect structure, where voids, grain boundaries and second phase particles were incorporated into the lattice. Shape, size and density of these defects could be varied to investigate their effect on material deformation and removal under nano indentation, nano scratching and single grit grinding conditions. The simulation provided the nature of plastic deformation, material removal, chip formation, material defects and frictional forces. The models could be refined using the experimental data obtained from the tests. The MD simulation could also be used to study the abrasive wear at nanometric scales by considering diamond tip or abrasive grit as a non-rigid body. The limitation for MD simulation is that it can only model the deformation layer of several or at maximum several tens of nanometres due to its requirement for extremely large computation capacity.

10.2.3 Grinding Approaches for Micro Grinding

In general, approaches used for grinding features of conventional sizes can be employed for features of micrometric geometry. Nevertheless, for profile grinding, while most components of axisymmetrical profiles are ground with the conventional cross grinding technique, the parallel grinding method [16, 29, 35, 48] had some advantages, such as improved surface finish and tool life. The advantages are important as wheel wear is critical for micro grinding and the improved surface finish can significantly enhance the chances of making micro grinding as the final finishing process.

A schematic illustration of the two grinding methods for profile fabrication is shown in Figure 10.8. Figure 10.8a shows the cross grinding set-up, where a grinding wheel is in point-contact mode with a workpiece. At the contact point, the direction of the wheel velocity is perpendicular to that of the workpiece. For the parallel grinding method in Figure 10.8b, a grinding wheel is tilted horizontally (X-Z plane) by 45 degrees with respect to the work rotational axis. As the wheel traverses inward against the work rotational axis, the rotational direction of the workpiece and the cutting direction of the wheel at the grinding point are parallel to each other. This in turn generates circular grinding marks and improves the overall surface roughness of the ground product. Furthermore, the grinding point moves along the cross sectional profile of the grinding wheel as it traverses towards the centre of the workpiece, increased its functional grinding area substantially. Wheel life is therefore improved.

10.3 Implementation Perspectives

10.3.1 Truing and Dressing

Wheels for micro grinding need to be prepared with special care. This is because the contact area between the wheel and work material is extremely small in micro grinding and the active cutting grits involved in the grinding functioning zone is thus very few. Therefore, special attention must be paid in the truing and dressing process to avoid unnecessary loss of the cutting edges. Truing and dressing of conventional grinding wheels have been extensively investigated, however, practical techniques suitable for micro grinding wheels are limited.

For profile grinding, grinding wheels are normally trued into spherical shapes. For truing the micro grinding wheels, Kuriyagawa *et al.* [29, 36] developed a wheel profile generating method, which used a stainless steel pipe as a truer to generate the spherical profile of resin bond micro wheels, and then employed silicon carbides slurry for a dressing to expose the cutting grits. The truing efficiency with the steel pipe as a truer may need to be improved for some applications. Alternatively, a metal bond cup wheel of diamond abrasives can be used as the truer to replace the steel pipe or diamond abrasive particles can be extroplated on the truer surface, as shown in Figure 10.9a. As shown in Figure 10.9b, the truing of a micro wheel was performed by moving the cup wheel truer slowly along the *Z*-direction towards the rotating micro wheel. A semi-spherical shaped micro wheel was thus generated based on the curve

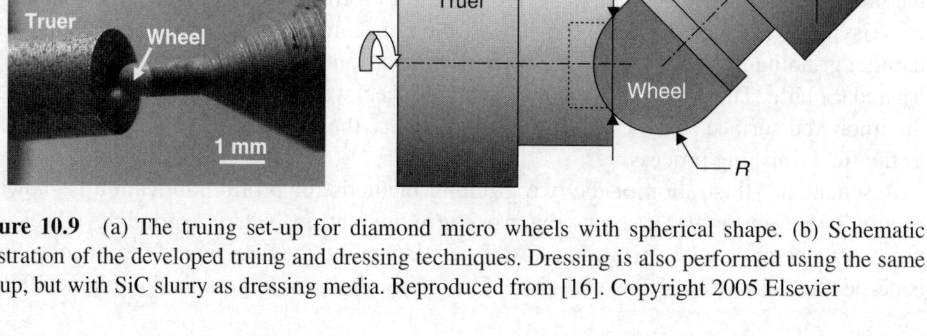

Figure 10.9 (a) The truing set-up for diamond micro wheels with spherical shape. (b) Schematic illustration of the developed truing and dressing techniques. Dressing is also performed using the same set-up, but with SiC slurry as dressing media. Reproduced from [16]. Copyright 2005 Elsevier

generating principle. The radius of the semi-sphere of the wheel can be described by the following equation:

$$R = D/2\cos\theta \qquad (10.7)$$

where R is the generated profile radius, D the truer diameter and θ the wheel inclination angle. The truer has much larger diamond grains and is more strongly bonded than the wheel to be trued to increase the efficiency in which the bonding material and diamond grains of the micro wheel are removed.

Wheels for micro grinding of profiles can be trued using profiled templates, such as a iron lapping plate shown in Figure 10.10a, where a resin bond wheel (top) was trued using a profiled cast iron plate that was machined to have a mirrored image of the required wheel shape. Diamond pastes of various particles sizes were employed on the iron plate for adjusting the truing intensities. If needed, a small amount of oil drops could also be added during truing. Figure 10.10b shows the wheel profile after truing using this curve generating method. In order to achieve the required profile accuracy for the profiled wheel shape, the whole truing process was arranged as a grinding process. In other words, truing must be repeated until reaching the required wheel profile accuracy, as shown in Figure 10.11. With this method, the average profile error (peak to valley value) achieved was below 2 μm for the diamond abrasive wheels of grit sizes below 1 μm in diameter. Such accuracy was reasonably good for fabricating spherical or aspherical components of profile accuracy of below 0.5 μm in PV value.

For exposing the cutting grits on the micro grinding wheels, dressing should be carried out using less agressive approaches than those for conventional grinding wheels. This is because the cutting grits in the grinding contact zone is much less than conventional grinding, dressing must be carried out to reduce the lose of abrasive grits to a minimum. For this purpose, the free abrasive lapping method was developed [15, 35]. In this method, as shown in Figure 10.12, fine conventional abrasive particles such as silicon carbide or alumina slurries, can be used to erode the bond materials and expose diamond grits on the wheel. Oil can also be used as lubricants to relieve the dressing intensity and reduce the grit pull-out. The dressing process normally takes only a few minutes. It should be noted that during dressing the gap between the wheel and the lapping plate remained unchanged. The loose abrasive dressing method was also used for the on-machine dressing of micro grinding wheels, as shown in Figure 10.13a, where the set-up was the same as that illustrated in Figure 10.9. Again, a small gap between the wheel and the truer was maintained during dressing and alumina pastes were employed.

10.3.2 Characterization of Wheel Topography and Cutting Edge Distribution

During micro grinding, the abrasive grits will be worn and eventually dropped from the wheel surface and new cutting grits will be exposed, which in turn changes wheel surface topography. Detailed information on wheel topography, such as grit distribution, cutting edge height variation and wheel shape, is required if grinding needs to be accurately controlled in ductile regime. As a micro grinding wheel normally has only a small number of cutting edges involved in material removal, the understanding of wheel surface characteristics and grit size distribution on the wheel surface after truing and dressing become more critical than the conventional

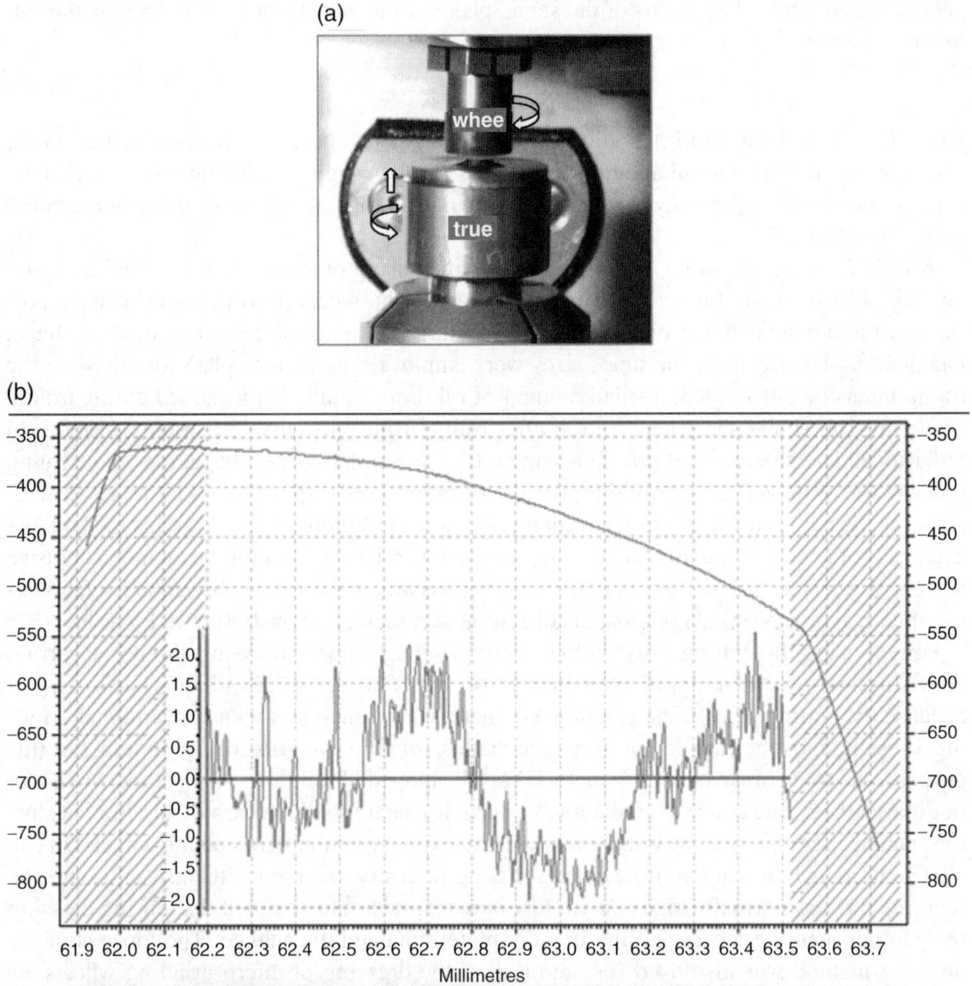

Radius 5.2000 mm			
Pa 0.8003 µm			
	Pv 2.0554 µm		
	Pt 4.3570 µm		
	Pz 4.3570 µm		

Figure 10.10 (a) Trueing set-up for micro spherical wheel. (b) Trued profile of a grinding wheel of 1 mm in diameter. The wheel profile was measured using a Talysurf profilometer (b) Reproduced with permission from [25]. Copyright 2004 Elsevier

grinding. Improper dressing would lead to the reduction of functional cutting grits and increase the wheel wear during grinding. Scanning electron microscopy (SEM) is a powerful tool to characterize wheel surfaces, with which detailed topographies after truing and dressing can be examined. Figure 10.14a shows the SEM image of a micro wheel after truing, showing the unopened wheel surface. If dressing was not properly conducted, the exposure of cutting

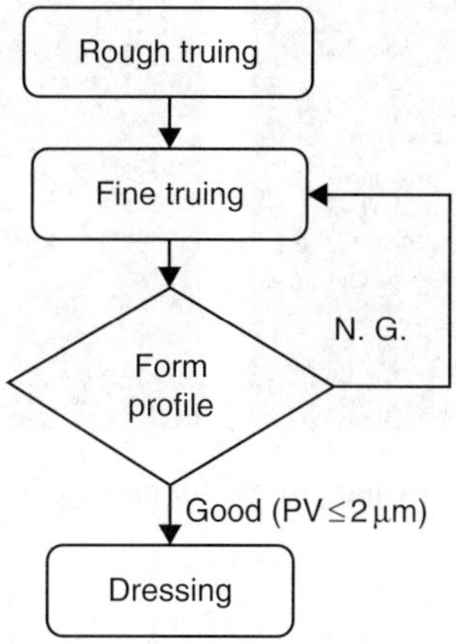

Figure 10.11 A flow chart showing the truing and dressing procedure for micro grinding wheels

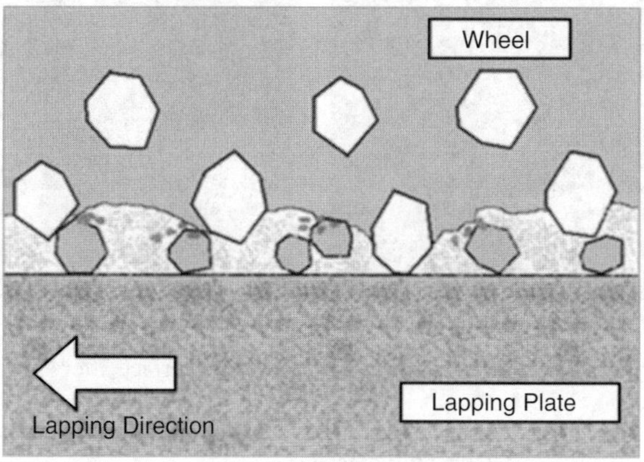

Figure 10.12 Illustration of the free abrasive lapping process for dressing micro grinding wheels Shaded particles are dressing abrasives and unshaded particles arediamond grits on the wheel. Reproduced from [15]. Copyright 2005 Elsevier

edges on the wheel could be very poor. A properly dressed wheel should have the abrasive grains well exposed and few grits pulled out, as shown in Figure 10.14b.

The grit distribution on a dressed wheel can be carefully examined using SEM and the exposed and pull-out grits can then be measured. Figure 10.15 shows the densities of exposed and pull-out grits of a resin bond diamond wheel of grit size of 4 µm in diameter after dressing using different conditions. In comparison, the theoretical grit density (grit per mm^2), N, was also calculated using the following formula [41]:

$$N = 6V_g / \pi d_g^{\,2} \tag{10.8}$$

where V_g is the volume fracture of abrasives in the wheel and d_g the average grit diameter. As can be seen in Figure 10.15, the measured density was less than 80% of the theoretical density. The dressing condition had substantial influence on the grit distribution in the grinding wheel.

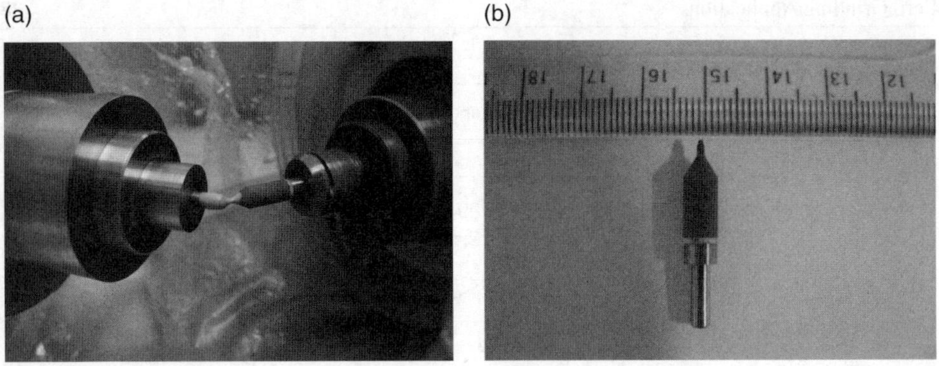

Figure 10.13 (a) Illustration of a dressing process for micro grinding wheels. (b) The micro grinding wheel after trued and dressed

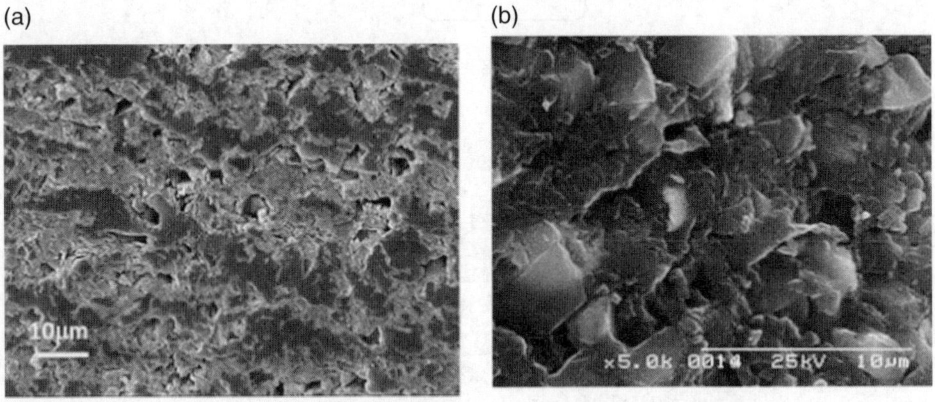

Figure 10.14 (a) Wheel surface after truing. (b) Wheel surface after appropriate dressing

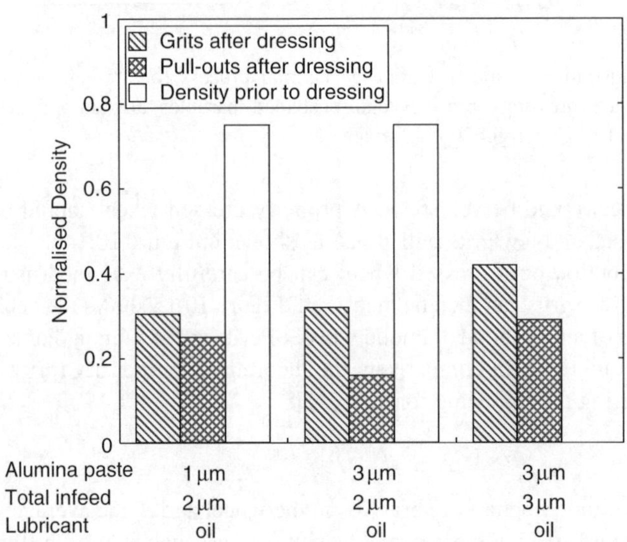

Figure 10.15 Densities of exposed and pull-out grits on the wheel surfaces after dressing using lapping method with alumina pastes of different grit sizes. The resin bond wheel has a grit size of 10 μm and a volume fraction of 25%

The grit pull-out was significant though the dressing conditions were selected with special care. Balance of efficiency and intensity in dressing would be always an important issue in the design of micro grinding process.

10.3.3 Measurement of Grit Height Distribution

Little information on the quantitative measurement of grit height distribution and its evolution during micro grinding is available. It is thus of great significance to develop technologies for obtaining such information. Both stylus profilometry and confocal microscopy can be used to assess the grit height distribution of a grinding wheel.

In the profilometry based method, the wheel topography information, shown in Figure 10.16a, can be obtained by scanning the wheel surface using a stylus profilometer. The SEM image in Figure 10.16b demonstrated that the grit distribution pattern obtained

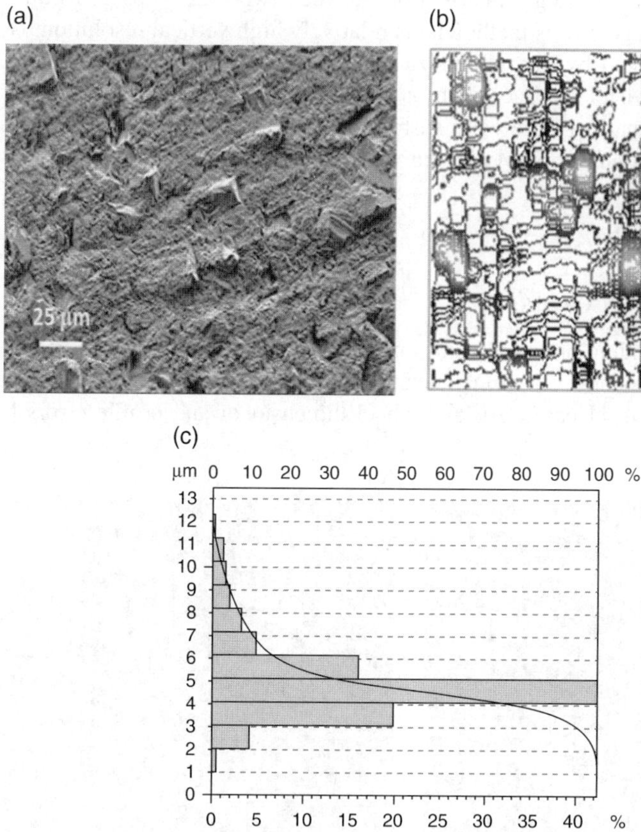

Figure 10.16 (a) SEM image and (b) profilometer-scanned topography of wheel surface with a grit size of 20 μm after truing and dressing, showing good exposure of diamond abrasive grits. (c) Distribution of grit heights. Reproduced from [15]. Copyright 2005 Elsevier

from the stylus scanning was identical to that observed via SEM. By examining the overall bearing ratio of the wheel surface provided by the profilometer, size and shape of active abrasive grits and their height distribution were obtained, as shown in Figure 10.16c.

10.3.4 Characterization of Abrasive Wear

Diamond abrasive grits on a micro wheel are being worn during grinding of brittle materials because abrasive wear occurs whenever the grits are loaded against a solid object that has equal or even lower hardness. The abrasive wear is expected to be more significant in micro/nano grinding as much less active abrasive grits are involved compared with conventional grinding. Figure 10.17 shows the comparison of the grinding wheel surfaces prior to and after grinding of cemented tungsten carbide for about 20 minutes. Significant wheel wear occurred. Diamond grits were worn and pulled out. It is thus important to investigate the wear rate of abrasive grits, and how the being worn and worn grits affect the material removal and the critical point for ductile to brittle transition. Using the method for measuring grit height, the difference of grit height distribution between two time steps gives quantitative wheel wear. The advantage of this method is its relatively high vertical resolution. However, lateral resolution is limited by the probe size and the reliability can be influenced by the contamination of the wheel surface, such as adherence of dusts and grinding chips on the wheel surface. Alternatively, the worn surface characteristics of the abrasive grit can be examined using high resolution SEM and the wear volume can also be measured using the confocal microscopy.

10.3.5 Compensation Grinding

In an ultra precision grinding process, such as micro grinding, major error sources that affect the profile accuracy of ground surfaces include machine positioning error, workpiece and wheel setting error, grinding wheel path error and wheel dimensional and profile errors [28, 38].

Figure 10.17 Comparison of grinding wheel surfaces (a) prior to and (b) after grinding for a certain period of time. Resin bond wheel of a grit size of 15 µm and abrasive volume fraction of 25%

Among them, tool path and positioning errors associated with a modern precision machine tool are negligible, compared with the other error sources. Tool setting error can be satisfactorily reduced by using the on-machine truing and dressing technology [14, 15]. The geometric (or dimension and profile) error of a grinding wheel is usually caused by the truing and dressing process and the grinding wear. The error develops with the progress of grinding and affects the ground profile accuracy. Thus, this error must be determined accurately and compensated in the coming cycle of grinding.

Figure 10.18 shows a flow chart that is often used in ultraprecision profile grinding processes, which is especially important for profile grinding of micro components. After the first grinding cycle, the profile of the part being ground is measured using a profilometer. If the accuracy of the ground profile does not meet the required accuracy, grinding has to be continued. The tool path for the next grinding procedure must be taken into consideration of the error that was produced in the last grinding cycle through a powerful compensation algorithm. Therefore, the grinding process is also named as compensation grinding.

The efficiency of the compensation grinding depends on the compensation approach that is employed. There are two important issues that significantly affect the compensation accuracy and efficiency in the process. One is the way that the residual profile error is compensated for in the modified tool path. The other is how to reduce the effect of wheel geometric error on the ground profile in the initial grinding cycle. Figure 10.19 shows the effect of the different compensation approaches. The conventional method for compensation is to translate the tool path along the Z-axis with an equal distance, ΔE_i, as illustrated in Figure 10.19a. The distance between a point on the previous profile and the corresponding point on the new profile is defined as ΔE_i' and $\Delta E_i \neq \Delta E_i$'. Moreover, the value of ΔE_i' in Figure 10.19a slightly varies with the grinding location and the compensation is thus inaccurate. This affects the convergence to the required profile accuracy. Figure 10.19b demonstrates an improved approach. Instead of translating the tool path along the Z-axis, the error, ΔE_i, is compensated for by shifting it along the radial direction or the direction of its normal vector at the grinding point. The normal vector is determined by connecting the grinding point with the reference centre of the target profile. The offset

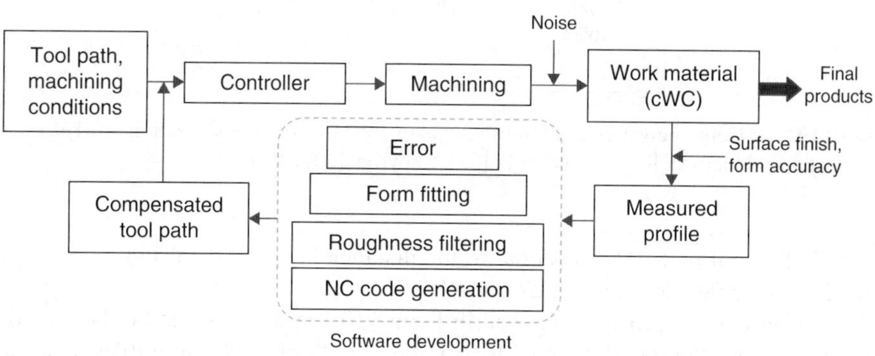

Figure 10.18 A flow chart of micro profile grinding, indicating the role of error compensation in the grinding process

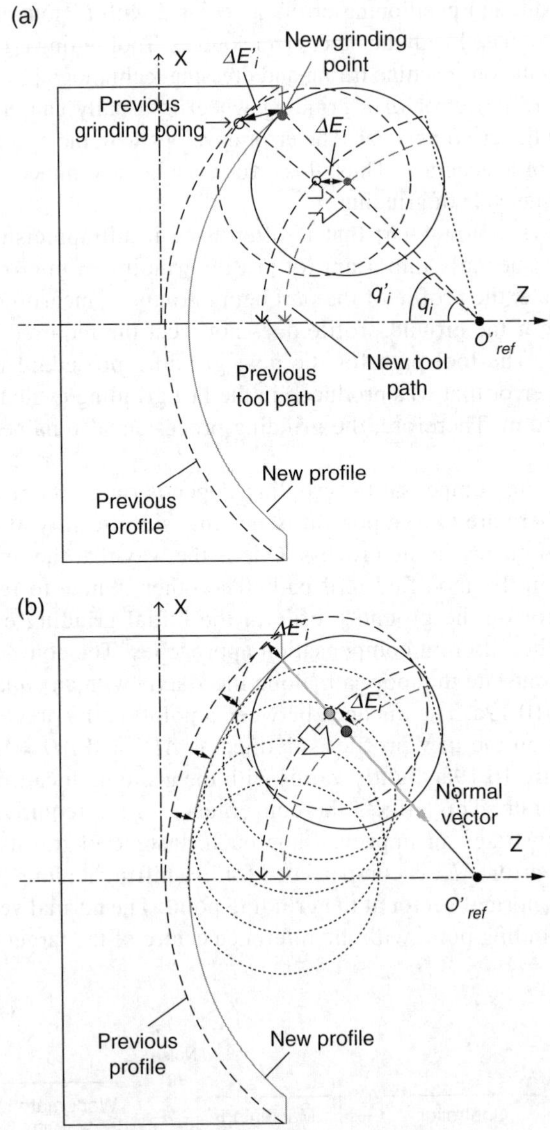

Figure 10.19 (a) Commonly used axial translational method for error compensation. (b) The improved radial translational method. Reproduced from [28]. Copyright 2007 Elsevier

distance of the tool path, ΔE_i', is equal to the distance between the grinding points prior to and after compensation, that is, $\Delta E_i = \Delta E_i$'.

The compensation approach described in Figure 10.19a demonstrated to be effective in the grinding of micro aspherical moulds. A micro mould of ~600 μm in diameter made of cemented tungsten carbide was ground using the approach. As shown in Figure 10.20, the starting profile error after primary grinding (i.e. grinding without compensation) was 1.02 mm

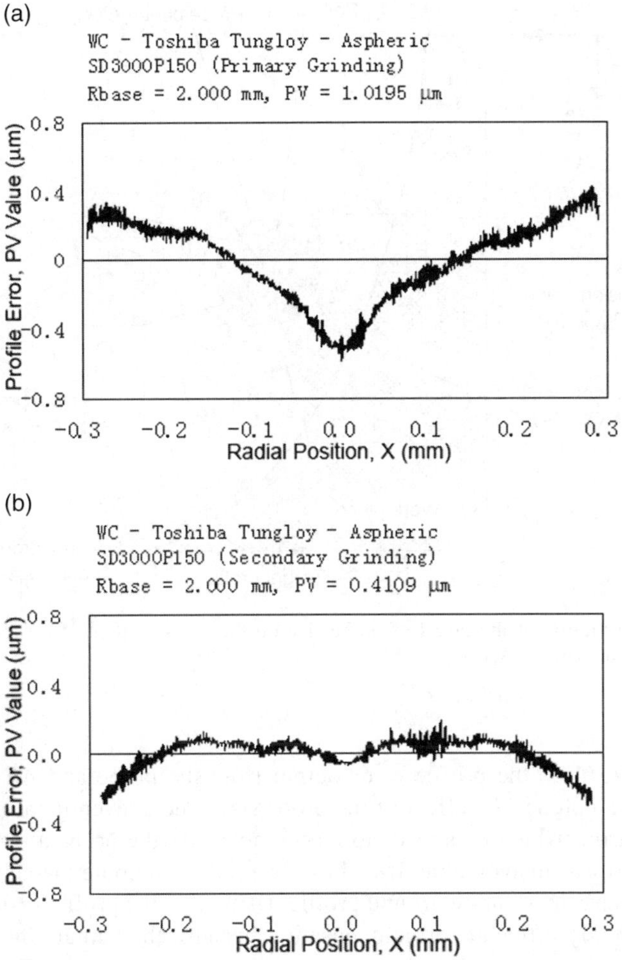

Figure 10.20 Progress of profile error with compensated grinding of mould insert of 600 μm in diameter. Reproduced from [28]. Copyright 2007 Elsevier

in PV. After only one cycle of compensation grinding, the profile error was significantly reduced to 0.41 μm.

The employment of the on-machine measurement of the ground profile can effectively eliminate the error resulting from the re-setting of the workpiece and substantially improve the machining efficiency. When using the probe to measure the ground profile, an error is induced because the radius of the probe affects the contact point. Figure 10.21 schematically shows the effect of the probe size on the measurement. A point on the target profile (or the designed profile), that is, $D(X_{di}, Z_{di})$, corresponds to the point F_i on the actual ground profile. The actual residual error, F_iD_i or δ, along Z-axis at point Di can be expressed as

$$\delta = \Delta E_t - \Delta E_g \tag{10.9}$$

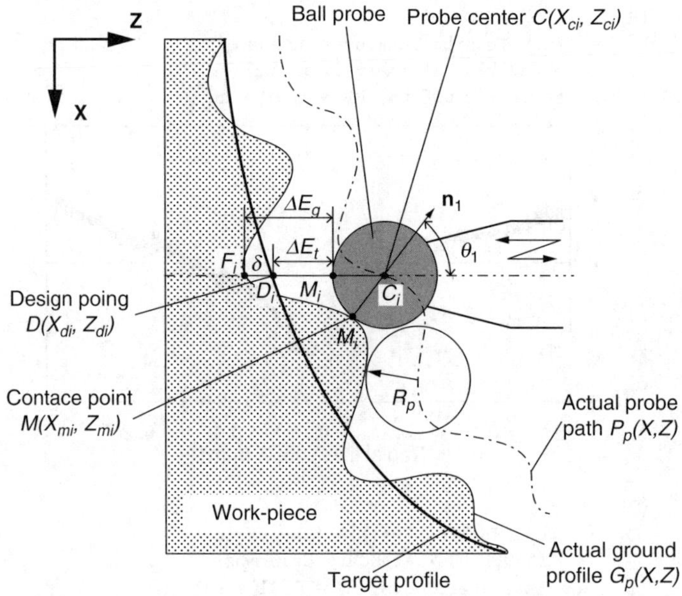

Figure 10.21 Illustration of the effect of probe size on the measurement of form error. Reproduced from [12]. Copyright 2010 Elsevier

where ΔE_t, or D_iM_i', is the profile error output from the probe and ΔE_g or F_iM_i' is the overall error that includes the effect of the probe size. The conventional method uses ΔE_t as the residual error, which thus leads to inaccuracies. As the probe always has a certain dimension, an error is unavoidable. It is thus essential to compute the actual profile error δ in order to achieve an accurate ground profile. Using δ as the profile residual error could still be insufficiently efficient for compensation because the actual contact point of the probe is at point M_i, instead of F_i, due to the size effect of the probe. Therefore, this size effect must be excluded when computing the compensated probe and tool paths for the next grinding cycle.

The information of the wheel geometry at the initial stage of grinding is important for improving the grinding efficiency. Such information is normally unknown, unless extra effort is paid to achieve it. In most grinding processes, direct use of a wheel for profile grinding would result in a significant profile error. This thus increased the difficulty for the subsequent compensation process. Therefore, an extra grinding cycle could be performed, not on the work material, but on a carbon disc, which was thus named as a preliminary grinding cycle [28]. The wheel working path for the preliminary cycle was generated using the target profile parameters. The profile of the ground carbon disc was then measured using a Talysurf profilometer. The profile error appeared on the carbon disc should be a direct reflection of the wheel geometric error from the truing and dressing process because the tool wear was negligible when grinding carbon. The radius and profile errors of the ground profile were thus obtained from the subsequent analysis in the profilometer. A compensated tool path was generated for grinding the actual work material. Therefore, the tool path used for the first grinding cycle on the work material had taken the wheel dimension and profile errors into account.

10.3.6 Pragmatic Aspects in Profile Grinding

In the micro grinding of brittle materials, the wheel wear is significant in most cases. Compensation can only be done after one grinding cycle, which does not help on the error caused during the grinding cycle. Therefore, pragmatic techniques that reduce the wear and prolong the effective life of the wheel are always beneficial. For instance, for generating convex profiles, the cup wheels can be trued to have profiles with a larger lip radius, as shown in Figure 10.22, compared with the conventional shape shown in Figure 10.9. For wheels with the same manufacturing dimensions, a larger lip radius has increased the wheel functional area, thus reduced the wear along the radius direction.

Swivelling the grinding wheel can vary the contact area of a wheel with the workpiece, as shown in Figure 10.23. This is equivalent to the increase in wheel functional area, which thus reduces the overall wheel wear. Unless the grinding machine used has a high precision accuracy in the wheel inclination axis, otherwise, special care must be taken as the swivelling of the wheel would introduce the positional error.

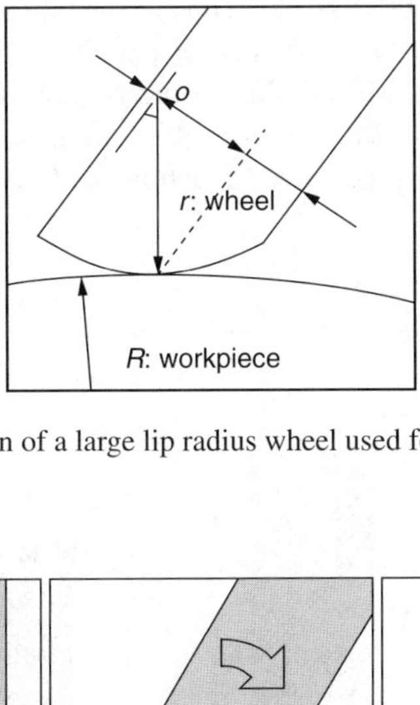

Figure 10.22 Illustration of a large lip radius wheel used for convex profile grinding

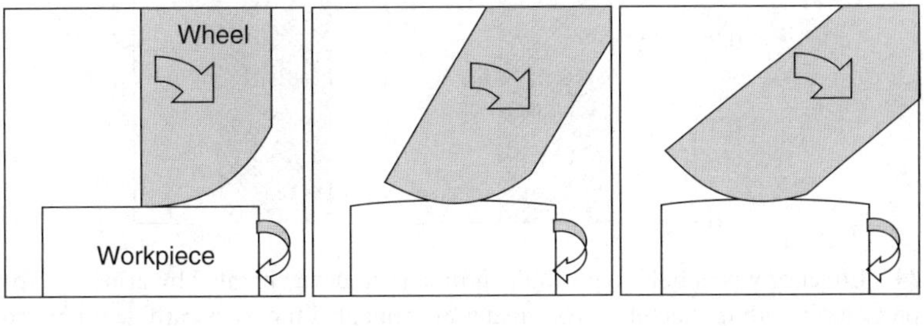

Figure 10.23 Schematic illustration of the swivel grinding process

10.3.7 Parametric Effects in Profile Grinding

10.3.7.1 Effect of Bond Type

For profile grinding, the wheel bond type has substantial effect on the ground surface finish and profile accuracy. Normally, a softer bond material will produce a better surface finish, but a greater profile error. Figure 10.24 shows the comparison of profile errors obtained from the grinding wheels of resin and metal bonds when the other grinding conditions remained the same. The profile error produced by the resin bond wheel is about 30% greater than that by the metal bond. This was because the metal bond wheel had less wear, thus leading to higher profile accuracy. However, the rubbing effect from the resin bond wheel had slight advantage in obtaining better surface finish over the metal bond wheel. As a result, the selection of wheel bond types must be based on the grinding requirements. In many cases, compromise must be made to achieve the balance of surface finish and profile accuracy.

10.3.7.2 Effects of Wheel Speed and Feed Rate

The effects of grinding wheel speed on the profile accuracy and surface roughness are shown in Figure 10.25. The increase in wheel speed resulted in improvements in both surface finish and profile accuracy. It should be noted that the increase in wheel speed is greatly hindered by the grinding wheel size used in micro grinding. For the Toshiba machine used, the maximum spindle speed is 20 000 rpm. Thus, if the wheel diameter of 1 mm is used, the maximum wheel speed is only 1 m/s. This is far from the optimal speed that is used in the conventional grinding. In this case, rubbing would be very severe and the removal amount of the grinding process was very little. This is also another compromise that has to be taken in micro grinding unless a very advanced machine with extremely high spindle speed is

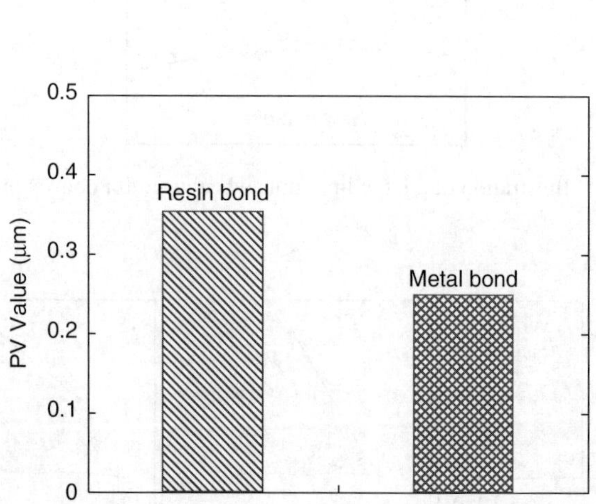

Figure 10.24 Effect of wheel bond type on the profile accuracy generated by grinding. Toshiba ULG-1002 ultra precision grinding machine, work material: cemented tungsten carbide, diamond wheel grit size: 4 µm, wheel speed: 8 m/s, feed rate: 0.2 mm/min., depth of cut: 0.2 µm. The staring profile error remained the same as 1.2 µm in peak-to-valley value.

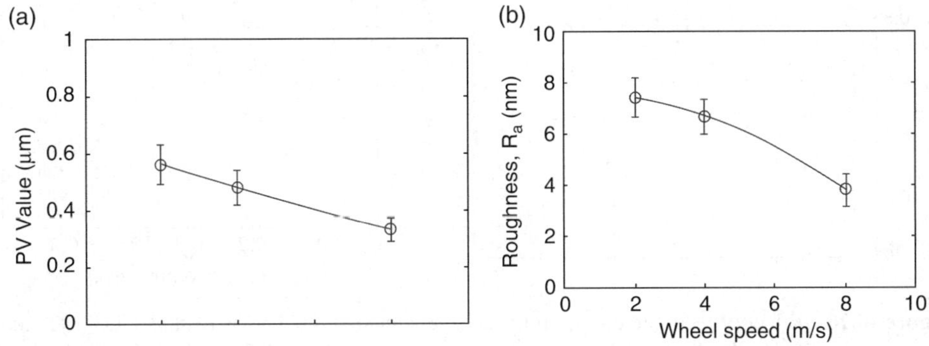

Figure 10.25 Effect of wheel speed on (a) profile accuracy and (b) roughness. Toshiba ULG-1002 ultra precision grinding machine, work material: cemented tungsten carbide, diamond resin bond wheel, grit size: 8 μm, feed rate: 0.2 mm/min., depth of cut: 0.2 μm

available. The effects of feed rate on the profile accuracy and surface roughness are insignificant. As shown in Figure 10.26, the increase in feed rate led to a slight increase in profile error, as well as surface roughness.

The complexity of profile being fabricated also has influence on the ground profile accuracy being achieved. Figure 10.27 demonstrated an example, in which both spherical and aspherical concave surfaces were ground under the same grinding conditions. The profile error obtained in the aspherical profile was found to be about 22% greater than that in the spherical profile.

10.4 Application Cases

Two micro grinding application cases are described in this section. The first example demonstrates the development of pragmatic technologies for fabrication of micro moulds made of cemented tungsten carbide. The second case introduces the process for grinding fibre optic connectors made of combination materials.

10.4.1 Micro Grinding of Aspherical Moulds

10.4.1.1 Grinding Set-Up and Conditions

The grinding experiments were performed on a Toshiba ULG-1002, ultra precision grinding machine. Figure 10.28a shows a full view of the machine and 10.28b demonstrated an example of the profile grinding set-up. Detailed machine specification is shown below.

- two-axis (x, z), resolution of 10 nm;
- spindle up to 20,000 rpm;
- maximum table speed 1000 mm/min;
- work spindle 100–3000 rpm.

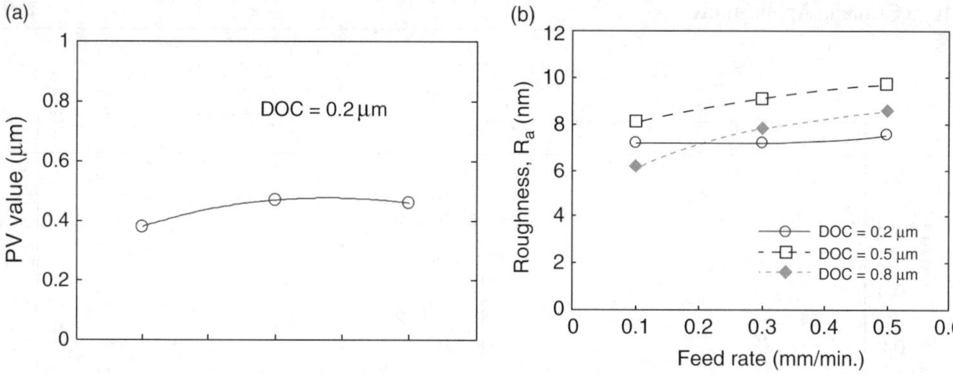

Figure 10.26 Effect of feed rate on (a) profile accuracy and (b) roughness. Toshiba ULG-1002 ultra precision grinding machine, work material: cemented tungsten carbide, diamond resin bond wheel, grit size: 8 µm, wheel speed: 8 m/s, depth of cut: 0.2 µm

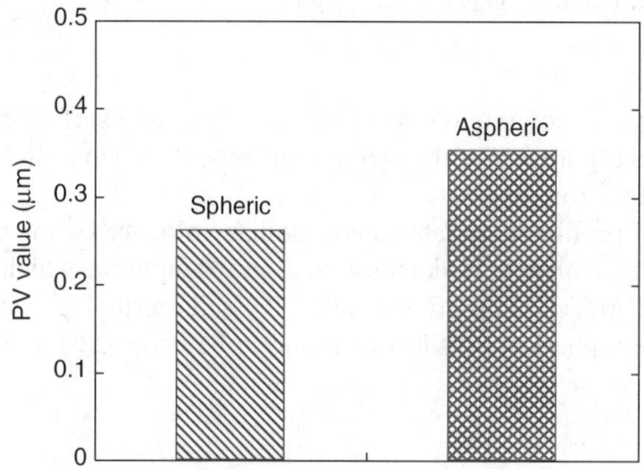

Figure 10.27 Effect of profile shapes on the ground profile accuracy. Toshiba ULG-1002 ultra precision grinding machine, work material: cemented tungsten carbide, diamond resin bond wheel, grit size: 8 µm, wheel speed: 8 m/s, feed rate: 0.2 mm/min., depth of cut: 0.8 µm

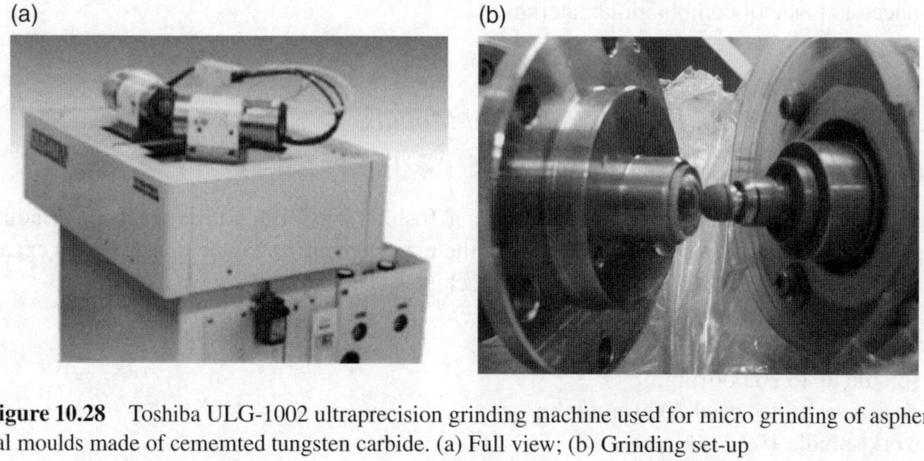

Figure 10.28 Toshiba ULG-1002 ultraprecision grinding machine used for micro grinding of aspherical moulds made of cememted tungsten carbide. (a) Full view; (b) Grinding set-up

Micro diamond wheels (manufactured by Noritake) used in the experiment had a mesh size of 3000 (an equivalent grit size of 8 μm in diameter) and bond strength of grade P. Cemented tungsten carbide (supplied by Toshiba Tungaloy) was used as the work material. The truer used was a metal bond diamond cup wheel with a mesh size of 400. The curve generation method introduced earlier was used for truing the grinding wheel. A very small amount of coolant was used in truing. The free abrasive lapping method was used for dressing the wheel. Dressing used the same tool as that in truing, but the cup wheel remained not in contact with the wheel, with a small gap of several micrometers. During dressing, a silicon carbide paste with mesh size of 6000 was employed. The entire dressing process lasted only three to five minutes. Detailed truing and dressing conditions are summarized in Table 10.2.

10.4.1.2 Grinding Process

Parallel grinding approach was used in the grinding of micro aspherical moulds, which aimed to achieve a good surface finish. In this work, based on the conventional compensation process, a new grinding protocol for the fabrication of aspherical surfaces was developed. The flow chart of the grinding protocol is shown in Figure 10.29. In the new protocol, a preliminary grinding cycle (i.e. grinding was performed not on work material, but on easy-to-grind carbon disc) was introduced, which reduced the effect of the initial wheel geometric error on the ground profile, hence decreased the repeated grinding cycles. In particular, the method for compensating the residual profile error was used.

Table 10.2 Truing, dressing and grinding conditions used. [16]

Wheel	SD3000P150B, 1 mm in diameter
Truer/dresser	SD400N75M, cup wheel type
Truing conditions	
Wheel speed (rpm)	10,000
Truer speed (rpm)	200
Feed rate (μm/min)	1
Coolant	Water soluble type
Dressing conditions	
Wheel speed (rpm)	2,000
Truer speed (rpm)	200
Feed rate (μm/min)	0
Coolant	none
Slurry for dressing	GC6000
Grinding conditions	
Wheel speed (rpm)	20,000
Truer speed (rpm)	300
Feed rate (mm/min)	0.1
Infeed per pass (μm)	0.8
Coolant	Water soluble type

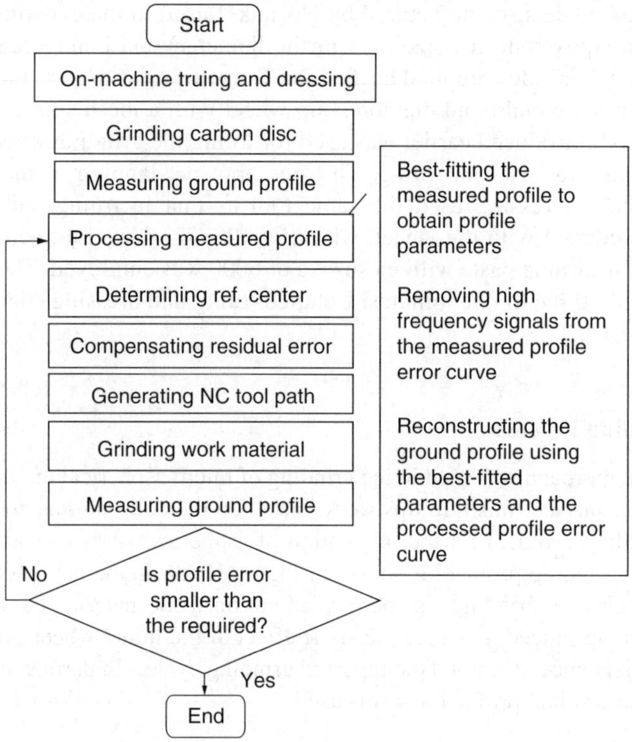

Figure 10.29 A flow chart of the grinding process. Reproduced from [28]. Copyright 2007 Elsevier

10.4.1.3 Ground Micro Moulds

Figure 10.30 shows a ground micro aspherical insert with a diameter of 1 mm. The profile error of 0.4 μm in PV was obtained after the first compensation grinding cycle. This value is generally acceptable for mould inserts used in production of glass lenses. To examine the compensation method, the second compensation grinding was performed, which further reduced the form error to 0.2 μm. Further improvement of profile accuracy by increasing the compensation cycles was found to be insignificant. The results indicated that the form error remained in the range of 0.2 μm after the third or fourth compensation cycles. A surface finish of 7 nm in R_a and 100 nm in R_{max} was obtained.

The capability of the developed micro grinding strategy was tested by making the smallest possible aspherical moulds. A micro aspherical insert of 200 μm in diameter was ground. A SEM micrograph of the ground insert is shown in Figure 10.31a. As can be seen in Figure 10.31b, the profile error of 0.44 μm was achieved using the same micro grinding strategy as that for the insert of 1 mm in diameter. The profile parameters for this aspherical mould are: R_{base}=0.567 mm, K=-1.685, A_4=0.890 and A_6=–0.276. The AFM analysis showed the average surface roughness of the ground surface was 3.821 nm.

The surface topography of the micro wheel observed under the SEM showed that the wheel was adequately self-dressed with the grinding condition used. About 56% of abrasive grits

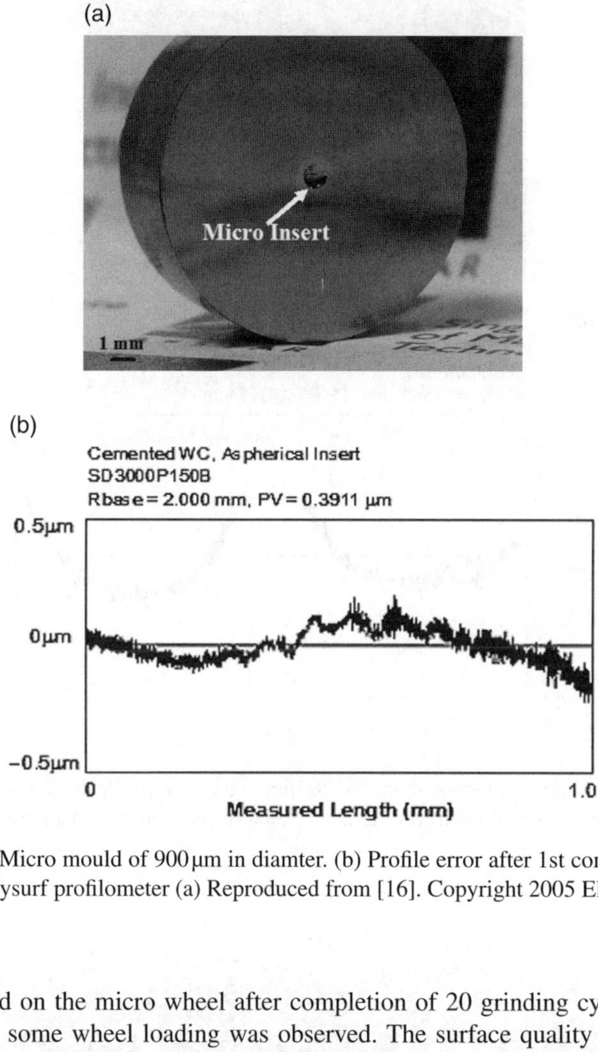

Figure 10.30 (a) Micro mould of 900 μm in diamter. (b) Profile error after 1st compensation grinding measured using Talysurf profilometer (a) Reproduced from [16]. Copyright 2005 Elsevier

was still remained on the micro wheel after completion of 20 grinding cycles. With further prolonged usage, some wheel loading was observed. The surface quality was not seriously affected by the wheel loading, but the form accuracy of the ground inserts had deteriorated due to increased tool wear. It should also be noted that though the average surface roughness reached several nanometres, some defects could still be observed under high magnification, as shown in Figure 10.32. The defects should be attributed to the existence of cobalt binders, which could not be avoided, even though the material removal was expected to be conducted in the ductile regime. This is the drawback of using such a composite material. In addition, grinding marks could also be observed on the magnified ground surface. For the fabrication of optical lenses, the marks in regular patterns are not favourable.

Several aspherical moulds of different diameters were fabricated using the same grinding conditions. The profile error in PV value after first compensation grinding was measured and plotted as a function of the size of mould. Size effect was observed. As shown in Figure 10.33, a greater profile error corresponded to a smaller size of mould ground. Possible reasons for this size effect include: (1) when mould size was decreased, the average grinding speed across

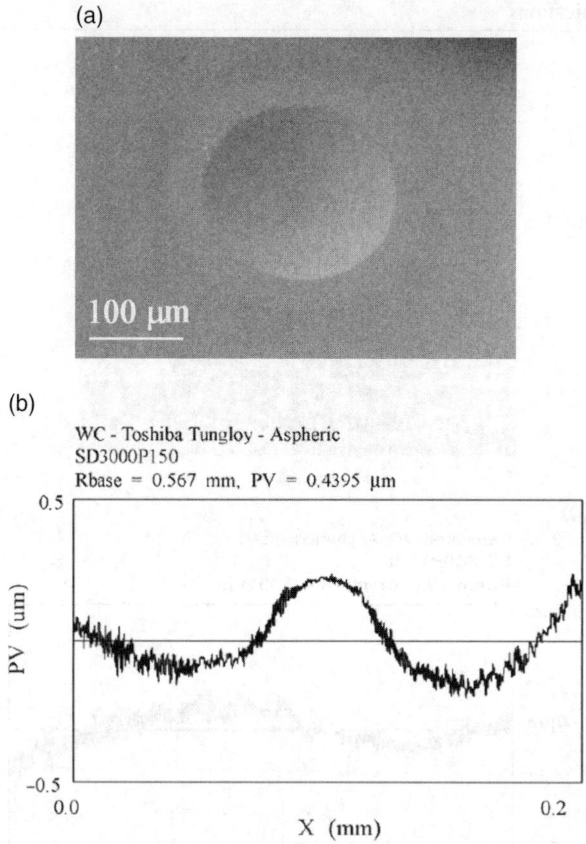

Figure 10.31 (a) A micro asphercial mould of 200 μm in diamter. (b) Profile error after 1st compensation grinding measured using Talysurf profilometer. Reproduced from [16]. Copyright 2005 Elsevier

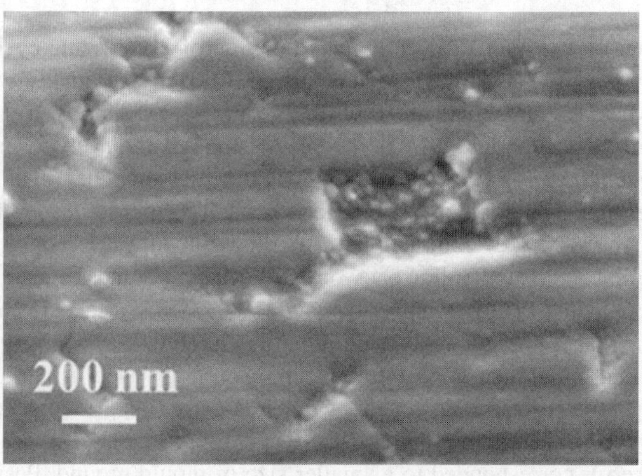

Figure 10.32 SEM images of ground surfaces

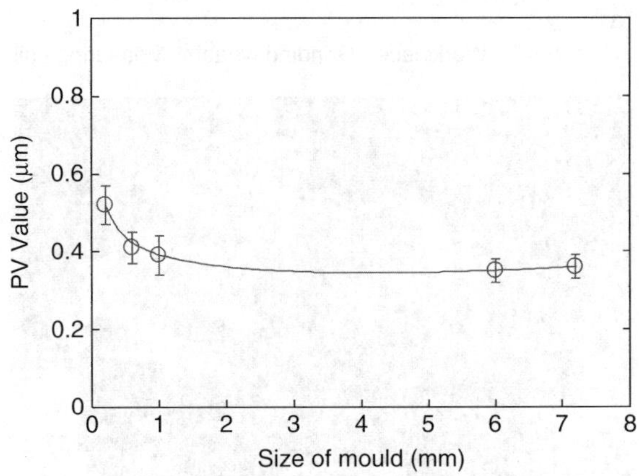

Figure 10.33 Size effect in the micro grinding of mould inserts

the entire ground area was reduced. As shown in Figure 10.25a, a smaller grinding speed would lead to a larger profile error. (2) When the size of mould shrank, the measured information of the profile error was reduced. This could affect the compensation accuracy. Apparently, future work should be directed towards this understanding.

The grinding of an aspheric mould of 6 mm in diameter was also performed on a 4-axis ultra-precision machine tool (NACHI, ASP005P), as shown in Figure 10.34a, using a resin bond wheel of mesh size of 2000. For this machine, the work spindle of this machine can linearly move along *Z*-axis at a positioning resolution of 1 nm and *Y*-axis at a resolution 100 nm. The grinding spindle is installed on a rotary table (*B*-axis). Its rotational movement has a resolution of 0.0001 degree. The grinding spindle can also move linearly along *X*-axis at a positioning resolution of 1 nm and reach a speed of 45 000 rpm. To avoid the grinding interference and achieve the best possible surface quality, a cylindrical wheel with arc nose radius of 1 mm was used and the inclined-axis grinding mode was employed. The grinding spindle axis had an inclined angle of 45 degrees relative to the work spindle axis. The profile error after three cycles of compensation grinding reached 0.2 μm, as shown in Figure 10.34b, which was better than the best profile accuracy (about 0.36 μm) achieved with the Toshiba machine. The increased wheel speed used, the on-machine profile measurement and the higher positioning accuracy of the machine should all be attributed to the improvement in profile accuracy.

10.4.2 Micro Grinding of Optical Fibre Connectors

A fibre optic connector consists of a zirconia ferrule of 2.5 mm in diameter and a glass fibre of 125 μm in diameter. The fibre is centred inside the ferrule which has a spherical face at its open end. The spherical end face is precisely ground prior to the fibre assembly. After assembly, the fibre protrudes from the spherical end face of the ferrule. The spherical end face, together with the fibre, has to be polished to achieve an extremely even and smooth surface. Currently in production, manual polishing is first used to remove the excess fibre and epoxy

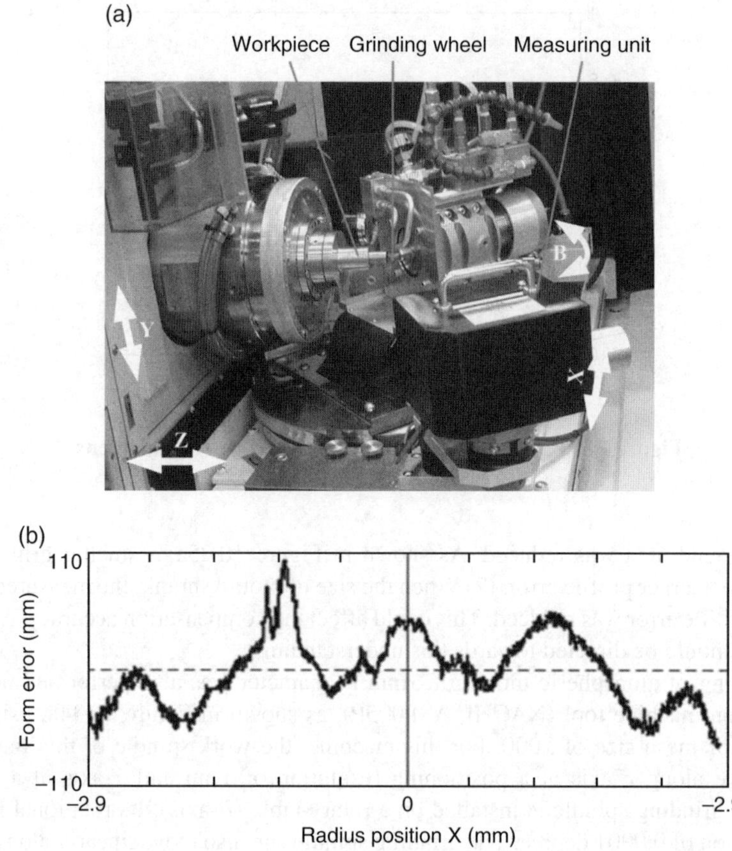

Figure 10.34 (a) NACHI, ASP005P, ultra precision grinding machine with on-machine measuring probe. (b) The profile error measured using Talysurf with a PV value of 185 nm, which is in good agreement with the value of 177 nm via on-machine measurement. Reproduced from [12]. Copyright 2010 Elsevier

(for holding the fibre inside the ferrule) bead. There exist some problems in the current machining processes. First, it is difficult to control end face profile accuracy during polishing because of the use of flat polishing pads. Second, four polishing stages (including the manual polishing) are required to reach the required surface finish and the machining efficiency is thus low.

10.4.2.1 Grinding Set-Up and Conditions

An ultra precision grinding technology was developed for machining the spherical end faces of fibre connectors using inclined resin bond diamond cup wheels, aiming at replacing the polishing process. The grinding was completed on a CNC grinding machine (Loh

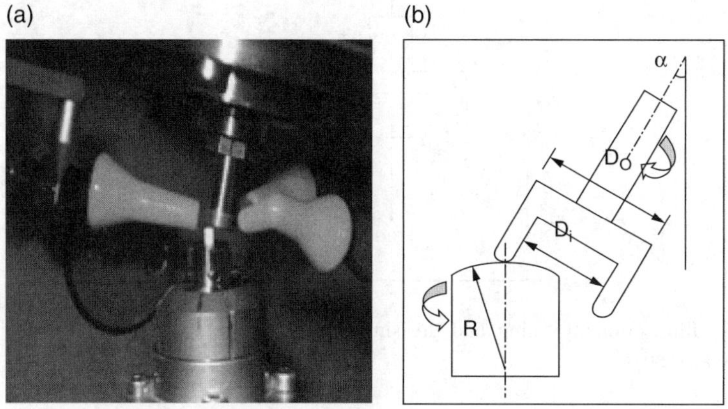

Figure 10.35 (a) Grinding set-up on the Loh Grinder. (b) Schematic illustration of convex spherical surface generation using a cup wheel Reproduced with permission from [25]. Copyright 2004 Elsevier

Spheromatic 25-2SL, Germany) shown in Figure 10.35a. A cup wheel grinding method was used to generate the spherical profile of end faces. In this method, the cup wheel was tilted and the top part of the tilted wheel contacted the workpiece during grinding, as shown in Figure 10.35b. The radius of the generated convex spherical surface, *R*, can be expressed as [25],

$$R = \frac{D_m - 2r\sin\alpha}{2\sin\alpha} \tag{10.10}$$

where *r* is the lip radius of the cup wheel illustrated in Figure 10.2c, the tilt angle and

$$D_m = \left(D_o + D_i\right)/2 \tag{10.11}$$

where D_i and D_o are the cup wheel inner and outer rim's diameters, respectively.

Resin bond diamond wheels of different grit sizes were used in the experiments. A medium grade of wheel hardness (Grade N) was selected in order to obtain a reasonably good self-dressing. This was taken the consideration that the combination materials of zirconia (relatively ductile) and silica (brittle) need to be ground simultaneously. Prior to grinding, the resin bond diamond wheels were trued and dressed using a cast iron lapping plate, as illustrated in Figure 10.36. The lapping plate has the same profile as the required shape of the trued wheels. Diamond pastes with particle size of 15 and 6 μm were selected for the 4 μm grit size wheel during rough and fine truing, respectively. For the 20 μm grit size wheel, diamond pastes with particle size of 50 and 30 μm were employed on the lapping plate. A small amount of oil was dropped into the truing zone. After truing, an alumina paste of grit size of 3 μm was used to dress the wheels until satisfactory wheel topography was achieved. Oil was

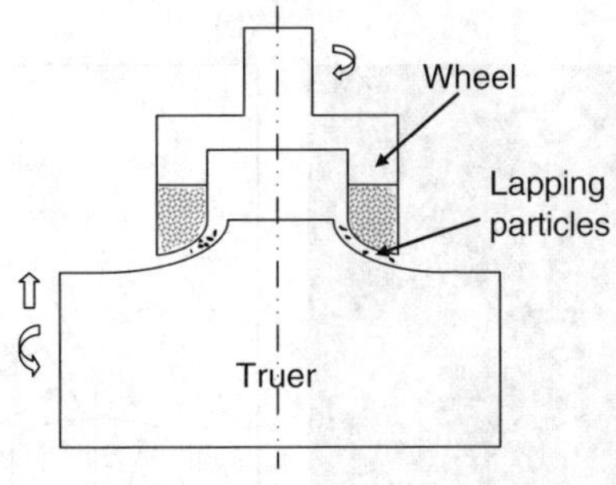

Figure 10.36 Illustration of truing and dressing set-up. Reproduced with permission from [25]. Copyright 2004 Elsevier

used as a lubricant for dressing to reduce abrasive pull-outs. It takes a few minutes for the whole dressing process.

Various wheel speed and feed rate were first used to examine their effects on ground surface quality and to find out the optimal grinding conditions. The total feed was fixed at 30 μm, which was relatively small because the ferrule end faces originally have a spherical shape. Under each grinding condition three connectors were ground. Spark-out was carried out at the end of each grinding cycle at the same grinding speed for ten seconds. Coolant at a relatively high flow rate (several litres per minute in this case) was supplied to the grinding area from three different directions. Detailed conditions are shown in Table 10.3.

10.4.2.2 Ground Fibre End Faces

As the grinding of connector end faces dealt with combination materials of both zirconia and silica, the response of the work materials to the grinding conditions was critical for the design of the grinding process. It should be noted that it was the fibre surface profile and finish which determined the final optical performance. Therefore, the optimization of the grinding process focused more on the achievement of better surface finish on the fibre surface.

Figure 10.37 shows the comparison of the connector end surfaces ground using two wheels with grit sizes of 4 μm and 20 μm, respectively. The fibre surfaces (circular areas at the centre) were much smoother than the surrounding zirconia surfaces. As expected, a smaller grit size produced a smaller surface roughness. The effect of grinding speed on the surface finishes on both fibre and ferrule was insignificant, as shown in Figure 10.38.

In the micro grinding process for fibre connector end faces, as the grinding wheel was trued into the spherical shape, the grinding wheel wear was directly reflected to the end face profile. No compensation grinding was employed as only single grinding cycle was allowed. In this case, attention was focused to reduce the grinding wheel wear and prolong the wheel life by using large lip radius. The profile error of the end surface reached an average PV value of 0.2 μm and surface roughness of 7 nm in Ra. The wheel wear had significant

Table 10.3 Truing, dressing and grinding conditions used. Reproduced with permission from [25]. Copyright 2004 Elsevier

Wheel	Diamond resin bond wheel, grade N, concentration 25% (corresponding to a volume fraction of 6.25%), grit sizes of 4 and 20 μm
Truer/dresser	Cast iron lapping plate
Truing conditions	
Wheel speed (rpm)	5,000
Lapping plate speed (rpm)	1,000
Feed rate (μm/min)	10
Slurry for truing	Diamond paste, 6–50 μm
Lubricant	Vegetable oil
Dressing conditions	
Wheel speed (rpm)	5,000
Lapping plate speed (rpm)	300
Feed rate (μm/min)	0
Slurry for dressing	Alumina paste, 3–10 μm
Lubricant	Vegetable oil
Grinding conditions	
Wheel speed (m/s)	26
Work speed (rpm)	0.023 times of the wheel speed
Feed rate (μm/min)	10
Total infeed (μm)	30
Coolant	Water soluble type, high flow rate

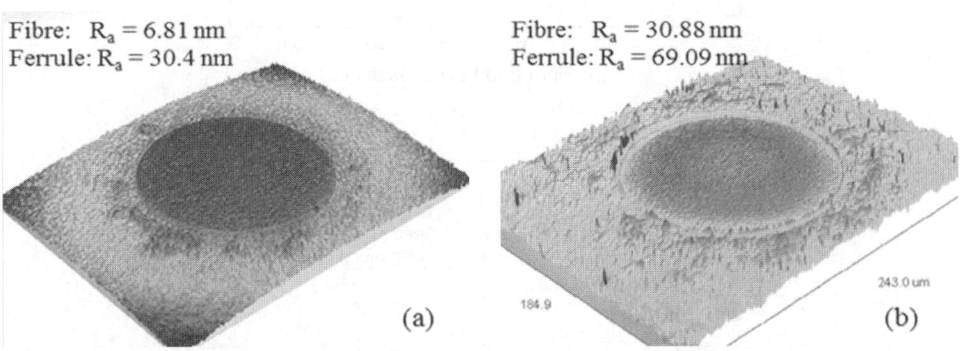

Figure 10.37 WYKO 3D images of specimens ground using wheels with grit sizes of (a) 4 μm and (b) 20 μm. Grinding speed and feed rate are 26 m/s and 10 μm/min., respectively. Reproduced with permission from [25]. Copyright 2004 Elsevier

influence on the profile accuracy when about 18 connectors were ground, as shown in Figure 10.39a. However, the wheel wear had little effect on the surface finish, as demonstrated in Figure 10.39b.

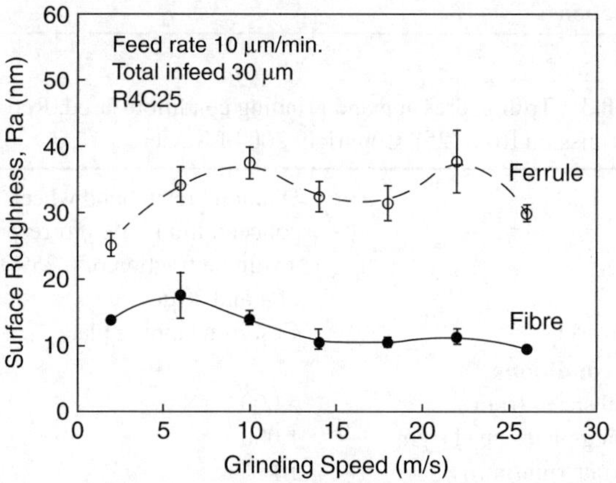

Figure 10.38 Effect of grinding speed on surface roughness. Reproduced with permission from [25]. Copyright 2004 Elsevier

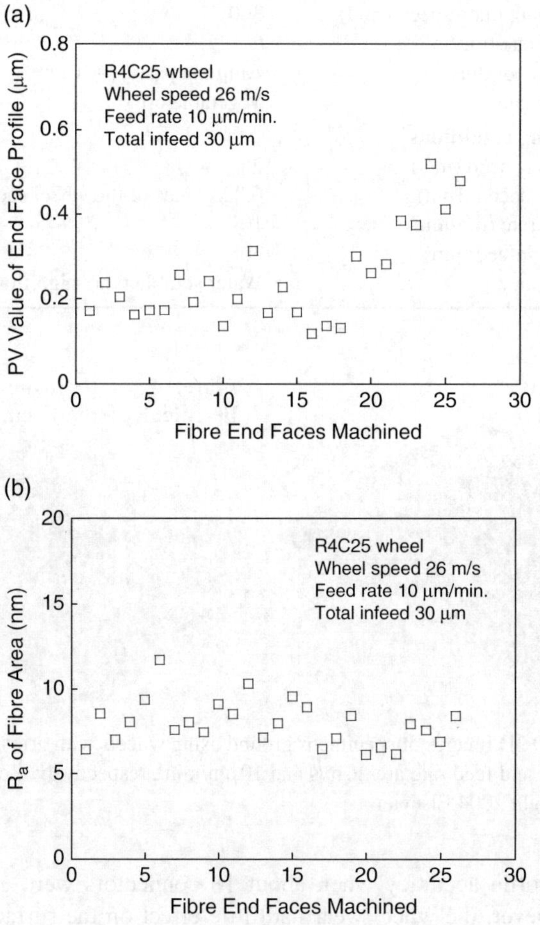

Figure 10.39 Effect of wheel wear on (a) profile error and (b) roughness (in fibre area) of the ground end faces. Reproduced with permission from [25]. Copyright 2004 Elsevier

Acknowledgements

The authors are grateful for valuable discussions with Professors T. Kuriyagawa and S.H. Yin, Drs. L. Yin, K. Ramesh and C.J. Xiong, and experimental assistances from W.K. Chen, Y.C. Liu, R. Irwan and Z.M. Gong. This work was financially supported by the Australian Research Council (ARC) under the Discovery Project Program, the Department of Industry, Innovation, Science and Research (DIISR) Australia under International Science Linkage Program.

References

[1] Adams, M.J., Allan, A.B., Briscoe, J., Doyle, P.J., Gorman, D.M., and Johnson, S. (2001) An experimental study of the nano-scratch behaviour on poly (methyl-methacrylate), *Wear* 251: 1579–1583.

[2] Aurich, J.C., Engmann, J., Schueler, G., and Haberland, R. (2009) Micro grinding tool for manufacture of complex structures in brittle materials. *Annals of the CIRP* 58(1): 311–314.

[3] Ball, M.J., Murphy, N., and Shore, P. (1991) Electrolytically assisted 'ductile' mode diamond grinding of BK7 and SF10 optical glasses. *Proceedings of the SPIE* 1573: 30–38.

[4] Bifano, T.G. (1988) *Ductile-regime grinding of brittle materials*, PhD Thesis, North Carolina State University, Raleigh, NC.

[5] Bifano, T.G., Dow, T., and Scattergood R.O. (1991) Ductile-regime grinding: a new technology for machining brittle materials,. *Trans. of ASME , J. of Eng Ind.* 113(5): 184–189.

[6] Blaedel, K.L., Taylor, J.S., and Evans C.J. (1999) Ductile-regime grinding of brittle materials in *Machining of Ceramics and Composites*, edited by S. Jahanmir, M. Ramulu and P. Koshy, Marcel Dekker, Inc., 139–176.

[7] Blake, P.N., and Scattergood, R.O. (1990) Ductile-regime machining of germanium and silicon. *Journal of the American Ceramic Society* 73(4): 949–957.

[8] Bridgman, P.W. (1947) The effect of hydrostatic pressure on the fracture of brittle substances. *Journal of Applied Science* 18: 246–258.

[9] Bridgman, P.W., and Simon, I. (1953) Effects of very high pressures on glass, *Journal of Applied Physics* 24: 405–413.

[10] Brinksmeier, E., Mutlugunes Y., Klocke, F., Aurich, J.C., Shore, P., and Ohmori, H. (2010) Ultra-precision grinding. *CIRP Annals – Manufacturing Technology* 59: 652–671.

[11] Brinksmeier, E., and Riemer, O. (2005) Metal cutting of microstructures. *Proceedings of the 1st International Conference on Multi-Material Micro Manufacture (4 M)*, Karlsruhe, Germany, 19–25.

[12] Chen, F.J., Yin, S.H., Huang, H. Ohmori, H., Wang, Y., Fan, Y.F., and Zhu, Y.J. (2010) Profile error compensation in ultra-precision grinding of aspheric surfaces with on-machine measurement, *International Journal of Machine Tools and Manufacture*, 50: 480–486.

[13] Chen, W.K., and Huang, H. (2003) Ultra precision grinding of spherical convex surfaces on combination brittle materials using resin and metal bond cup wheels, *J. Mater. Proc. Tech.* 140: 217–223.

[14] Chen, W.K., Kuriyagawa, T., Huang, H., Ono, H., Saeki, M., and Syoji, K. (2004) A novel form error compensation technique for tungsten carbide mould insert machining utilizing parallel grinding technology, *Key Eng. Mater.*, 257–258:141–146.

[15] Chen, W.K., Huang, H., and Yin, L. (2005) Loose abrasive truing and dressing of resin bond diamond wheels for grinding fibre optic connectors, *J. Mater. Proc. Tech.*, 159: 229–239.

[16] Chen, W.K., Kuriyagawa, T., Huang, H., and Yosihara, N. (2005) Machining of micro aspherical mould inserts, *Precision Engineering*, 29: 315–323.

[17] Cheng, K. (2002) Abrasive micromachining and microgrinding in *Micromachining of Engineering Materials*, edited by Joseph McGeough, Marcel Dekker, Inc., New York. Basel, 85–123.

[18] Corbett, J., McKeown, P.A., Peggs, G.N., and Whatmore, R. (2000) Nanotechnology: international developments and emerging products, *Annals of CIRP* 49(2):523–545.

[19] Engmann, J., Schuler, G.M., Haberland, R., Walk, M., and Aurich, J.C. (2009) Efficient technique for 3rd micro structuring of carbide and brittle materials with diamond micro-shaft grinding tools. *Proceedings of the 9th Euspen International Conference*, vol. 2, San Sebastian, Spain, 97–100.

[20] Dong, Z.G. Huang, H., and Kang, R.K. (2010) An investigation of the onset of elastoplastic deformation during nanoindentation in MgO single crystal (001) and (110) planes, *Materials Science and Engineering A*, 527: 4177–4184.

[21] Heinzel, C., and Rickens, K. (2009) Engineered wheels for grinding of optical glass. *Annals of the CIRP* 58(1):315–318.

[22] Huang, H. (2003) Machining characteristics and surface integrity of yttria partially stabilized zirconia in high speed deep grinding, *Mater. Sci. Eng. A* 345:155–163.

[23] Huang, H., Yin, L., and Zhou, L. (2003) High speed grinding of silicon nitride with resin bond diamond wheels, *J. Mater. Proc. Tech.* 141:329–336.

[24] Huang, H., and Liu, Y.C. (2003) Experimental investigations of machining characteristics and removal mechanisms of advanced ceramics in high speed deep grinding, *International Journal of Machine Tool and Manufacture* 43:811–823.

[25] Huang, H., Chen, W.K., Yin, L., Xiong, Z.J., Liu, Y.C., and Teo, P.L. (2004) Micro/Meso ultra precision grinding of fibre optic connectors and its influence on optic performance, *Preci. Eng.* 28: 95–105.

[26] Huang, H., and Kuriyagawa, T. (2007) Nanometric grinding of axisymmetric aspherical mould inserts for optic/photonic applications, *International Journal of Machining and Machinability of Materials*, 2:71–84.

[27] Huang, H., Irwan, R., and Kuriyagawa, T. (2007) A study on deformation and removal mechanisms of tungsten carbide, *Key Engineering Materials*, 329:385–390.

[28] Huang, H., Chen, W.K., and Kuriyagawa, T. (2007) Profile error compensation techniques in parallel nanogrinding of tungsten carbide aspherical mould inserts, *International Journal of Machine Tool and Manufacture*, 47:2237–2245.

[29] Hwang, Y., Kuriyagawa, T., and Lee, S.K. (2006) Wheel curve generation error aspheric microgrinding in parallel grinding method. *International Journal of Machine Tools and Manufacture* 46:1929–1933.

[30] Inasaki, I. (1987) Grinding of hard and brittle materials, *Annals of CIRP* 36(2):463–471.

[31] Irwan, R., and Huang, H. (2008) Mechanical properties and fracture characteristics of cemented tungsten carbide with fine microstructure studied by nanoindentation, *International Journal of Surface Science and Engineering* 2:29–40.

[32] Jahanmir, S., Xu, H.H.K., and Ives, L.K. (1999) Mechanisms of material removal in abrasive machining of ceramics in *Machining of Ceramics and Composites*, edited by S. Jahanmir, M. Ramulu and P. Koshy, Marcel Dekker, Inc., 11–84.

[33] Koehler, U., Guber, A.E., Bier, W., Heckele, M., and Schaller, T. (1996) Fabrication of microlenses by combining silicon technology, mechanical micromachining and plastic molding. Proceedings of SPIE 2687:18–22.

[34] Komanduri, R., Chandrasekaran, N., and Raff, L.M. (1998) Effect of tool geometry in nanometric cutting: a molecular dynamics simulation approach, *Wear*, 219:84–97.

[35] Kuriyagawa, T., Sepasy, M.S., and Syoji, K. (1996) A new grinding method for aspheric ceramic mirrors, *J. of Mater. Proc. Tech.* 62:387–392.

[36] Kuriyagawa, T., Lee, J., Saeki, M., and Syoji, K. (2001) Micro truing/dressing for small-sized aspherical mirror grinding in *Advances in Abrasive Technology IV*, Seoul, Korea, 111–114.

[37] Lafaye, S., and Troyon M., (2006) On the friction behaviour in nanoscratch testing. *Wear*, 261:905–913.

[38] Lee, W.B., Cheung, C.F., Chiu, W.M., and Leung, T.P. (2000) An investigation of residual form error compensation in the ultra-precision machining of aspheric surfaces, *Journal of Materials Processing Technology*, 99:129–134.

[39] Li, K. and Liao, W. (1996) Surface/subsurface damage and the fracture strength of ground ceramics, *J. Mater. Proc. Tech.* 57:207–220.

[40] Liu, K., and Li, X.P. (2001) Ductile cutting of tungsten carbide, *J. Mater. Proc. Tech.*, 113:348–354.

[41] Malkin, S. (1989) *Grinding Technology: Theory and Application of Machining with Abrasives*, Ellis Horwood Limited, Chichester.

[42] Marinescu, I., Rowe, B., Yin, L., and Wobker, H.G. (2000) Abrasive processes in *Handbook of Ceramics Grinding and Polishing*, edited by I. Marinescu, H.K. Tönshoff and I. Inasaki, Noyes Publi., 94–189.

[43] Masuzawa, T. and Tönshoff, H.K. (1997) 'Three-dimensional micromachining by machine tools, *Annals of CIRP* 46(2):621–628.

[44] Ohmori, H., Ebizuka, N., Morita, S., and Yamagata, Y. (2001) Ultraprecision micro-grinding of germanium immersion grating element for mid-infrared super dispersion spectrograph, *Annals of CIRP* 50(1):221–224.

[45] Ohmori, H., Katahira, K., Uehara, Y., and Lin, W. (2002) ELID-grinding of microtool and applications to fabrication of microcomponents. *International Journal of Nanotechnology* 41(2):193–204.

[46] Ramesh, K., Huang, H., Yin, L., and Zhao, J. (2004) Microgrinding of deep micro grooves with high table reversal speed, *Int. J. Mach. Tool Manuf.* 44:39–49.

[47] Rapaport, D.C. (1995) *The Art of Molecular Dynamics Simulation*, Cambridge University Press.

[48] Saeki, M., Kuriyagawa, T., and Syoji, K. (2002) Machining of aspherical molding dies utilising parallel grinding method, *J. Jpn Soc. Preci. Eng.* 68:1067–1071.

[49] Stachowiak, G.W., and Batchelor, A.W. (1993) *Engineering Tribology*, Elsevier.
[50] Sun, H.Q., Irwan, R. Huang, H., and Stachowiak, G.W. (2010) Surface characteristics and removal mechanism of cemented tungsten carbides in nanoscratching, *Wear* 268:1400–1408.
[51] Suzuki, H., Kodera, S., and Hara, S. (1995) Grinding of micro aspherical surface. *Proceedings of the 10th Annual Meeting of the ASPE*, vol.12, Austin, USA, 183–186.
[52] Suzuki, H., Kuriyagawa, T., Syoji, K., Tanaka, K., Yan, J., and Wajima, N. (1998) Study of precision grinding of micro aspherical surface (3rd report) – feasibility study of micro aspherical surface by inclined rotational grinding, *J. of Japan Soc. Preci. Eng.* 64(9):1350–1354.
[53] Suzuki, K., Uematsu, T., and Nakagawa, T. (1987) On-machine trueing/dressing of metal bond grinding wheels by electro-discharge machining. *Annals of the CIRP* 36(1):115–118.
[54] Swain, M.V. (1979) Microfracture about scratches in brittle solids. *Proceedings of the Royal Society* A366, 575–597.
[55] Tönshoff, H.K., Denkena, B., Friemuth, T., and Reichstein, M. (2003) Precision grinding of components for aerostatic micro guidance, *Preci. Eng.* 27:185–188.
[56] Upadhyaya, G.S. (2005) *Cemented Tungsten Carbides: Production, Properties and Testing*, Culinary and Hospitality Industry Publications Services, Texas.
[57] Venkatesh, V.C., Izman, S., Vichare, P.S., Mon, T.T., and Murugan, S. (2005) The novel bondless wheel, spherical glass chips and a new method of aspheric generation. *Journal of Materials Processing Technology* 167:184–190.
[58] Waida, T. (2002) Micro grinding. *AIST (Japan) Special Report* (AIST01-C00018).
[59] Weck, M., McKeown, P.A., Bonse, R., and Herbst, U. (1995) Reduction and compensation of thermal errors in machine tools. *Annals of the CIRP* 44(2):589–598.
[60] Xu, H.H.K., and Jahanmir, S. (1995) Scratching and grinding of a machinable glass-ceramic with weak interfaces and rising **T**-curve, *J. Ameri. Ceram. Soc.* 78:497–500.
[61] Yamamoto, Y., Suzuki, H., Okino, T., Hijikata, Y., Moriwaki, T., Fukuta, M., Nishioka, M., and Kojima, Y. (2004) Ultra precision grinding of micro aspherical surface – Development of a three-axes controlled single point inclined grinding method. *Proceedings of the 13th Annual Meeting of the ASPE*, vol.34, St. Louis, USA, 558–561.
[62] Yin, L., Pickering, J.P., Ramesh, K., Huang, H., Vancoille, E., and Sponwage, A. (2004) Ultraprecision grinding of tungsten carbide for spherical mirrors, *Proc. Inst. Mech. Eng., Part B: J. Eng. Manuf.*, 218: 419–429.
[63] Yin, L., Huang, H., Chen, W.K., Xiong, Z.J., Liu, Y.C., and Teo, P.L. (2004) Polishing of fibre optic connectors, *International Journal of Machine Tool and Manufacture*, 44:659–668.
[64] Yin, L., Ramesh, K., Wan, S., Liu, X.D., Huang, H., and Liu, Y.C. (2004) Abrasive flow polishing of micro bores, *Materials and Manufacturing Processes*, 19(2):187–207.
[65] Yin, L., Sponwage, A., Ramesh, K., Huang, H., Pickering, J.P., and Vancoille, E. (2004) Influence of microstructure on ultraprecision grinding of cemented carbides, *International Journal of Machine Tool and Manufacture*, 44:533–543.
[66] Yin, L., Vancoille, E., Lee, L.C., Huang, H., Ramesh, K., and Liu, X.D. (2004) High-quality grinding of polycrystalline silicon carbide spherical surfaces, *Wear* 256:197–207.
[67] Yin, L., Pickering, J.P., Ramesh, K., Huang, H., Sponwage, A., and Vancoille, E. (2005) Planar nanogrinding of a fine grained WC-composite for optical surface finish, *International Journal of Advanced Manufacturing Technology*, 26:766–773.
[68] Yin, L. and Huang, H. (2008) Brittle materials in nano-abrasive fabrication of optical mirror-surfaces, *Precision Engineering*, 32:336–341.
[69] Zhang, B., and Howes, T.D. (1994) *Material-removal mechanisms in grinding ceramics Annals of CIRP* 43(1):305–308.
[70] Zhang, B., and Liu, X. (2003) Grinding of nanostructural ceramic coatings: damage evaluation, *Int. J. Mach. Tool Manuf.* 43:161–167.
[71] Zarudi, I., and Zhang, L.C. (1998) Effect of ultraprecision grinding on the microstructural change in silicon monocrystals, *J. Mater. Proc. Tech.* 84:148–158.

11

In-Process Micro/Nano Measurement for Micro Cutting

Wei Gao, Kang-Won Lee, Young-Jin Noh, Yoshikazu Arai and Yuki Shimizu
Tohoku University, Japan

11.1 Introduction

Microoptics with micro structures such as micro-grooves, micro-prisms and micro-lenses are widely used for enhancement of functional performances of optical systems [1–3]. In general, the surface form and surface quality are two main quantities of an optical component [4–5]. With remarkable progress in micro-optics, the dimensions and the pitches of the micro structures are getting smaller, while the shapes are getting more complicated and the required accuracies are getting higher. Some applications, such as FPD (Flat Panel Display), also require micro structures to be fabricated over larger areas.

Micro cutting based on single point diamond turning [6–14] is a promising machining method for fabricating such kinds of micro-optics. Diamond turning is a conventional ultra-precision machining technology for generation of smooth surfaces with relatively long spatial wavelengths such as polygon mirrors and parabolic reflectors. The accurate shape and sharp edge of the single point diamond cutting tool together with the accurate tool path generated by the slides and the spindle of the diamond turning machine make the machined surface have good form accuracy and a high surface quality [15–19]. The fast tool servo or fast tool control system has also been developed to improve the bandwidth of diamond turning from several tens of Hz, which is limited by the heavy mass of the machine slide, to several kHz so that micro structures can be fabricated [20–21]. A fast tool servo system generally consists of a PZT actuator for driving the cutting tool, a capacitive displacement sensor for measuring the actual depth of cut and a controller for feedback control of tool motion [22–28]. The authors have demonstratedthat 3D sinusoidal micro structures with pitches of 10 μm can be fabricated by fast tool servo in combination with a micro cutting tool [29, 30].

On the other hand, it is important to monitor the long-time micro-cutting process when 3D micro-structures are fabricated over a large area because the process is influenced by a lot of factors such as tool damage and tool wear. It has long been recognised that the cutting force is

Micro-Cutting: Fundamentals and Applications, First Edition. Edited by Kai Cheng and Dehong Huo.

a good parameter to reflect the cutting process [31, 32]. In-process force monitoring systems have also been developed for the ordinary cutting process by using various force sensors such as strain gauges [33] and dynamometers [34]. For the cutting process of micro structures by fast tool servo, it is necessary to detect the small amplitude of the dynamic micro cutting force. For this purpose, it is important to choose a sensitive force sensor with a high bandwidth and a high stiffness. It is also desired to set the position of the force sensor to be close to the cutting point.

This chapter presents a hybrid instrument, which consists of a fast tool servo and a piezoelectric force sensor [35–37]. In Section 11.2, a design strategy is presented for integration of the force sensor into the fast tool servo so that the instrument can have a high sensitivity for in-process force detection as well as a high stiffness and bandwidth for cutting of micro structures. In-processing force measurement by using the hybrid instrument is described in Section 11.3. The in-process force data are employed to estimate the tool wear in Section 11.4. A new technique is presented in Section 11.5 for on-machine surface profile measurement by using the hybrid instrument as a force controlled stylus probing instrument followed by the summary of the chapter in Section 11.6.

11.2 The Hybrid Instrument for Micro Cutting and In-process Measurement

Figure 11.1 shows a schematic of the hybrid instrument for micro cutting and in-process measurement on a diamond turning machine. The instrument is composed of a fast tool servo system (FTS) and a force sensor, which is integrated for detection of the contact force or cutting force generated at the interface between the workpiece surface and the cutting tool. The fast tool servo consists of a single crystal diamond cutting tool, a ring type PZT actuator, a wave washer spring and a capacitive displacement sensor. The cutting tool is mounted at the front end of tool holder and moved by the PZT actuator. The wave washer spring, which is supported by the base, plays the important role of the motion guide for translation of the motion generated by the PZT actuator to the cutting tool. The spring mechanism also improves the dynamic performance by applying a pre-load against the PZT actuator. The displacement sensor, which is mounted at the end of shaft, measures the actual displacement of the centre of the head part for feedback control of the cutting tool motion. When the gap between the sensor and the spring is adjusted, the axis of the displacement sensor is aligned perpendicular to the plane of the spring by using the bolts so that the measurement errors caused by the out-of-perpendicularity can be reduced.

Figure 11.2 shows a schematic of the micro-cutting process for fabricating micro structures on a cylinder workpiece by the hybrid instrument on a diamond turning machine. The hybrid instrument is mounted at the Z-slide of the machine tool. The workpiece is mounted by the vacuum chuck on the spindle of the machine tool. The profile of the micro structures is generally determined with the geometry of the cutting tool and the tool path (tool motion) generated by the PZT actuator and the machine spindle. In the micro-cutting process shown in Figure 11.2, the profile of the micro structures along the axial direction is generated by the cutting tool shape and that along the circumferential direction is generated by the motion of cutting tool. The motion data for the cutting tool can be expressed by the depth of cut applied by the PZT actuator of the fast tool servo and the position on the circumferential direction

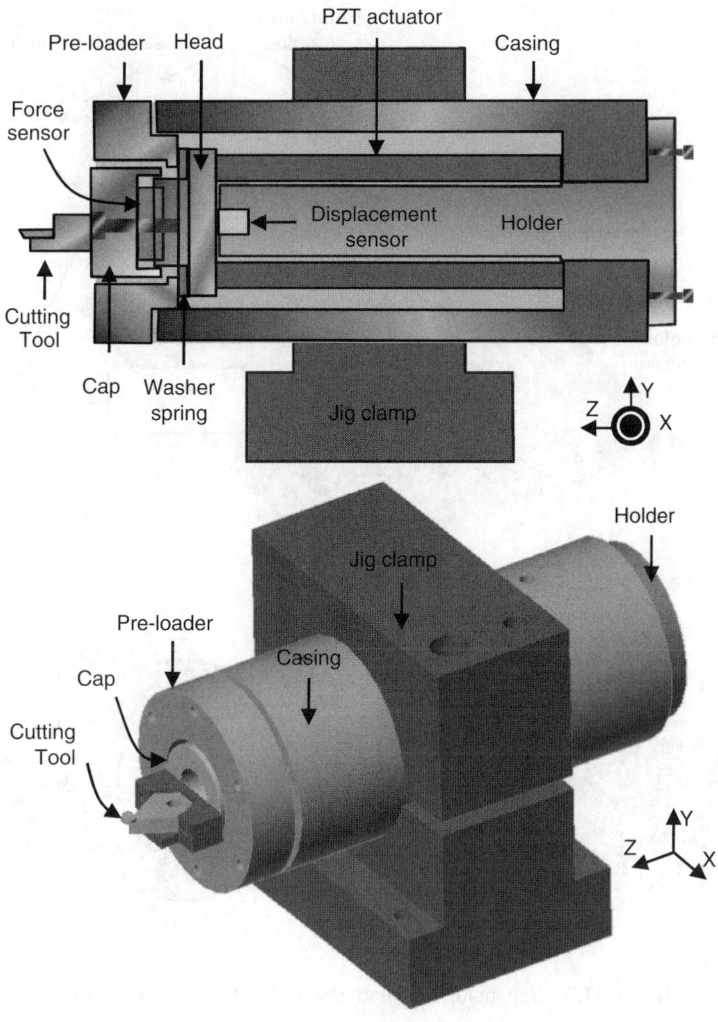

Figure 11.1 Schematic of the hybrid instrument

provided by the machine spindle. Assume that the length of the desired micro structures is equal to the pitch along the circumferential direction, the cutting position on the circumferential direction can be expressed by:

$$\theta(j) = \frac{2\pi}{S} j \qquad j = 0, 1, 2, \ldots N \tag{11.1}$$

where, S is the pulse number of the rotary encoder of the spindle per each revolution. j is the jth rotary encoder pulse. Assume that the profile of the desired micro structure is a parabola

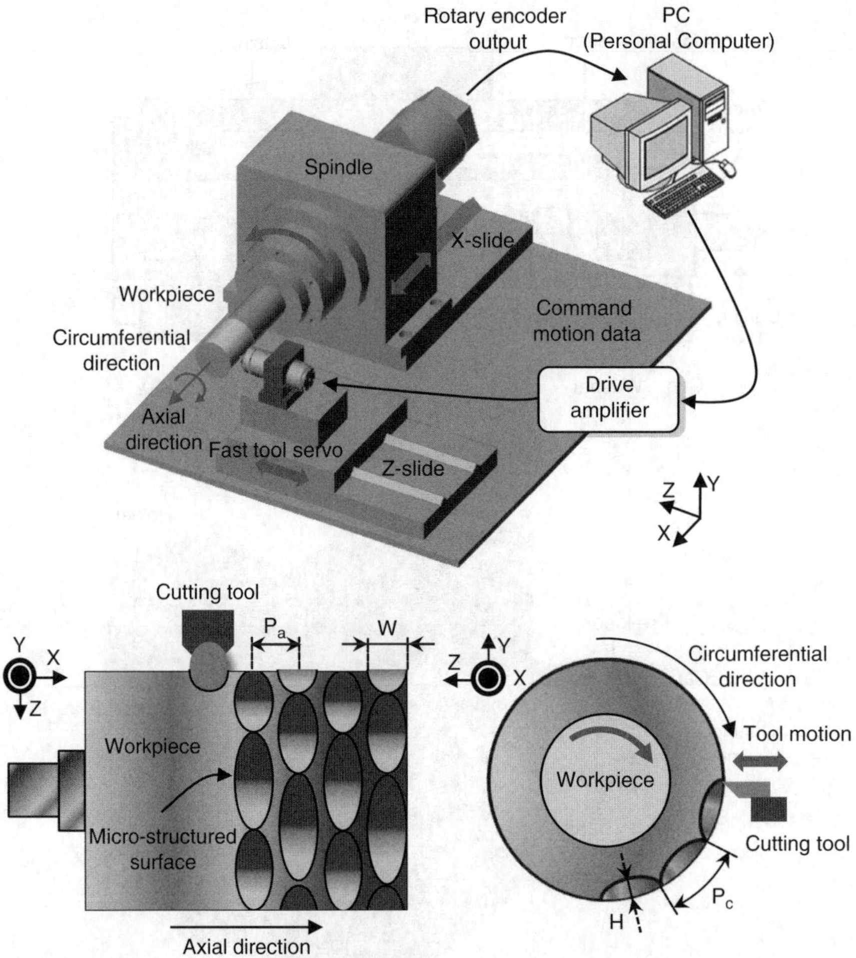

Figure 11.2 Schematic for micro cutting by the hybrid instrument

along the circumference direction, the cutting motion data for the cutting tool at position j, which corresponds to the depth of cut generated by the PZT actuator, can be calculated as follows:

$$\begin{aligned} f(j) &= A\left\{(r\theta(j) - nP_c) - \frac{P_c}{2}\right\}^2 \\ &= A\left\{\left(\frac{2\pi r}{S}j - nP_c\right) - \frac{P_c}{2}\right\}^2 \end{aligned} \tag{11.2}$$

where, A is the constant of the parabolic profile, P_c is the pitch of the micro structure, n is the number of the micro structure along the circumference direction, and r is the radius of the workpiece. The motion data are automatically generated based on the profile of micro structures. The generated motion data is stored in the personal computer (PC) and transferred to the PZT actuator of the hybrid instrument. During the fabrication, the motion of the PZT actuator and that of the spindle are synchronised by the trigger signal, which is transferred from the rotary encoder.

Figure 11.3 shows a schematic of in-process measurement for monitoring the fabrication process and evaluating the machined surface form by the hybrid instrument on a diamond turning machine. The in-process measurement is basically carried out to utilise the contact force between the workpiece and the cutting tool detected by the force sensor, which is integrated with the instrument. At first, the hybrid instrument can be used for detecting the initial machining point between the cutting tool and the pre-cut surface before the fabrication. The process of detecting the initial machining point is important to prevent the diamond cutting tool from being broken due to unexpected impact. It is also desired to detect the cutting force during the machining process because the factors influencing the surface quality and the form accuracy of the machined surface such as machine motion errors, cutting tool form errors, fast tool servo errors, thermal drifts, vibrations, and so on, can be monitored through the cutting force. The in-process cutting force can also be used for monitoring the cutting tool wear.

After the fabrication, the hybrid instrument can be applied for measuring the surface form of the machined micro structure on the machine tool, which will be addressed in detail in Section 11.5. As can be seen in Figure 11.3, the workpiece is rotated by the spindle or translated by the X-slide so that the micro structures can be scanned by the cutting tool along the axial direction (X-direction) and the circumferential, respectively. In this process, the cutting tool is employed as a stylus probe to detect the machined surface. During the scanning, the contact force between the cutting tool and the measured surface is kept as a small value through servo control of the Z-directional tool position by the PZT actuator based on the force sensor output of the hybrid instrument. The tool motion, which can be measured by the rotary encoder of the spindle, the linear encoder of the X-slide and the displacement sensor of the hybrid instrument, represents the surface form of machined surface.

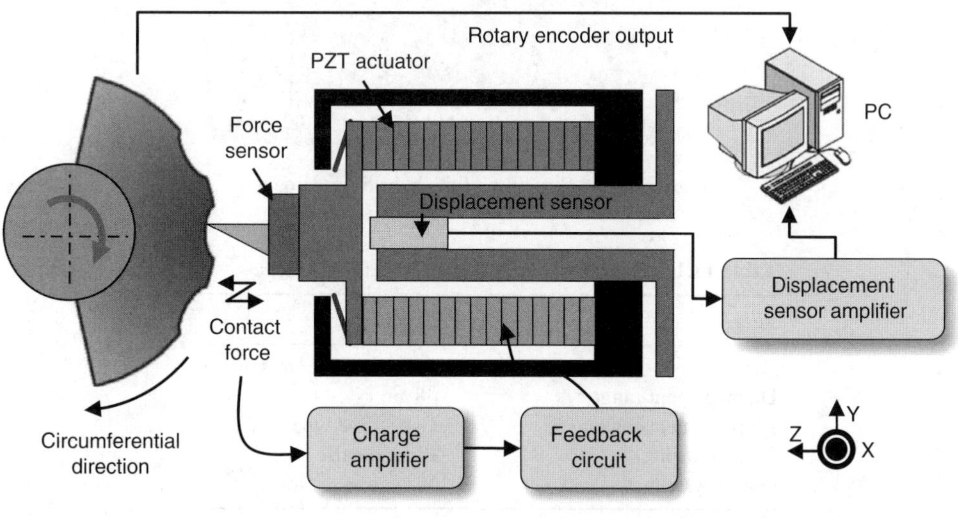

Figure 11.3 Schematic for in-process measurement by the hybrid instrument

Figure 11.4 shows the designed and constructed hybrid instrument. The size of the overall system of the hybrid instrument was 110 mm x 70 mm x 70 mm. Table 11.1 shows the specification of the instrument. A ring type PZT actuator was mounted in the casing under a pre-load applied by a wave washer spring. The PZT actuator, which could generate a nominal displacement of 48 μm, had an outer diameter of 20 mm, an inner diameter of 14 mm, a length of 60 mm and a stiffness of 117.5 N/μm. The capacitance of the PZT actuator was 15 μF and its elastic constant was 44 GPa.

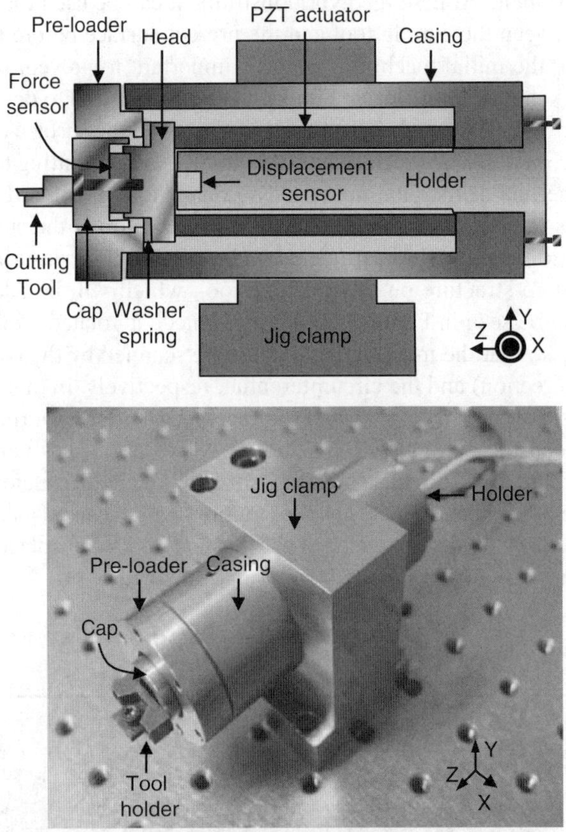

Figure 11.4 Structure of the hybrid instrument

Table 11.1 The specification of the hybrid instrument

Item	Specification
Dimension	110 mm x 70 mm x 70 mm
Displacement range	48 μm
Displacement sensor	Capacitive type
Force sensor	Piezoelectric type
Guide system	Wave washer spring

The PZT actuator was controlled by a personal computer via a PZT amplifier with an amplification of 15. Table 11.2 shows the specification of the PZT actuator [38]. The PZT actuator was pre-loaded with a force of 280 N by the wave washer spring in order to improve the dynamic performance of the instrument and to prevent the influence by external disturbances such as the shear force. The PZT actuator amplifier had an output range from 0 to 150 V, a maximum peak current of 0.9 A and a maximum average current of 2.5 A. The capacitive displacement sensor had a measurement range of 125 μm, a bandwidth of 20 kHz and a resolution of approximately 12.5 nm. Table 11.3 shows the specification of the capacitive displacement sensor [39].

A piezoelectric force sensor, which had the characteristics of high stiffness and high sensitivity, was employed as the force sensor. The force sensor, which was mounted at the back of the tool holder, was adjusted by a bolt. The force sensor had an outer diameter of 12 mm, an inner diameter of 5 mm and a stiffness of 953.3 N/μm. Table 11.4 shows the specification of the force sensor [40].

Table 11.2 The specification of the PZT actuator

Item	Specification
Dimension	ϕ 20 mm × L 60 mm (inner diameter: 14 mm)
Displacement range	48 μm
Stiffness	117.5 N/μm
Capacitance	15.0 μF
Elastic constant	44 GPa

Table 11.3 The specification of the capacitive displacement sensor

Item	Specification
Measurement range	125 μm
Diameter of the probe	3.94 mm
Resolution	12.5 nm
Bandwidth	20 kHz
Linearity	±0.1%

Table 11.4 The specification of the force sensor

Item	Specification
Type	Piezoelectric type
Dimension	ϕ 12 mm x 5 mm (Inner diameter: 5 mm)
Capacitance	1.24 nF

The stiffness of the instrument, which is the capacity of resistance to deformation against the external force, directly affects the tool motion accuracy. In particular, the axial stiffness of the instrument has to have enough stiffness for cutting micro structures. The stiffness of the instrument is basically determined by those of the elements in the instrument. Figure 11.5 shows the theoretical model for calculating the stiffness of the hybrid instrument at the axial direction (Z-direction). Assuming that each part of the instrument is an elastic body, the stiffness of the instrument along the Z-direction can be expressed as:

$$k_z = \frac{1}{\frac{1}{\mathrm{k_c}} + \frac{1}{k_f} + \frac{1}{k_h} + \frac{1}{k_p} + \frac{1}{k_f}} \tag{11.3}$$

Substituting the calculated stiffness data of the elements in the instrument in Table 11.5, the stiffness of the instrument along the Z-direction is evaluated to be approximately 30.13 N/μm.

Figure 11.6 shows the experimental setup for evaluating the stiffness of the hybrid instrument. A load was applied to the hybrid instrument through pulling the tool holder by a reference weight. Figure 11.7 shows the stiffness of the instrument along the Z-direction by experiment and theoretical calculation. The stiffness by the experiment was evaluated to be 34.67 N/um over a range of 5 N, which is close to the previously calculated value by the theoretical stiffness model.

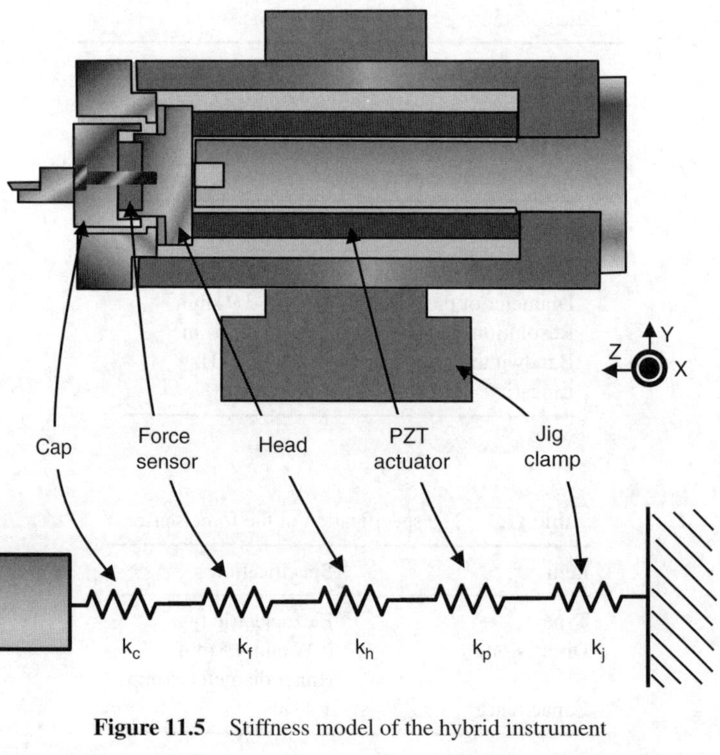

Figure 11.5 Stiffness model of the hybrid instrument

Table 11.5 The calculated stiffness of each element of the hybrid instrument

Item	Specification
Cap	3506.5 N/μm
Piezoelectric force sensor	953.3 N/μm
Head	7319.4 N/μm
PZT actuator	117.5 N/μm
Jig clamp	30.58 N/μm

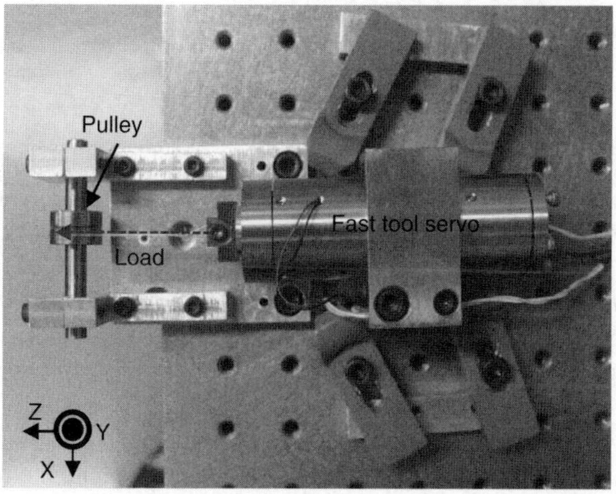

Figure 11.6 Experimental setup for evaluating stiffness

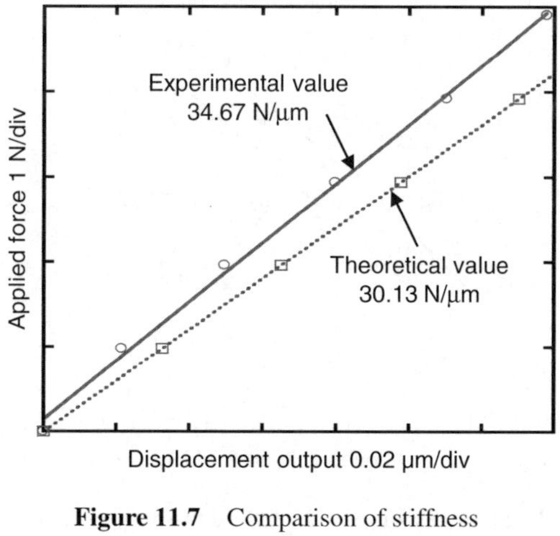

Figure 11.7 Comparison of stiffness

A high bandwidth of the hybrid instrument is required for fabricating complex micro structures. The dynamic response of the hybrid instrument is dominated by both the mechanical factors and the electronic factors. The mechanical factors include the stiffness and the moving mass of the instrument, which determines the resonance frequency of the instrument in the Z-direction as follows:

$$f_z = \frac{1}{2\pi}\sqrt{\frac{k_z}{m}} \tag{11.4}$$

where, m is the effective mass of the moving elements, which is approximately 0.045 kg. The resonance frequency is calculated to be 7509.4 Hz. The electronic factors depend on the capacitance of the PZT actuator, the applied signal and the capacity of the amplifier. The maximum frequency determined by the electronic factors can be expressed as [41]:

$$f_{\max} = \frac{I_{\max}}{\pi C V_{p-p}} \tag{11.5}$$

where, I_{max} is the maximum current of the amplifier, C is the capacitance of the PZT actuator and $V_{p\text{-}p}$ is the peak to peak voltage of a sine wave signal. The PZT actuator employed in the hybrid instrument has a stroke of 50 μm with respect to a peak to peak voltage of 150 V. The capacitance of the PZT actuator, which is proportional to the stroke of the PZT actuator, is 15 μF. Take into consideration the maximum current of the amplifier is 0.9 A, the allowable maximum frequency is calculated to be 127 Hz when the PZT actuator is oscillated with the amplitude of full stroke. It can be seen that the frequency response of the hybrid instrument is much more dependent on the electronic factors.

Figure 11.8 shows a static response of the hybrid instrument when the system was in the open-loop mode. After fitting the output curves of the displacement, the residual error is also

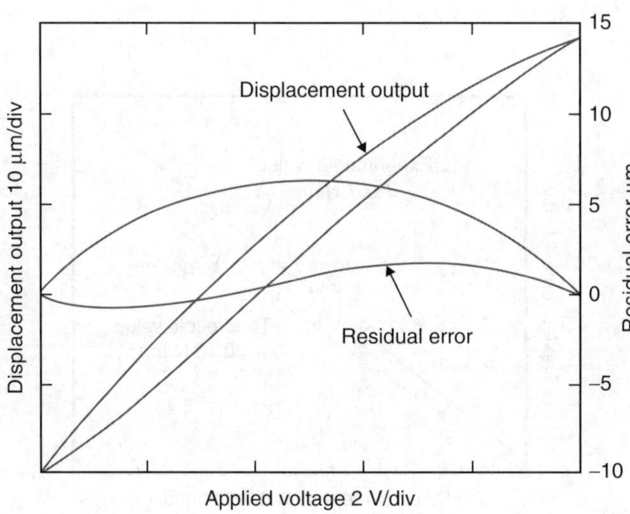

Figure 11.8 Static response of the hybrid instrument in open-loop mode

indicated in Figure 11.8. The hybrid instrument has a residual error of approximately 7 μm over a range of 48.5 μm. Figure 11.9 shows the static response when the system was under the closed-loop control. A PID controller was employed in the experiment. The PID gains were adjusted by trial and error. The residual error was reduced to be approximately 80 nm compared with that in Figure 11.8.

The dynamic response of the hybrid instrument was measured by using a FFT analyzer with a bandwidth of 40 kHz. The PZT actuator was oscillated by a swept sine wave with an amplitude of 500 nm. Figure 11.10 shows the dynamic response of the hybrid instrument along the axial direction. As can be seen in Figure 11.10, the hybrid instrument had a resonance frequency of approximately 7837 Hz, which is quite close to the previously calculated value. The measured cut-off frequency (-3 dB) is evaluated to be approximately 975 Hz.

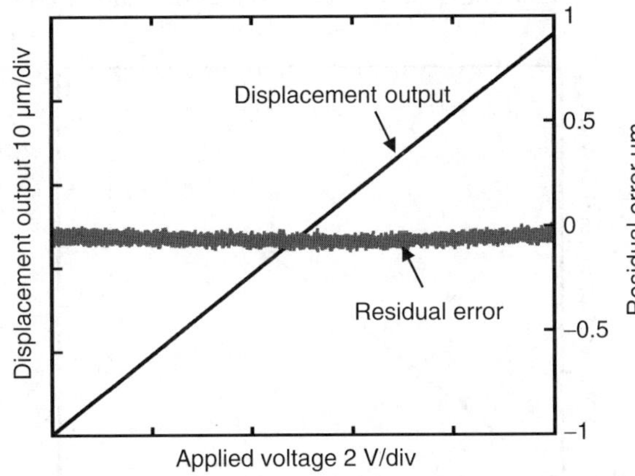

Figure 11.9 Static response of the hybrid instrument in closed-loop mode

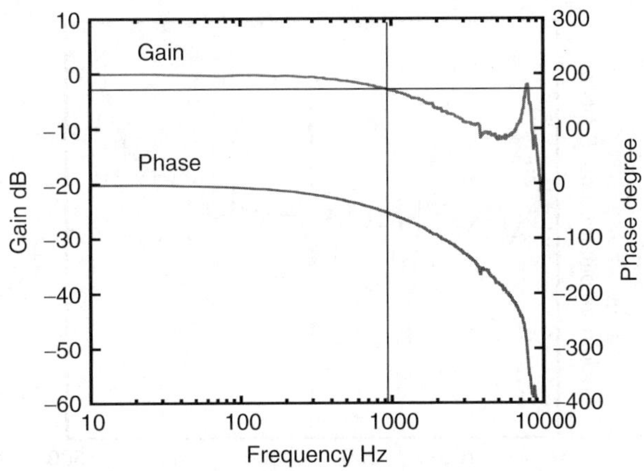

Figure 11.10 Dynamic response of the hybrid instrument

11.3 In-process Measurement of Micro Cutting Force

Figure 11.11 shows the force sensor output when a force was applied along the axial direction over a range of 0.5 N. It can be seen that the sensor output increased linearly with respect to the increased force. The sensitivity of force sensor was evaluated to be approximately 282 mV/N. It can be seen that the piezoelectric force sensor has a good linearity.

A modulation method was employed to improve the sensitivity of the force sensor. The cutting tool was oscillated by the PZT actuator with a certain frequency and a small amplitude of less than several nanometers so that the output of the force sensor could be modulated to an AC signal with the same oscillation frequency. A lock-in amplifier [42–44] was employed to detect the amplitude of the modulated force sensor output with a high sensitivity without the influence of electronics noises. When using the modulation technique, it is necessary to select a suitable oscillation frequency for the PZT actuator. Figure 11.12 shows the result of the stability of the lock-in amplifier output with respect to the oscillation frequency. The oscillation

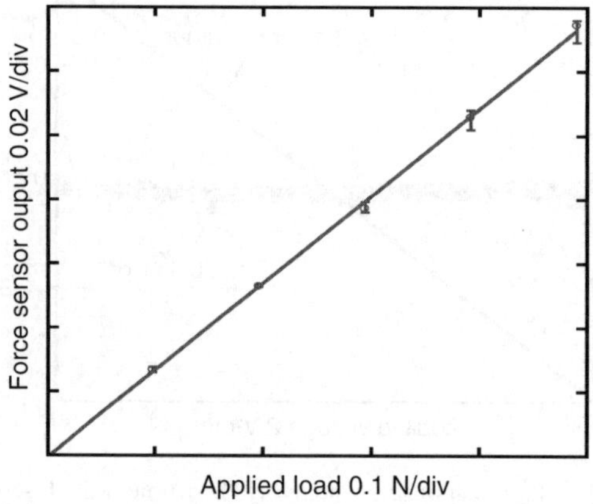

Figure 11.11 Sensitivity of the force sensor of the hybrid instrument

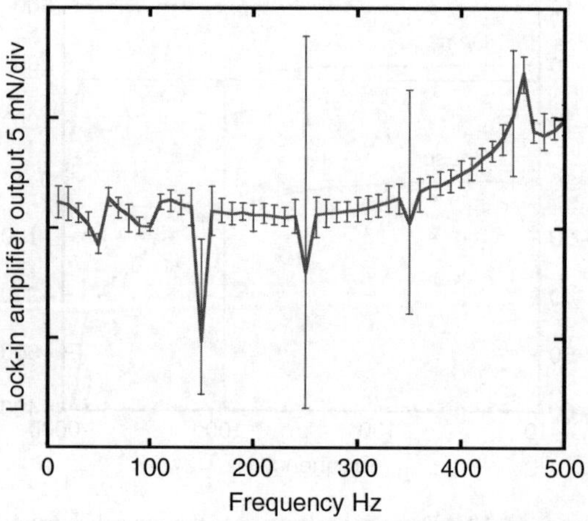

Figure 11.12 Stability of the force sensor output of the hybrid instrument

amplitude was 3 nm. As can be seen in Figure 11.12, the output was most stable when the oscillation frequency was in a range from 160 Hz to 230 Hz. The modulation method based on the oscillated PZT actuator and the lock-in amplifier was applied to the initial contact experiment and the measurement of machined surface form.

An experiment was conducted on a diamond turning machine [45] to evaluate the capability of detecting the contact force between the workpiece and the cutting tool by the hybrid instrument. An oxygen free copper cylinder workpiece and an aluminium cylinder workpiece with Ni-P plating were employed. Each of the workpieces was pre-cut by a cutting tool with a nose radius of 2 mm. After the pre-cutting, the cutting tool was replaced by a cutting tool with a nose radius of 0.2 mm for the contact experiment. The diamond cutting tool, which was oscillated with an amplitude of 3 nm and a frequency of 380 Hz, was moved toward the workpiece with steps of 10 nm by the Z-slide of the diamond turning machine. The threshold value of the force sensor for detecting the tool-workpiece contact was determined to be 0.5 mN from the

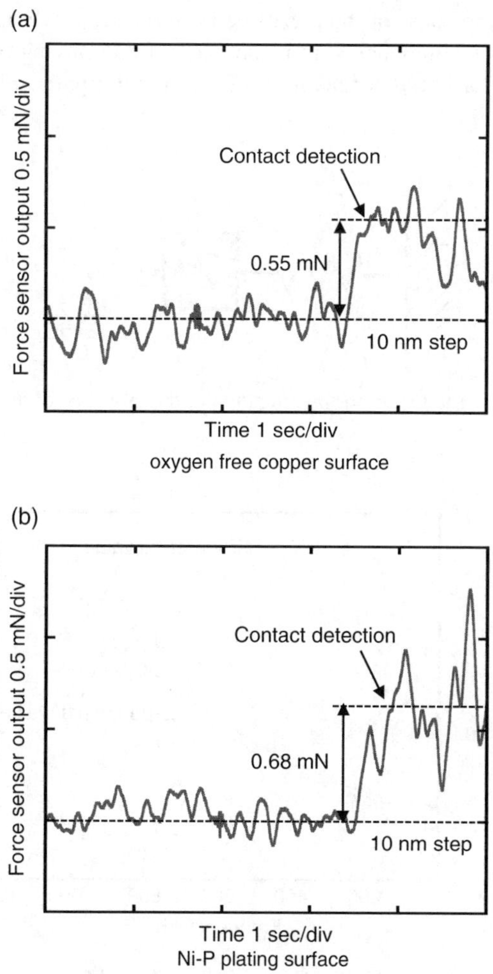

Figure 11.13 Detection of contact force

variation of the amplitude of the force sensor output from the lock-in amplifier. The force sensor outputs with respect to a 10 nm contact depth are shown in Figure 11.13. The contact forces corresponding to the 10 nm contact depth were 0.55 mN and 0.68 mN for the oxygen free copper surface and the Ni-P plating surface, respectively. The stability of the piezoelectric force sensor output was evaluated to be approximately 0.2 mN. It can be seen that the hybrid instrument has a capability of detecting the contact between the cutting tool and the workpiece with a resolution of less than 10 nm for the given conditions.

On the other hand, the piezoelectric force sensor is influenced by the inertial force generated by the moving part, which include the tool holder and the cap. Measurement of the inertial force is important for accurate cutting force measurement. A theoretical model for calculation of the inertial force is shown in Figure 11.14. From this model, F_t can be expressed as:

$$F_t = F_c + ma \tag{11.6}$$

where, k represents the stiffness of the piezoelectric force sensor, F_c is the generated cutting force, F_t is the total force components, m is the mass of the moving parts including the tool holder and the cap and a is the acceleration of the moving parts. Figure 11.15 shows the

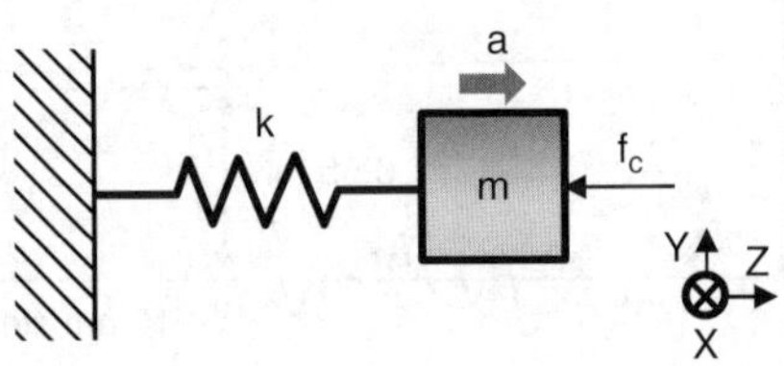

Figure 11.14 The model for investigation of the influence of the inertia force

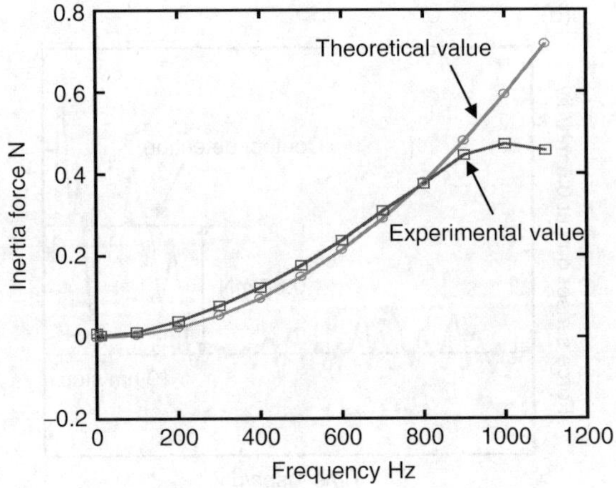

Figure 11.15 Comparison of inertial forces

calculated theoretical value and the measured value of the inertial force on the condition without any external forces. The amplitude of oscillation was 1 μm and the mass of the moving part was 0.015 kg. To reduce the influence of electronic noises, a low pass filter with a cut-off frequency of 1 kHz was applied to the charge amplifier. It can be seen that the measured inertial force is close to the calculated theoretical value.

Micro-lens arrays were then fabricated on the oxygen free copper cylinder workpiece and the aluminium cylinder workpiece with Ni-P plating by the hybrid instrument, respectively. Figure 11.16 shows the photograph of the experimental setup. The experiment was carried out on the diamond turning machine. A diamond cutting tool with a nose radius of 0.2 mm was employed for the fabrication. The designed length of the micro-lens array along the circumferential direction and the depth along the Z-direction were 200 μm and 5.2 μm, respectively. The pitch of the micro-lens array along the axial direction was set to be 100 μm.

A 3D profile measuring microscope [46] was employed for evaluating the machined surface form of the micro-lens array. Figure 11.17 shows the images and the measured profiles of the machined micro-lens arrays on the oxygen free copper surface and the Ni-P plating surface. The depths were evaluated to be 5.29 μm and 5.26 μm in Figure 11.17a and 11.17b, respectively. The deviations of the actual depths from the designed depth were 0.09 μm and 0.06 μm, respectively.

Experiments of in-process cutting force measurement were also carried out when micro-lens arrays were fabricated on the surfaces. The designed length of the micro-lens array along the circumferential direction and the depth along the Z-direction were 200 μm and 5.2 μm, respectively, which were the same as those shown in Figure 11.17. The pitch of the micro-lens array along the axial direction was changed from 100 μm to 50 μm. Figure 11.18 shows the fabricated micro-lens images obtained by the optical microscope. As can be seen in

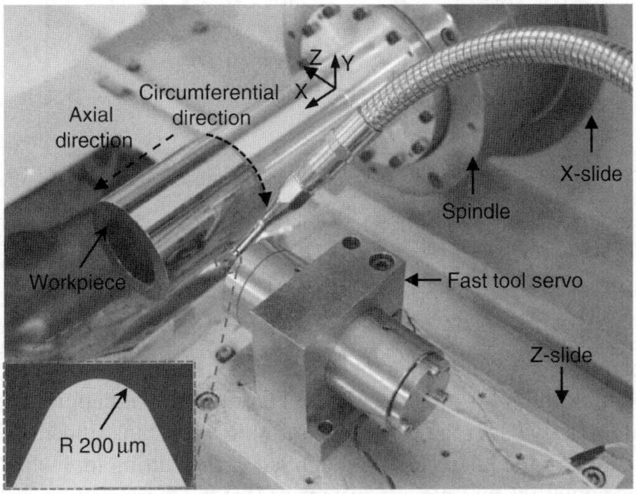

Figure 11.16 Setup for fabrication of the micro-lens array

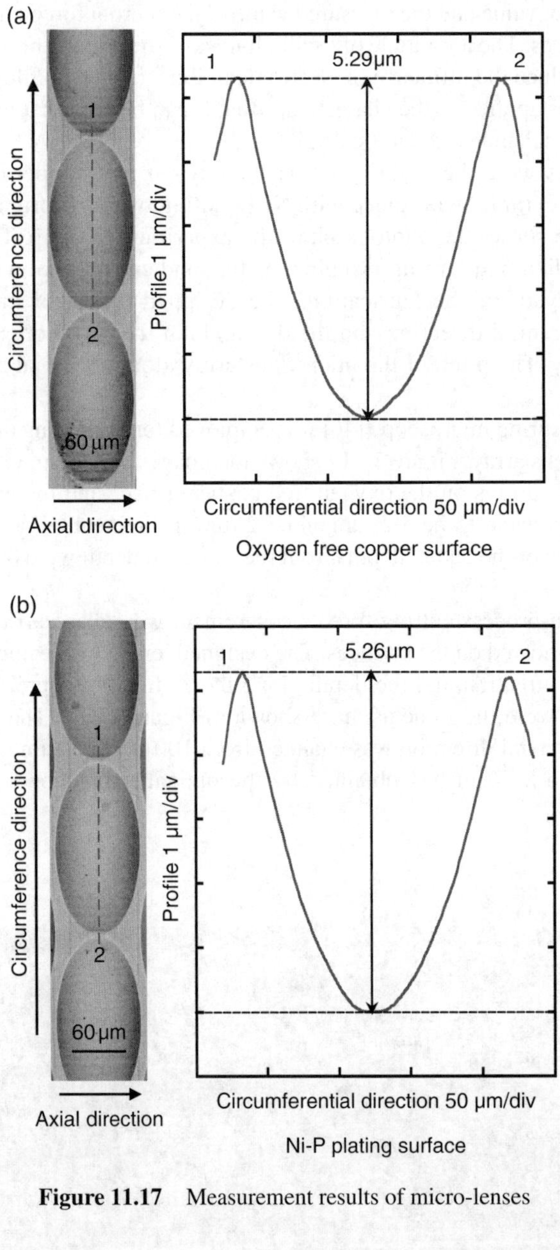

Figure 11.17 Measurement results of micro-lenses

Figure 11.18, there are burrs and chips on the oxygen free copper surface but no burrs and chips are observed on the Ni-P plating surface. During the fabrication process, the cutting forces were monitored by using the force sensor in the hybrid instrument. The same cutting motion for fabrication of the micro-lens was applied to the cutting tool when the cutting tool did not make contact with the workpiece surface so that the inertial force caused by the moving part could be detected for compensation of the in-process cutting force data.

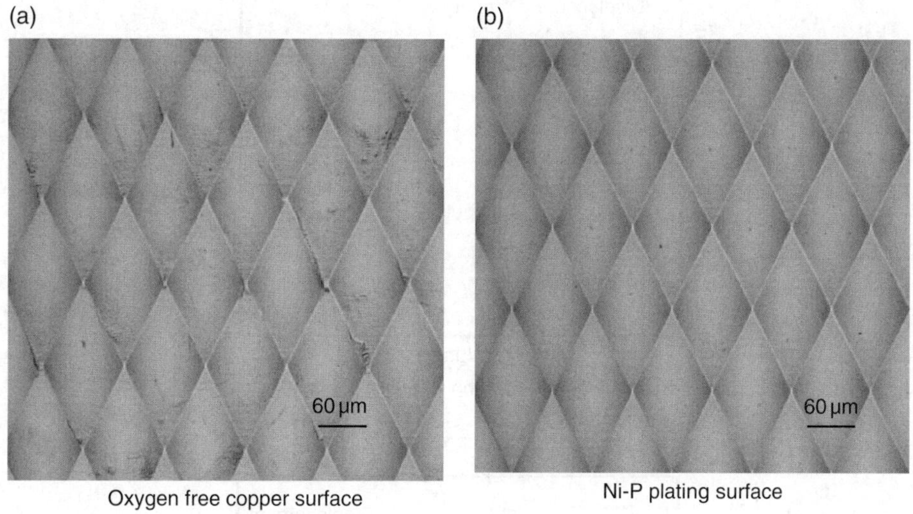

Figure 11.18 Image of fabricated micro-lens

Figure 11.19a shows the applied cutting motion and the corresponding inertial force measured in the experiment, in which the workpiece was not cut by the cutting tool. It can be seen that the amplitude of the inertial force was approximately 0.06 N. Figure 11.19b shows the force sensor output during cutting the micro-lens on the Ni-P plating surface. The force sensor output shown in Figure 11.20b, which was a sum of the inertial force and the actual cutting force, was approximately 1.03 N. The actual cutting force can thus be evaluated by taking the difference between the force sensor output shown in Figure 11.19b and that shown in Figure 11.19a.

Figure 11.20 shows the evaluated actual cutting forces during the micro-lens array fabrications on the oxygen free copper surface and the Ni-P plating surface, respectively. The actual cutting forces were 0.57 N and 0.97 N, respectively. The Ni-P surface was harder than the oxygen free copper, resulting in a larger cutting force. These results show that the hybrid instrument has the capability of measuring the cutting force in real time during the fabrication.

11.4 In-process Measurement of Micro Wear of Cutting Tool

Wear of the cutting tool is one of the factors to reduce both the surface quality and form accuracy of the fabricated micro structures. With the increase of the fabrication area on the micro-structured surface, it is important to monitor the micro wear of the cutting tool. The cutting force generated at the interference between the cutting tool and the workpiece is a good barometer to reflect the tool wear.

This section presents in-process measurement experiment of micro wear of the cutting tool by using the developed hybrid instrument. In the experiment, micro-lens arrays were fabricated on a Ni-P plating surface with a hardness of Hv 540 by using a new tool and a used tool with micro wear, respectively, to identify the relationship between the tool wear and the

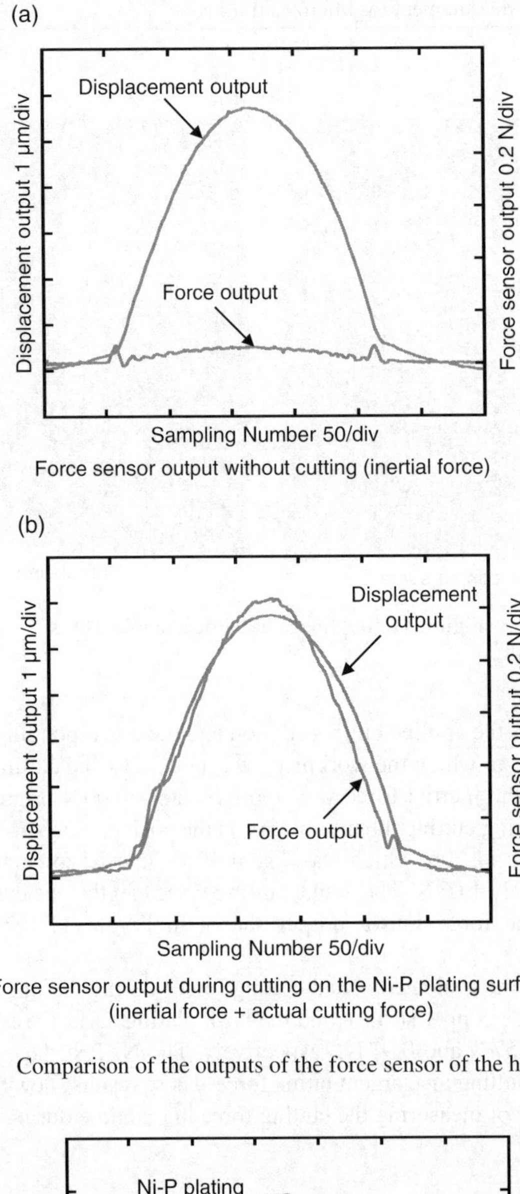

Figure 11.19 Comparison of the outputs of the force sensor of the hybrid instrument

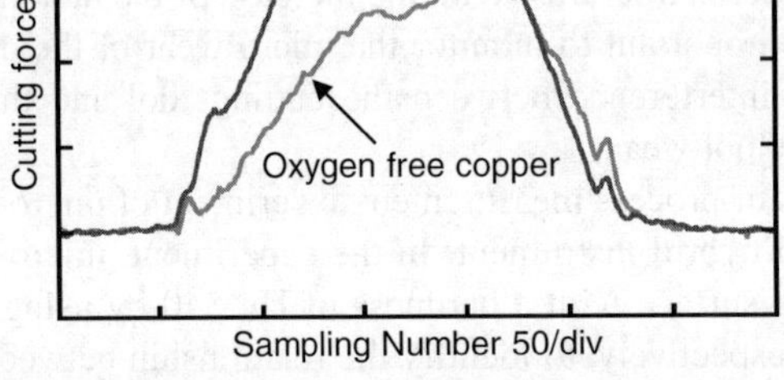

Figure 11.20 Comparison of actual cutting forces measured by the hybrid instrument for fabricating the micro-lens array on different surfaces

variation of cutting force. The nose radius of the tools was 0.2 mm. Figure 11.21 shows the profiles of the rake surfaces of the cutting tools. The optical microscope shown before was employed for the profile measurement. It can be seen that the used cutting tool has a wear with a depth of approximately 150 nm on the rake face.

Micro-lens arrays, which were designed to have a depth of 5.2 μm and a pitch of 200 μm, was fabricated on the Ni-P plating surface by using the two cutting tools, respectively. Figure 11.22 shows the measured cutting forces during the fabrication. The amplitude of the actual cutting force by the new tool was measured to be 0.91 N and that by the used cutting tool was measured to be 1.12 N. It can be seen that the cutting force by the new tool was

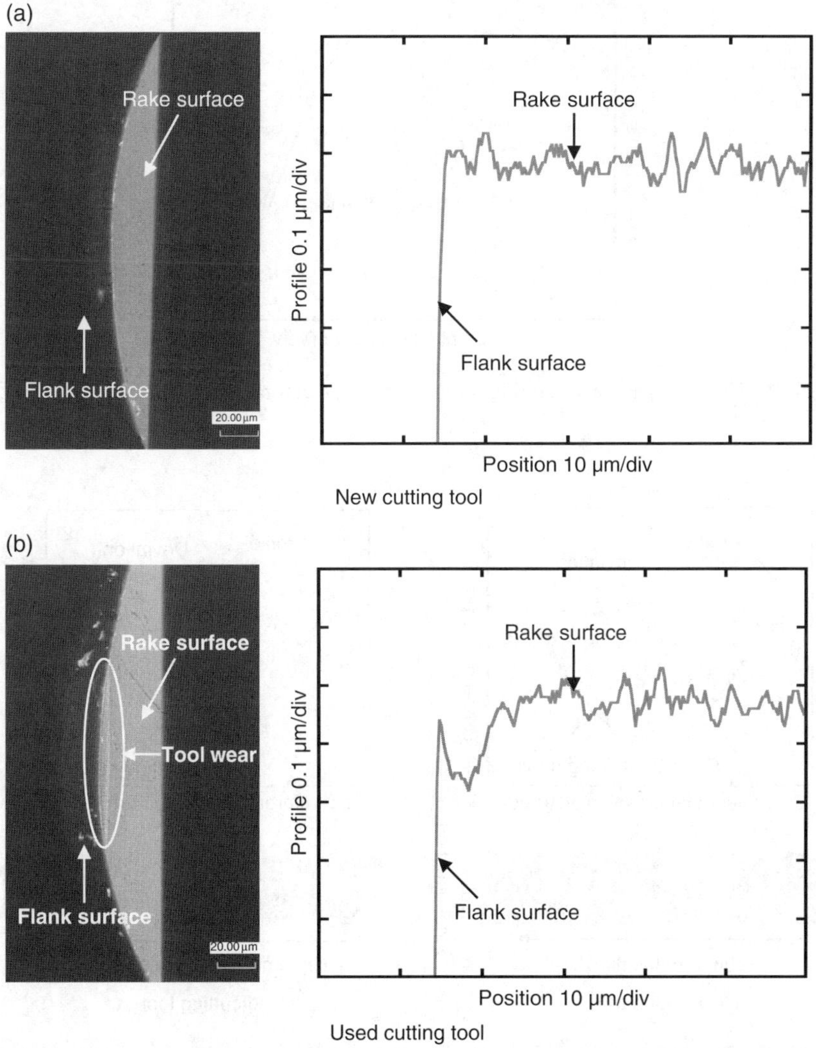

Figure 11.21 Measurement results of cutting tool profile

smaller than that by the used tool. The results confirmed that the cutting force is increased by the tool wear and it is possible to monitor the micro wear of the cutting tool from the variation of the force sensor output of the hybrid instrument during the fabrication. Figure 11.23 shows a comparison of surface qualities of micro-lenses fabricated by the two tools. The optical microscope was employed to measure the surface profiles. The deviation of the profile of the micro-lens surface from a best fitting curve is shown in each of the data graph. It can be seen that the micro-lens generated by the new tool had better surface quality than that of the use cutting tool. The results shown in Figures 11.22 and 11.23 are consistent with the

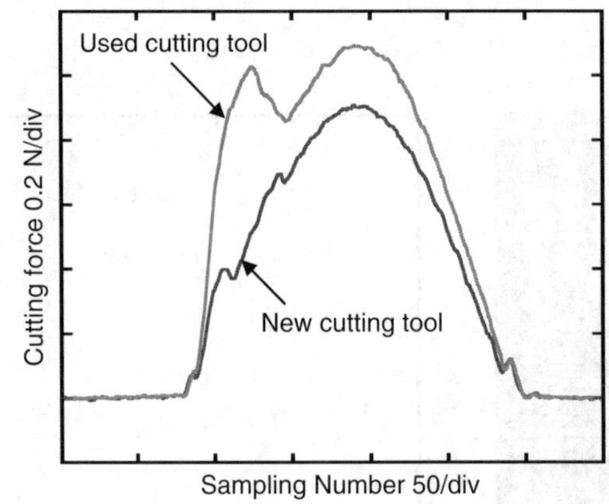

Figure 11.22 Comparison of cutting forces by the new cutting tool and the used cutting tool

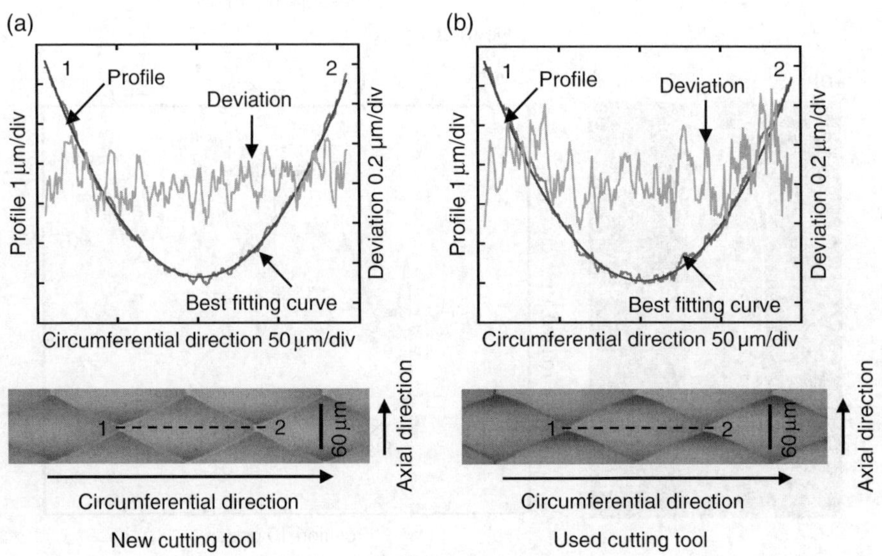

Figure 11.23 Comparison of surface qualities of micro-lenses by the new cutting and the used cutting tool

fact that the tool wear will increase the cutting force and decrease the surface quality. It should be noted that the high frequency components shown in Figure 11.22 were not accurate because it is difficult for the optical microscope to measure the micro-lens with small dimensions.

Experiments were also carried out to investigate the relationship between tool wear and workpiece hardness. Micro-lens arrays were fabricated on workpieces with different hardness by using a new cutting tool. A soft Ni-P plating workpiece with a hardness of Hv 540 and a hard Ni-P plating workpiece with a hardness of Hv 700 were employed. The micro-lens array was designed to have a pitch of 200 μm along the circumferential, a pitch of 50 μm along the axial direction, and a depth of 5.2 μm. The variation of the cutting force was measured when the fabrication area was increased.

Figure 11.24 shows the measured cutting forces for different lines of the micro-lens arrays along the axial direction. It can be seen that the harder the workpiece, the larger the cutting force was. The cutting force also increased with the increase of the number of the micro-lens array line. In the case of the soft workpiece with a hardness of Hv 540, the cutting force slightly increased from 0.94 N to 0.99 N when the line number changed from 1 to 400. On the other hand, the cutting force for the hard material with a hardness of Hv 700 increased from 2.12 N to 2.34 N when the line number changed from 1 to 100. The increase of the cutting force was caused by the growth of the tool wear and more tool wear was generated by the harder material. It is thus possible to estimate the growth of the tool wear from the increase of the cutting force.

Figure 11.25 shows a comparison between the surface qualities of the micro-lenses fabricated on the workpieces with different hardness. One of the micro-lenses on the first line of the micro-lens array and that on the last line of the micro-lens array were measured. The measured sectional profiles of the micro-lenses along the axial direction are shown in the data graphs. A commercial stylus surface form instrument [47], instead of the optical microscope, was employed for more accurate measurement of the micro-lens. In the case of the soft workpiece

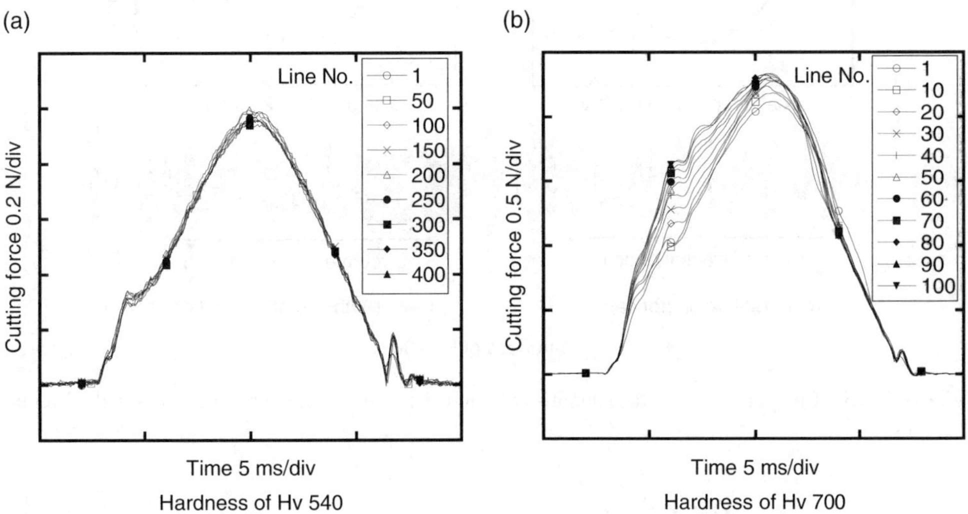

Figure 11.24 Comparison of cutting forces when the micro-lens array was fabricated on workpieces with different hardness

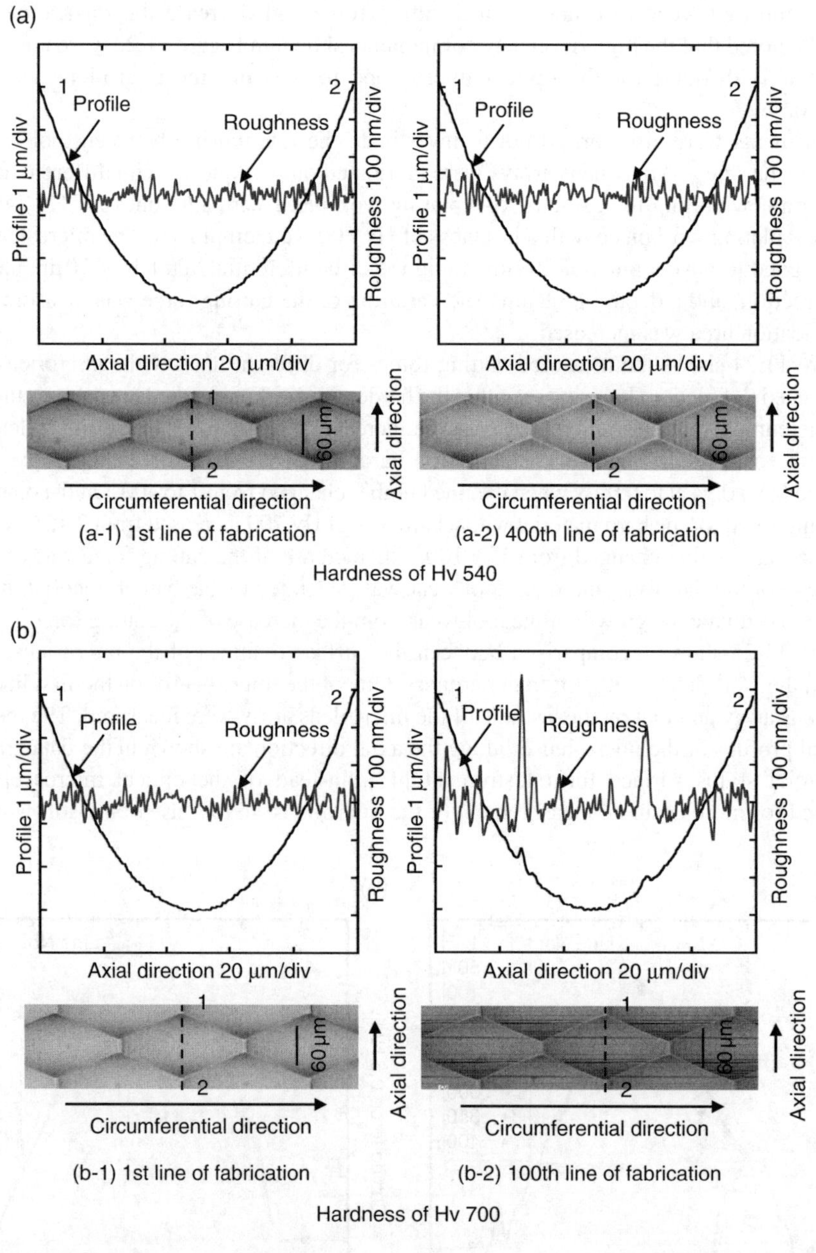

Figure 11.25 Comparison of surface qualities of micro-lenses fabricated on workpieces with different hardness

with a hardness of Hv 540, the surface roughness had little change, which was 0.08 µm for the 1st line and 0.11 µm for the 400th line. On the other hand, however, the surface roughness for the hard material with a hardness of Hv 700 increased significantly from 0.1 µm (1st line) to 0.4 µm (100th line). Some scratches can be observed on the surface of the micro-lens at the

400th line shown in Figure 11.25(b-2). The scratches were caused by the tool wear. The experimental results confirmed the fact that the micro wear of the cutting tool could significantly reduce the surface quality of micro structures. It has also verified that the hybrid instrument has the capability of estimating the growth of tool wear during the fabrication from the in-process cutting force data.

11.5 In-process Measurement of Micro Surface Form

After the fabricating process of micro structures, it is necessary to carry out in-process measurement of the micro surface form from the viewpoints of measurement efficiency and error compensation. As described in Section 11.2, the hybrid instrument can be employed as a stylus probing instrument for in-process measurement of micro surface form. The performance of the hybrid instrument as a stylus probing instrument mainly depends on the measurement of the cutting tool displacement as well as the detection and control of the contact force between the tool and the workpiece surface. In particular, the ability of detecting the small contact force is critical for the hybrid instrument because it is necessary to control the contact force to be as small as possible when the cutting tool scans over the workpiece surface so that the damage on the workpiece surface caused by the scanning operation can be reduced.

Experiments were carried out for identifying the basic performance of the instrument for in process surface form measurement. To detect the contact force with a high sensitivity, the cutting tool was oscillated by the PZT actuator of the hybrid instrument with a certain frequency and a small amplitude in the order of several nanometers so that the output of the force sensor can be modulated to an AC signal with the same oscillation frequency as shown in Figure 11.26. A lock-in amplifier was used to detect the amplitude of the modulated force sensor output with a high sensitivity without the influence of electronic noises. A PID controller was employed

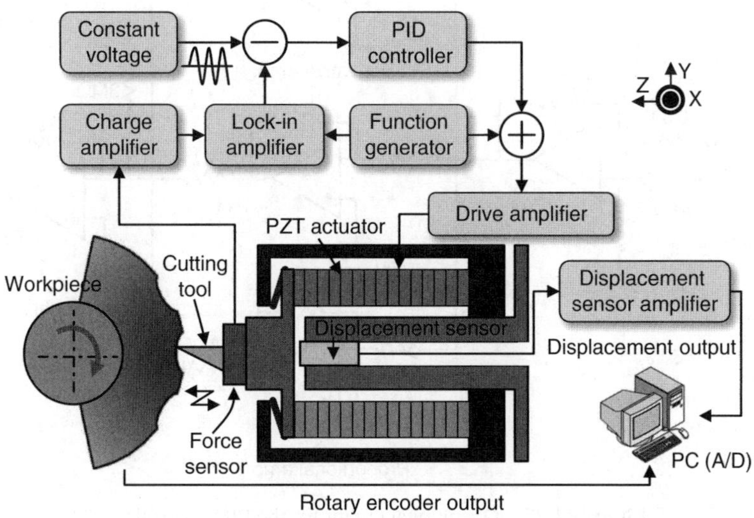

Figure 11.26 Principle of in-process surface form measurement by the hybrid instrument

for servo control of the contact force as shown in Figure 11.27. The controller was tuned by a trial-and-error method and the proportional gain Kp, the integral gain Ki and the derivative gain Kd of the PID controller were set to be 0.000042, 1.42 and 0.00552, respectively.

The ability of the system to detect the contact force between the cutting tool and the workpiece was investigated. The hybrid instrument, which was mounted on the Z-slide of the diamond turning machine, was moved by the Z-slide toward the workpiece surface along the Z-direction. A cylinder workpiece of Ni-P plating and a diamond cutting tool with a radius of 0.2 mm were employed in the experiment. The cutting tool was oscillated with an amplitude of 3 nm and a frequency of 220 Hz.

Figure 11.28 shows the relationship between the displacement of the Z-slide and the corresponding force sensor output when the tool made contact with the workpiece surface. As can be seen in Figure 11.28, the amplitude of the force sensor output was evaluated to be approximately 0.2 mN before the cutting tool contacted the workpiece surface. Once the cutting tool contacted the workpiece surface, the Z-slide was stopped so that the state of contact could be kept. It can be seen that the force sensor output, which included the influence due to the vibrations, increased to 0.5 mN once the cutting tool contacted with the workpiece surface. These results show that the force sensor had a resolution of approximately 0.2 mN and the contact force between the cutting tool and the workpiece surface could be controlled at a small amplitude of 0.5 mN.

After the contact between the cutting tool and the workpiece surface was established, an experiment was carried out to indentify the performance of the hybrid instrument for detection of the workpiece displacement. Figure 11.29 shows a schematic of the experiment. The hybrid instrument was moved by the Z-slide of the diamond turning machine with a step of 30 nm. During the movement, the cutting tool was moved by the PZT actuator of the hybrid instrument to track the movement of the Z-slide through keeping the cutting tool in contact with the workpiece surface at a contact force of 0.5 mN based on the feedback control of the PID

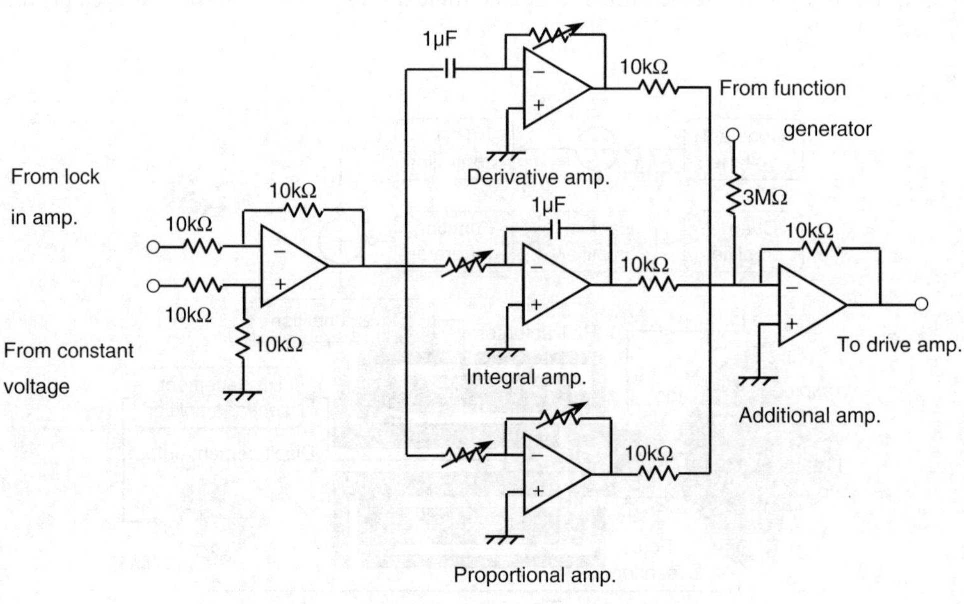

Figure 11.27 Operational circuit for the PID controller

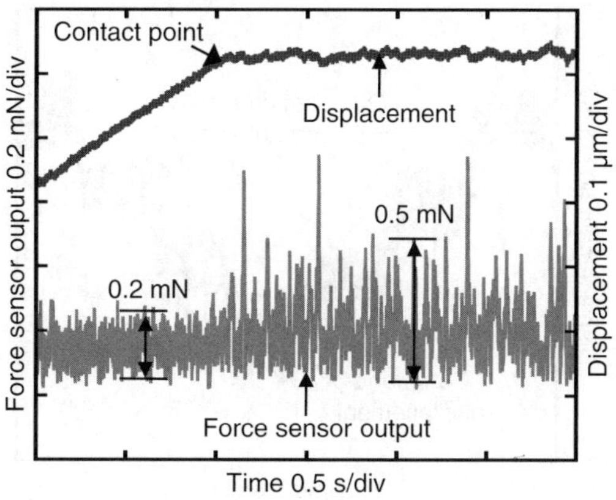

Figure 11.28 Experiment result of contact force

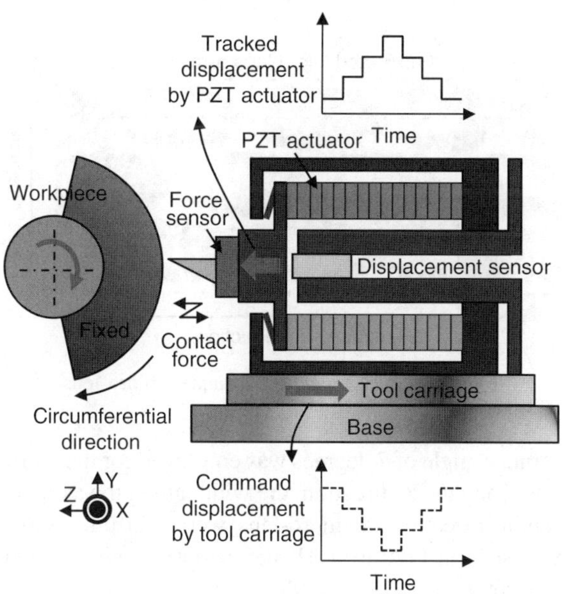

Figure 11.29 Schematic of tracking experiment by the hybrid instrument

controller. Figure 11.30 shows the relationship between the output of the capacitive displacement sensor of the hybrid instrument and the command displacement of the Z-slide. It can be seen that the hybrid instrument could successfully track and measure the displacement of the cutting tool relative to the workpiece surface and the cutting tool.

The hybrid instrument was then employed for experiments of fabrication and in-process measurement of micro-lens surface forms. Micro-lenses were fabricated on the cylinder workpiece of Ni-P plating on the diamond tuning machine by the hybrid instrument. A diamond cutting tool with a nose radius of 200 µm, a rake angle of 0 degree, an included angle of

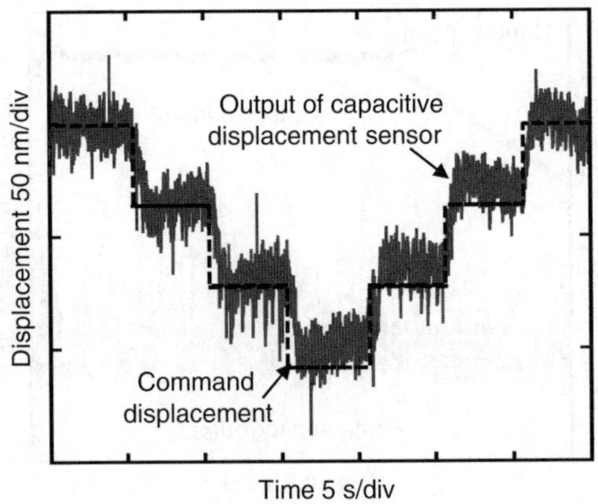

Figure 11.30 Experiment result for the hybrid instrument to track the displacement of the cutting tool relative to the workpiece surface

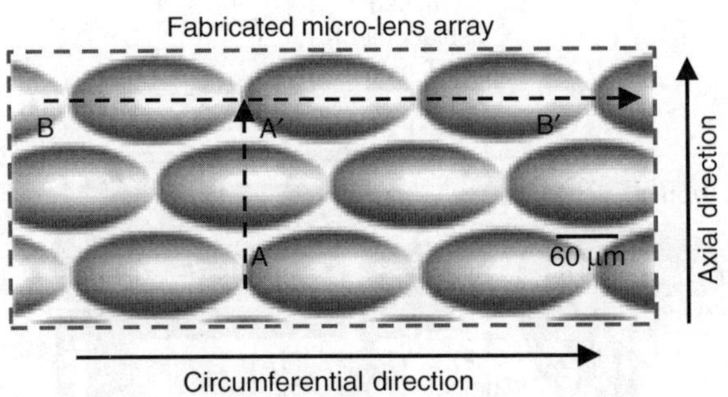

Figure 11.31 Image of fabricated micro-lens

55 degrees, and a clearance angle of 7 degrees was employed for the fabrication. The designed depth of the micro-lens along the Z-direction, the width along the axial direction and the pitch along the circumferential direction of micro-lens were 5.20 μm, 90.60 μm and 200.00 μm, respectively. As can be seen in Figure 11.31, there were no visual defects and burrs on the machined micro-lens surface.

Figure 11.32 shows a schematic of the scanning traces for in-process surface form measurement by the hybrid instrument. As can be seen in the figure, the scan was carried out across the centre point of the micro-lens along the axial direction (X-direction) and the circumferential direction of the cylinder workpiece, respectively. Before the scan, the cutting tool was first brought to make contact with the surface. In this process, the cutting tool was moved toward the workpiece surface step by step along the Z-direction by using the Z-slide of the diamond turning machine with visual feedback till the gap between the tool edge and the surface was within several micrometers. With the PID controller turned on, the tool was then moved by the hybrid instrument to contact the workpiece surface automatically with the feedback of the force sensor output. The contact was judged to be made when the force sensor output was larger than a threshold value, which was determined from the stability of the force sensor

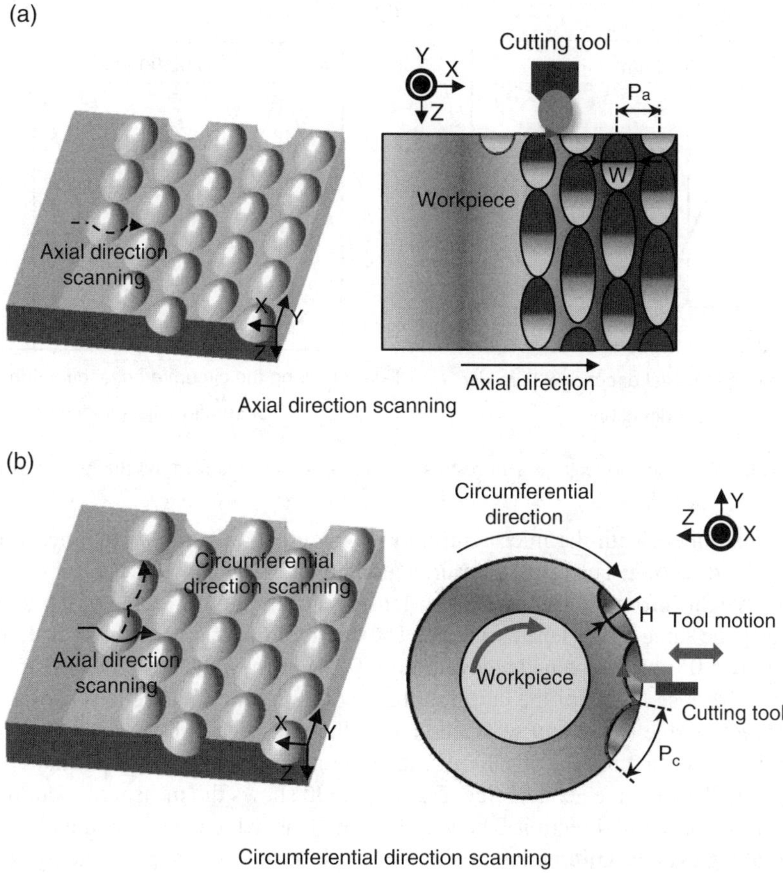

Figure 11.32 Scanning traces of the micro-lens by the hybrid instrument

output shown in Figure 11.28. Once the contact between the cutting tool and the workpiece surface was made, the workpiece surface was scanned along the axial direction and the circumferential direction, respectively, with the contact force being kept constant. It should be noted that the measurement result by using the cutting tool along the axial direction (X-direction) only contained the information of the lens height because of the nose radius of the cutting tool was comparable to the curvature radius of the sectional profile of the lens along this direction, as shown in the right of Figure 11.32a. On the other hand, the measurement result by using the cutting tool could provide accurate information along the circumferential direction because the cutting edge radius of the tool was less than 100 nm [17, 19], which was very small compared with the sectional profile of the lens along this direction as shown in Figure 11.32b.

Two lines AA′ and BB′ shown in Figure 11.31 were scanned. The scan speeds along the axial direction and circumferential direction were set to be 0.025 mm/min and 0.1 degree/min, respectively. The sampling intervals were set to be 40 nm and 0.002 degrees, respectively. Figure 11.33 shows the in-process measurement results of the micro-lens surface profile by

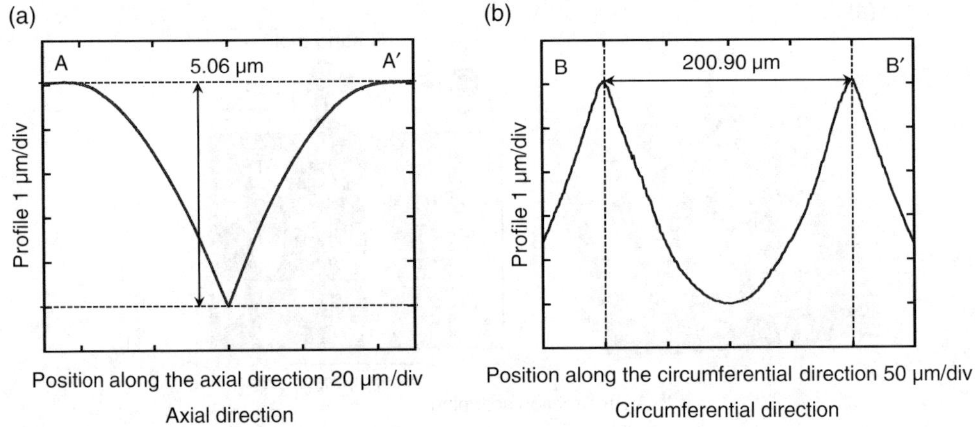

Figure 11.33 In-process measurement results of micro-lens surface form by the hybrid instrument

the hybrid instrument with the micro cutting tool. Figure 11.33a shows the measured result along the axial direction. The result measured along this direction by the cutting tool can only provide the height information of the micro-lens because the nose radius of the cutting tool was basically the same as the lens curvature. The depth of the micro-lens was evaluated to be approximately 5.06 μm, which had a deviation of 0.14 μm from the design value. It should be noted that if the nose radius of the cutting tool becomes larger than the lens curvature along the axial direction due to the tool wear, the top of the tool edge will not be able to touch the deepest point of the micro-lens. This will make the measurement result of the depth of the micro-lens smaller than the actual value. Figure 11.33b shows the measured sectional profile along the circumferential direction. The length of the lens, which was also the pitch between two lenses along the circumferential direction, was evaluated to be approximately 200.90 μm, which had a deviation of 0.90 μm from the design value. Although the edge radius of the cutting tool would increase due to the tool wear, the edge radius of the worn tool was still much smaller than the curvature of the micro-lens along the circumferential direction. Therefore, the tool wear did not have significant affects on the result shown in Figure 11.33b.

11.6 Summary

A hybrid instrument has been presented as an example of in-process micro/nano measurement for micro cutting. The hybrid instrument, which is combined with a fast tool servo and a piezoelectric force sensor, can be employed not only for fabrication of microstructures on a diamond turning machine but also for in-process measurement. A careful design has been made to integrate the force sensor into the fast tool servo so that the micro cutting force can be accurately detected without reducing the cutting performance of the fast tool servo. A number of in-process measuring technologies based on the force sensor output of the instrument have been presented. At first, the contact between the cutting tool and the workpiece surface, which is an essential process before starting the fabrication of micro structures, can be automatically and accurately established by in-process monitoring the contact force by the force sensor of the instrument. This technology is especially helpful to present damages of the fragile micro cutting tools with small nose sizes. Secondly, the dynamic cutting force during fabrication of micro structures can be detected in real time, which is important for monitoring the time-consuming process of fabricating large area micro-structured surfaces. Experimental results have also demonstrated that the growth of micro wear of the cutting tool during the fabrication process can be estimated in-process by using the force sensor output of the instrument. Finally, a novel in-process surface form measuring technology of the fabricated micro structures has been presented. The hybrid instrument, which has been employed to fabricate the micro structures, is employed as a force-controlled stylus probing instrument for in-process measurement of the surface forms of the micro structures. The cutting tool is scanned over the surface of the micro structure by the slide and the spindle of the diamond turning machine while the position of the cutting tool is servo-controlled by the instrument in such a way that contact force between the tool and the surface is maintained to be a small value. As a result, the surface form can be accurately obtained from the locus of the cutting tool motion.

References

[1] Huang, Y.P., Shieh, H.P.D., and Wu, S.T. (2004) Applications of multidirectional asymmetrical micro-lens array light-control films on reflective liquid-crystal displays for image quality enhancement. *Applied Optics*. 43(18):3656–3663.

[2] Pan, C.T., and Su, C.H. (2007) Fabrication of gapless triangular micro-lens array. *Sensors and Actuators A : Physical*, 134(2): 631–640.

[3] He, M., Yuan, X.C., Ngo, N.Q., Bu, J., and Tao, S.H. (2004) Single-step fabrication of a microlens array in sol-gel material by direct laser writing and its application in optical coupling. *Journal of Optics A: Pure and Applied Optics*, 6(1): 94–97.

[4] Sherrington, I., and Smith, E H. (1986) The significance of surface topography in engineering, *Precision Engineering*, 8(2):79–87.

[5] Bruzzone, A.A.G., Coasta, H.L., Lonardo, P.M., and Lucca, D.A. (2008) Advances in engineered surfaces for functional performance. *Annals of the CIRP*, 57(2):750–769.

[6] Fawcett, S.C., and Engelhaupt, D. (1995) Development of Wolter I x-ray optics by diamond turning and electro-chemical replication. *Precision Engineering*, 17(4):290–297.

[7] Lucca, D.A., Rhorer, R.L., and Komanduri, R. (1991) Energy dissipation in the ultraprecision machining of copper, *Annals of CIRP*, 40(1):69–72.

[8] Ikawa, N., Donaldson, R.R., Komanduri, R., Konig, W., Aachen, T.H., McKeown, P.A., Moriwaki, T., And Stowers, I.F. (1991) Ultraprecision metal cutting – the past, the present and the future, *Annals of CIRP*, 40(2):587–594.

[9] Taniguchi, N. (1994) The state of the art of nanotechnology for precessing of ultraprecision and ultrafine products. *Precision Engineering*, 16(1):5–24.

[10] Saito, TT. (1978) Diamond turning of optics: the past, the present, and the exciting future. *Optical Engineering*, 17(6):570–573.

[11] Keauskopf, B. (1984) Diamond- turning: reflecting demands for precision. *Manufacturing Engineering*: 90–100.

[12] Moriwaki, T., and Okuda, K. (1989) Machinability of copper in ultra-precision micro diamond cutting. *Annals of the CIRP*, 38(1):115–118.

[13] Ikawa, N., Shimada, S., and Tanaka, H. (1992) Minimum thickness of cut in micromachining. *Nanotechnology*, 3(6):6–9.

[14] Kim, D.S., Chang, I.C., and Kim, S.W. (2002) Microscopic topographical analysis of tool vibration effects on diamond turned optical surfaces. *Precision Engineering*, 26(2):168–174.

[15] Chen, C-C., Chen, C-M., and Chen, J-R. (2007) Toolpath generation for diamond shaping of aspheric lens array, *Journal of Materials Processing Technology*, 192–193(1):194–199.

[16] Childs, T.H.C., Sekiya, K., Tezuka, R., Yamane, Y., Dornfeld, D., Lee, D.E., Min, S., and Wright, P.K. (2008) Surface finishes form turning and facing with round nosed tools, *Annals of the CIRP*, 57(1):89–92.

[17] Gao, W., Motoki, T., and Kiyono, S. (2006) Nanometer edge profile measurement of diamond cutting tools by atomic force microscope with optical alignment sensor. *Precision Engineering*, 30(4):396–405.

[18] Asai, S., Taguchi, Y., Horio, K., Kasai, T., and Kobayashi, A. (1990) Measuring the very small cutting-edge radius for a diamond tool using a new kind of SEM having two detectors, *Annals of CIRP*, 39(1):85–88.

[19] Gao, W., Asai, T., and Arai, Y. (2009) Precision and fast measurement of 3D cutting edge profiles of single point diamond micro-tools. *Annals of the CIRP*, 58(1):451–454.

[20] Dow, T.A., Miller, M.H., and Falter, P.J. (1991) Application of a fast tool servo for diamond turning of nonrotationally symmetric surfaces. *Precision Engineering*, 13(4):243–250.

[21] Ludwick, S.J., Chargin, D.A., Calzaretta, J.A., and Trumper, D.L. (1999) Design of a rotary fast tool servo for ophthalmic lens fabrication. *Precision Engineering*, 23(4):253–259.

[22] Patterson, S.R., and Magrab, E.B. (1985) Design and testing of a fast tool servo for diamond turning, *Precision Engineering*, 7(3):123–128.

[23] Woronko, A., Huang, J., and Altintas, Y. (2003) Piezoelectric tool actuator for precision machining on conventional CNC turning centers, *Precision Engineering*, 27(4):335–345.

[24] Yang, Y., Chen, S., Huo, D., and Cheng, K. (2008) Performance analysis and optimal design of fast tool servo used for machining microstructured surfaces. *Journal of Mechanical Engineering Science*, 222(8):1541–1546.

[25] Miller, M.H., Garrard, K.P., Dow, T.A., and Taylor, L.W. (1994) A controller architecture for integrating a fast tool servo into a diamond into a diamond turning machine. *Precision Engineering*, 16(1):42–28.

[26] Okazaki, Y. (1990) A micro-positioning tool post using a piezoelectric actuator for diamond turning machines. *Precision Engineering*, 12(3):151–156.

[27] Gao, W., Araki, T., Kiyono, S., Okazaki, Y., and Yamanaka, M. (2003) Precision nano-fabrication and evaluation of a large area sinusoidal grid surface for a surface encoder. *Precision Engineering*, 27(3):289–298.

[28] Gao, W., Tano, M., Sato, S., and Kiyono, S., (2006) On-machine measurement of a cylindrical surface with sinusoidal micro-structures by an optical slope sensor. *Precision Engineering*, 30(3):274–279.

[29] Kimura, A., Gao, W., and Kiyono, S. (2007) Design and construction of a surface encoder with dual sine-grids. *International Journal of Precision Engineering and Manufacturing*, 8(2):20–25.

[30] Gao, W., and Kimura, A. (2007) A three-axis displacement sensor with nanometric resolution. *Annals of the CIRP*, 56(1):529–532.

[31] Sanjanwala, A., Choudhury, S.K., and Jain, V.K. (1990) On-line tool wear sensing and compensation during turning operation, *Precision Engineering*, 12(2):81–84.

[32] Byrne, G., Dornfeld, D., Inasaki, I., Ketteler, G., Konig, W., and Teti, R. (1995) Tool condition monitoring (TCM)-the status of research and industrial application, *Annals of the CIRP*, 44(2):541–567.

[33] Santochi, M., Dini, G., Tantussi, G., and Beghini, M. (1997) A sensor-integrated tool for cutting force monitoring, *Annals of the CIRP*, 46(1):49–52.

[34] Drescher, J.D., and Dow, T.A. (1990) Tool force model development for diamond turning, *Precision Engineering*, 12(1):29–35.

[35] Noh, Y.J., Arai, Y., Tano, M., and Gao, W. (2008) Fabrication of large-area micro-lens arrays with fast tool control. *International Journal of Precision Engineering and Manufacturing*, 9(4):32–38.

[36] Noh, Y.J., Arai, Y., and Gao, W. (2009) Improvement of a fast tool control unit for cutting force measurement in diamond turning of micro-lens array. *International Journal of Science and Engineering*, 3(3):227–241.

[37] Noh, Y.J., Lee, K.W., Arai, Y., and Gao, W. (in press) A force sensor integrated fast tool control system, *J of JSPE*.

[38] http://www.nec-tokin.com

[39] http://www.mtiinstruments.com

[40] http://www.fujicera.co.jp

[41] Holman, A.E., Scholte, P.M.L.O, Heerens, W. Chr., and Tuinstra, F. (1995) Analysis of piezo actuator in translation constructions. *Review of Scientific Instruments*. 66(5):3208–3215.

[42] Gao, W., Hocken, R.J., Patten, J.A., and Lovingood, J. (2000) Force measurement in a nanomachining instrument. *Review of Scientific Instruments*, 71(11):4325–4329.

[43] Gao, W., Hocken, R.J., Patten, J.A., Lovingood, J., and Lucca, D.A. (2000) Construction and testing of a nanomachining instrument. *Precision Engineering*, 24(4):320–328.

[44] http://www.thinksrs.com

[45] http://www.toshiba-machine.co.jp

[46] http://www.keyence.co.jp

[47] http://www.taylor-hobson.com

Index

Micro-Cutting: Fundamentals and Applications, First Edition. Edited by Kai Cheng and Dehong Huo.